OPTICAL FIBER SENSOR TECHNOLOGY

Optical Fiber Sensor Technology

Advanced Applications – Bragg Gratings and Distributed Sensors

Edited by

K. T. V. Grattan

and

B.T. Meggitt

City University, London, U.K.

KLUWER ACADEMIC PUBLISHERS

BOSTON / DORDRECHT / LONDON

A C.I.P. Catalogue record for this book is available from the Library of Congress.

ISBN 0-7923-7946-2

Published by Kluwer Academic Publishers,
P.O. Box 17, 3300 AA Dordrecht, The Netherlands.

Sold and distributed in North, Central and South America
by Kluwer Academic Publishers,
101 Philip Drive, Norwell, MA 02061, U.S.A.

In all other countries, sold and distributed
by Kluwer Academic Publishers,
P.O. Box 322, 3300 AH Dordrecht, The Netherlands.

Printed on acid-free paper

Printed in the Netherlands.

Contents

List of Contributors

K. T. V. Gratttan
Department of Electrical, Electronic & Information Engineering
City University
London, UK

G. R. Jones
Center for Intelligent Monitoring Systems
Department of Electrical Engineering and Electronics
University of Liverpool
Liverpool, UK

R. E. Jones
ReniShaw plc. - Metrology Division
Wotton-under-Edge
Gloucestershire, UK

R. Jones
Cambridge Consultants Ltd.
Science Park
Cambridge, UK

A. Othonos
Department of Natural Science: Physics
University of Cyprus
Nicosia, Cyprus

A. J. Rogers
Department of Electronic Engineering
School of Physical Sciences & Engineering
King's College
London, UK

A. Hartog
York Sensors
Chandlers Ford
Hampshire, UK

G. Murtaza
Center for Communication Networks Research
Department of Electrical and Electronic Engineering
The Manchester Metropolitan University
Manchester, UK

J. M. Senior
Faculty of Engineering and Information Sciences
University of Hertfordshire
Hertfordshire, UK

J. O. W. Norris
AEA Technology, Harwell,
Didcot,
Oxfordshire, UK

Preface

Advanced Applications in Optical Fiber Sensor Technology

The maturity in the subject of fiber optic sensors, seen in work presented at recent Conferences in the field and in the previous four volumes of this series on *Optical Fiber Sensor Technology* is reflected in the number of new, or enhanced applications of sophisticated fiber-based sensor systems which are continuing to appear. Indeed it is fiber sensor *systems*, rather than individual, relatively unsophisticated devices which have made the difference and opened up applications in the fields of environmental monitoring, measurement in civil engineering situations and in major industries such as petroleum and energy production. This has been brought about by combining the use of the best and most suited of the essential optical interactions concerned, coupled with sophisticated signal processing and data handling to make better measurements with optical fiber sensor systems than with conventional technology, often in special situations In this, the fifth text in the current series, the aim is to focus strongly on these application aspects, and to build upon the foundation of work reported in previous volumes.

The text comprises a series of commissioned chapters from leading experts in the field, the first of which discusses in some detail both the principles and applications of multimode optical fiber sensors. It draws upon the experience of three authors, coincidentally all having the same surname. Gordon Jones of the University of Liverpool, UK has co-ordinated the input of his colleagues in industry, Robert Jones and Roger Jones to include in the discussion a wide variety of multimode fiber systems. Following that Andreas Othonos draws upon his wide experience, including time spent with Ray Measures' group in Canada which had pioneered the application of fiber Bragg gratings in structural monitoring, in writing a chapter on fundamentals and applications of what are now key components of most optical fiber sensor systems. The chapter gives the essential groundwork for the subject, and then discusses a number of interesting applications which show the versatility of Bragg grating-based sensor systems. Alan Rogers' expertise in the fundamentals and applications of non-linear optics in fiber sensor systems has been known since his foundation work in the mid-1970s and he brings this to bear upon a chapter discussing both the basics and a range of topical uses of non-linear optical effects, enhanced by modern signal processing techniques. Arthur Hartog of York Technology was one of the pioneers of the use of non-linear optics in temperature measurement and in his chapter, again following a discussion of the key fundamentals of the subject, he emphasises the wide range of new applications of this sensor technology which has opened up in recent years. These devices, commercialized in the 1980s, represent some of the most successful of optical fiber sensor systems, and his discussion of the breadth of uses in industry is both fascinating and informative. Intensity-based fiber optical systems have required effective referencing to compete with other schemes, and Ghulam Murtaza and John Senior discuss in their chapter a wide range of such referencing schemes which are applicable to a number of different types of sensor systems and measurands. The key principles underpinning the wide range of optical fiber chemical sensors are discussed by John Norris, and seminal examples of the methods are discussed. The subject has grown rapidly over the years and European, American and other International Conferences regularly bring reports of the latest developments and applications which rely upon the key principles outlined herein.

The reputation of the authors as publishers in the leading international journals and key presenters at major Conferences gives then the basis upon which their contributions have been made and their authority to write as they do. They are the developers and users of the technology – in industry and from academia. The coverage of the subject is wide, and the material discussed will have a lasting impact upon both the fundamental understanding and applications of this sensor technology.

The editors are very grateful to Dr Tong Sun for her tireless efforts in typesetting this manuscript from the authors' original material and in preparing the diagrams from a wide range of sources for publication.

The editors hope that the readers of this fifth volume in the *Optical Fiber Sensor Technology* series will find, together with the companion four volumes, a valuable source of reference and information in what is a comprehensive series on devices, applications and systems.

1

Multimode Optical Fiber Sensors

G. R. Jones, R. E. Jones and R. Jones

1.1 INTRODUCTION

Multimode optical fiber is widely used in a range of sensor systems. Such multimode optical fiber sensors have advantages of:

- operating with substantial optical power over moderate distances inexpensively;
- utilizing the multiplicity of propagation modes within the fiber for sensing purposes;
- providing a means of sensing spectral signature changes over considerable wavelength ranges;
- relatively large dimensions so improving tolerances with respect to end effects and interconnections.

Thus whereas multimodal and polychromatic effects may lead to dispersion-based limitations to the high data rate demands of telecommunications, they may be used to advantage for sensing systems operating at lower data rates ($\leq$ MHz).

Since phase and polarization are not easily maintained in multimode fibers, the transmission of sensor information through the fiber needs to be by light intensity variations (although the sensing elements themselves may provide phase or polarization information). As a result, the evolution of multimode fiber sensing is concerned with:

- producing various intensity modulation sensor outputs for transmission (including phase and polarization modulating elements);
- overcoming the problems associated with the lack of intensity conservation in optical fiber systems.

Arguably multimode optical fiber sensors are capable of deployment for measuring a greater variety of measurands than other fiber systems, encompassing not only physical parameters (e.g. pressure, temperature etc.) but also chemical (e.g. impurity contamination etc.) parameters and non specific parameters (e.g. color, acoustical vibrations etc.). Both analog and

K.T.V. Grattan and B.T. Meggitt (eds.), Optical Fiber Sensor Technology, 1–77.

digital systems are available and their cost effectiveness for bulk applications has long been recognized [1].

The approach taken here is to establish the performance criteria and to formalize a general systems description within which framework transducer, data acquisition and signal processing requirements can be considered. These are discussed in section 1.2. Section 1.3 considers the implications of the system requirements upon the optical source and fiber transmission. Some important optical modulation principles are considered in section 1.4. Section 1.5 relates to signal processing and system architecture matters, whilst the present state of the technology and its future potential are is summarized in the conclusion

1.2 FORMAL SYSTEMS APPROACH

The systems approach to fiber sensing needs to establish how the system performance is to be judged and how the various system components interact to affect the overall performance. The former aspects may be addressed through specific performance criteria, whereas the latter require the formulation of a mathematical model from which the coupling of various terms representing the different components can be identified.

1.2.1 Performance criteria

The performance criteria for a measurement system may be identified from a consideration of the characteristic which relates the output of the measurement system to the measurand value (fig. 1.1). These are the sensitivity, noise, signal-to-noise ratio, resolution, dynamic range and accuracy. Consideration is also needed of transient response.

1.2.1.1 Sensitivity

The sensitivity (or scale factor), s, is the proportionality between the input (measurand) and output of the measurement system, i.e.

$$V_0 = sx \tag{1.1}$$

where V_0 is the output of the measurement system (e.g. volts) and x is the measurand value (e.g. pressure in pascals). Ideally, s should remain constant over the entire operating range of the transducer and should be independent of external conditions such as environmental temperature. Variations in s

for a given system are known as 'fading'. However, the value of s may be changed for different systems by varying either the transducer designs or system operation (section 1.2.2).

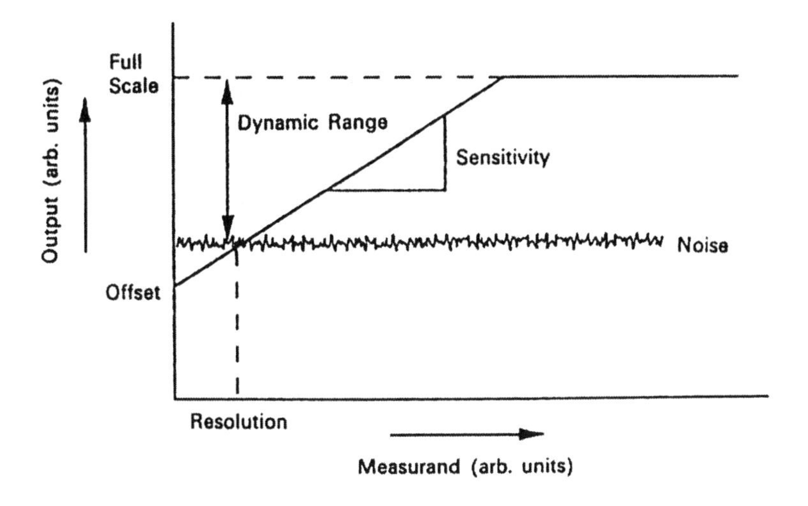

Figure 1.1. Relationship between sensor parameters

1.2.1.2 Noise

In electrically-based measurement systems, a fundamental limitation is due to electronic noise produced either in resistors or active devices, or by electromagnetic pick-up via connecting leads.

A noise is characterized by its frequency spectrum $V_n(f)$ (units $\mathrm{VHz}^{-1/2}$). Two important cases are Johnston noise (produced by random electron motion in resistive elements) and flicker noise. The former is independent of frequency whereas the latter is inversely proportional to frequency (fig. 1.2). $V_n(f)$ at a given frequency is the sum of all noise components $\sum V_n(f)$. Thus the total r.m.s output noise voltage from a system is

$$V_n = \int \left[\sum V_n(f) \right] df \tag{1.2}$$

which corresponds to the area under the curve of fig. 1.2.

This is merely a formalization of the fact that the r.m.s noise voltage depends upon the bandwidth (frequency range) of the measurement. For instance, for a system with only Johnston noise at 10 $\mathrm{nVHz}^{-1/2}$ (fig. 1.2) the r.m.s. output noise for a bandwidth up to 100 Hz would be 100 nV and would increase to 316 nV for a bandwidth up to 1 kHz.

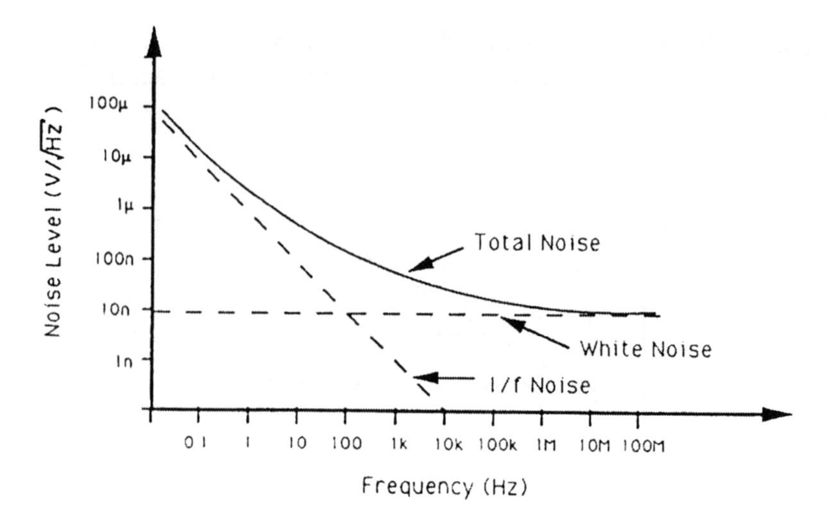

Figure 1.2. Spectra of white and 1/f noise

The discrimination of a signal against a noise background is quantified by the signal-to-noise ratio (S/N)

$$S/N = \frac{V_0}{V_N} = \frac{sx}{V_N} \tag{1.3}$$

1.2.1.3 Resolution

Clearly, the above noise considerations govern the smallest change in a measurand which a system can discern or resolve.

Thus the system resolution (R_s) is the value of the measurand which produces an output voltage equal to the noise voltage. From equation (1.3)

$$R_s = \frac{s}{V_N} \tag{1.4}$$

Since V_N is bandwidth dependent the implication of equation (1.4) is that for a fixed sensitivity s, resolution can only be improved at the expense of transient response.

1.2.1.4 Dynamic range

At the other extreme of the measurement range is the highest output signal that may be limited either by the transducer (e.g. length of travel of a displacement transducer), the system (e.g. output voltage reaching the supply rail) or the user requirement (e.g. unacceptable departure from linearity). This represents the full scale (FS) of the system. Thus the scale length which is available for measurement is the ratio between the full-scale and the noise voltage (fig. 1.2). This is known as the **dynamic range** and may be written as:

$$\Delta R = \frac{(FS)}{V_N} \tag{1.5}$$

1.2.1.5 Accuracy

The accuracy of a measurement system is the extent to which the output deviates from that of a calibrated standard. Thus although accuracy is related to the resolution of the system, the accuracy will in general be poorer than the resolution.

1.2.2 Formal representation of a fiber system

To relate the performance criteria described above to an optical fiber measurement system it is necessary to establish a formal theoretical description of the system. The general structure of such a system is shown in fig. 1.3. It consists of an optical source, optical fibers, a modulator element (which transduces the measurand to an optical signal), an optical detector and processing electronics. The output voltage, V_o, of the system depends upon the optical properties of each system component combined according to the mathematical expression

$$V_o = q\left\{\sum_{l,m} \int_\lambda \left[\int_l p(\lambda)F(\lambda)M_2(\lambda)dl \right] M_l(\lambda)R(\lambda)d\lambda \right\}^P \tag{1.6}$$

where $P(\lambda)$ is the spectral power distribution of the source, $F(\lambda)$ is the spectral transmission of the optical fiber, $M_l(\lambda)$ is the spectral modulation produced by the sensor element, and $R(\lambda)$ is the spectral responsivity of the detector. The optical signal may be polychromatic in nature, hence the need for integration with respect to wavelength λ. Propagation may occur over variable lengths of transmitting fiber, so integration of fiber-related aspects needs to be over the fiber length, l. In addition, the multimode nature of the fibers requires summation over all propagation modes designated by l, m (section 1.3.3). Intermodal power exchange caused by system components (connectors, modulator etc.) is taken into account by the factor $M_2(\lambda)$. The parameter q represents electronic signal processing effects of the circuitry which provides the voltage output V_o. The proportionality between voltage output and received optical power (which is the term in curly brackets) may be nonlinear, which leads to the exponent P.

Equation (1.6) provides an insight into several aspects of fiber monitoring. It embodies not only power conservation considerations but also spectral information which can be used for optimizing the spectral matching of components.

A special case which leads to a simplified description and which corresponds to an optical system which is most closely analogous to an electronic system involves intensity modulation with monochromatic light ($\lambda=\lambda_1$) and monomode fibers ($m = 0, l = 1$).

Here the amplitude of $M_1(\lambda_1)$ is proportional to the measurand X_i, i.e.

$$X_i = q_m M_1(\lambda_1) \tag{1.7}$$

with the additional assumption that $P = 1$, equation (1.6) reduces to a power budget expression from which the system sensitivity (eq.(1.1)) is more easily determined:

$$s = qP(\lambda_1)F(\lambda_1)M_2(\lambda_1)q_m R(\lambda_1) \tag{1.8}$$

This expression has several implications with regard to system limitations.

The ultimate resolution (section 1.2.1.3) is determined by additive electronic noise in the receiver via the parameter q. For a photodetector sensitivity of $0.5AW^{-1}$, a transimpedance amplifier with a feedback resistor of $1M\Omega$, a bandwidth of 1 kHz and an input current noise density of 0.2 $pAHz^{-1/2}$ (section 1.2.1.2), the minimum detectable change in optical power is 1.2 pW. Hence, for a typical received optical power of 1.2 μW the intensity resolution predicted is 1 in 10^6. In practice, it is difficult to achieve this level of resolution with intensity modulation. In the shot noise limit, increasing the intensity of the received signal is not advantageous since the shot noise is proportional to intensity.

However, in practice the dominant limitations are due not so much to q but to the fiber transmission, $F(\lambda)$, and fiber-related effects $M_2(\lambda)$. Firstly, changes can occur to these parameters due to aging or environmental temperature variations which produce fading (section 1.2.1). Secondly, although fiber systems are immune to electromagnetic interference, they are instead susceptible to mechanical noise and fluctuations caused by fiber microbending. It is the fading and fiber noise susceptibility which constitute the greatest barrier to the commercialization of optical fiber sensors and much research is concerned with overcoming these deficiencies.

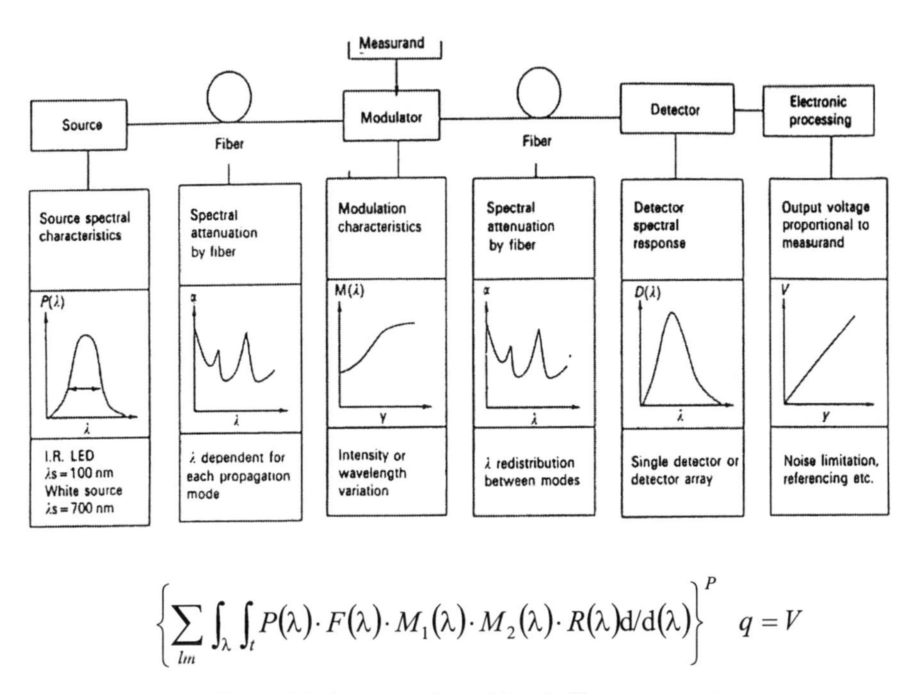

$$\left\{\sum_{lm}\int_{\lambda}\int_{t} P(\lambda)\cdot F(\lambda)\cdot M_1(\lambda)\cdot M_2(\lambda)\cdot R(\lambda)\mathrm{d}l\mathrm{d}(\lambda)\right\}^{P} q = V$$

Figure 1.3. Structure of a multimode fiber sensor system

In the case of the more general formulation (equation (1.6)) the situation is made more complicated not only by the complex integrated interdependence of the component parameters ($P(\lambda)$, $F(\lambda)$ etc.) but also because of the intermodal coupling via $\sum_{lm}$. The implication of this is that it is impossible to obtain a universally applicable simple analytical expression for the sensitivity, s. Instead equation (1.6) needs to be evaluated for each individual case and this involves detailed knowledge of not only the modulator characteristics but also each system component. It is for this reason that a rigorous systems description is essential for considering multimode fiber sensors.

Such an approach provides a powerful basis for exploring possible methods for overcoming system limitations of the type indicated above. It enables system components to be better optimized with regard to matching the spectral transmission windows of the interconnected optical elements of the system. It also allows various signal multiplexing (e.g. wavelength based) and system architecture possibilities to be assessed. The remainder of this chapter is based upon the implications of equation (1.6) for such considerations. The approach taken is to consider the mathematical form of each of the components representing parameters in equation (1.6) separately.

1.3 SOURCE AND FIBER EFFECTS

1.3.1 Spectral emission of source (P(λ))

The use of multimode sensors is, in general, less restrictive with regard to the type of optical source used than in the single mode case, so the mathematical form of the parameter $P(\lambda)$ may differ significantly depending upon systems and sensor requirements. Two simplifying extreme cases may be identified which correspond to a purely monochromatic source and an ideal white light source respectively. In the former case

$$P(\lambda) = P(\lambda_1) \tag{1.9}$$

so the wavelength integration in equation (1.6) becomes redundant, leading, for monomode propagation, to equation (1.8).

At the opposite extreme, corresponding to an ideal white light source,

$$P(\lambda) = p \text{ constant} \tag{1.10}$$

so equation (1.6) reduces to

$$V = q\left[\sum_{l,m} P\int_{\lambda}\left(\int_{l} F(\lambda)M_2(\lambda)dl\right)M_1(\lambda)R(\lambda)d\lambda\right]^p \tag{1.11}$$

In practice equation (1.8) is a good approximation for systems activated by laser sources, whilst there are situations in which equation (1.11) can apply to broadband sources. This latter category includes tungsten halogen, conventional LED and 'white light' LED sources. Conventional LED sources refer to moderate spectral width sources up to about 100 nm. Tungsten halogen sources are extremely wideband ranging from about 450 nm to the near infrared at 1.1 μm; they have proved to be reliable for providing a very wide spectral output with high spectral stability and capable of energizing up to eight sensors economically. The most recent addition to this group of broad band sources is the white LED covering the intermediate spectral range from 400 - 600 μm with a pronounced emission at the shorter wavelengths (figure 1.4). Early indications are that such sources have good power output, spectral stability and aging characteristics without the infrared heating effects of the tungsten halogen source, making them good candidates for optical fiber spectral and chromatic modulation systems.

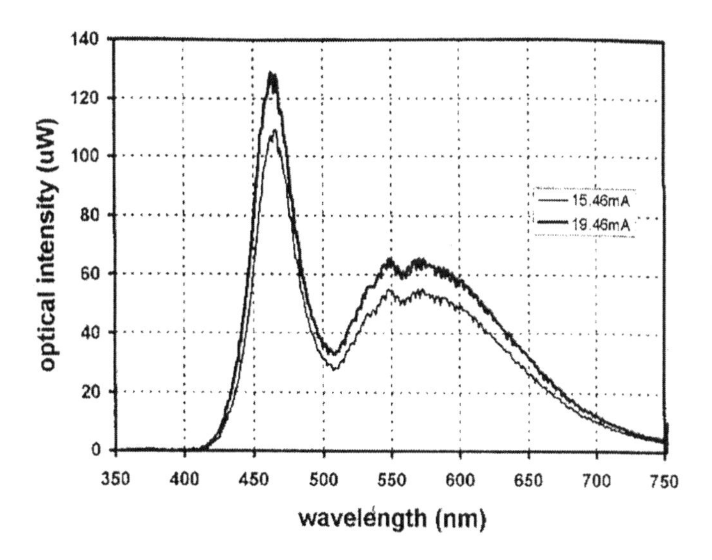

Figure 1.4. Change in spectrum of white LED with forward current

1.3.2 Wavelength-dependent fiber attenuation (F(λ))

The parameter F(λ) (equation (1.6)) takes account of both the attenuating and optical filtering action of the optical fiber and as such embodies the influence of several physical processes. These include the effect of Rayleigh scattering due to the structure of the optical fiber material, the optical absorption due to particular ionic/molecular impurities (such as OH) and residual effects such as losses associated with fiber bending (fig. 1.3). Conventionally these effects are incorporated via an attenuation coefficient and defined by (e.g. [2])

$$F(\lambda) = F_o \exp(-\alpha l) \qquad (1.12)$$

where F(λ), F_o are the optical powers of the signal after transmission along a length l of fiber and at launch respectively, The various loss effects are then incorporated into α as a summation, i.e.

$$\alpha = q_1\lambda^{-4} + q_2(\lambda) + q_3(\lambda) \tag{1.13}$$

where q_1 is the Rayleigh scattering coefficient (due to irregularities of size less than a tenth of the optical wavelength λ), $q_2(\lambda)$ is the absorption due to impurities and $q_3(\lambda)$ takes account of additional losses caused for instance by excessive bending of the optical fiber.

The complex dependence of $F(\lambda)$ upon optical wavelength is thus clearly apparent.

1.3.3 Fiber modal effects

The modal nature of signal propagation in optical fibers (i.e. propagation of stable radial and azimuthal distribution of the electromagnetic wave vectors) is governed by the solution to the scalar wave equation [2]

$$\Psi = U_0(r)\begin{Bmatrix} \cos L\phi \\ \hline \sin L\phi \end{Bmatrix} \exp(\omega t - \beta z) \tag{1.14}$$

where $U_0(r)$ represents the radial field distribution, ϕ is the angular cylindrical co-ordinate, ω the angular frequency, t time, z the axial co-ordinate and β the propagation constant defined by

$$n_2 k < \beta < n_1 k \tag{1.15}$$

with n_1, n_2 the core and cladding refractive indices respectively and $k = 2\pi / \lambda_0$ (λ_0 = free space wavelength). The field patterns associated with the various modes are designated by two integers *l*, *m* which give the number of field null points radially (via $U_0(r)$) and azimuthally (via cos$L\phi$, sin$L\phi$). As a result the optical intensity distribution across a section of a fiber is nonuniform and of a complex nature (fig. 1.5). (The fact that these modes propagate with different phase velocities, according to equation (1.15), and so cause signal dispersion, is of less consequence to multimode sensing than high data rate communications because less restrictive time scales are involved).

The number of such propagation modes sustained by an optical fiber is given by [2]

$$M = \frac{1}{2}\left(\frac{S_c}{S_c + 2}\right)\left(\frac{2\pi}{\lambda}a(NA)\right)^2 \tag{1.16}$$

which shows that the number of modes increases with fiber radius a and the numerical aperture NA. The parameter S_c governs the shape of the radial profile of the refractive index ($S_c \rightarrow \infty$ for step index, $S_c = 2$ for parabolic graded index). The significance of these relationships for multimode fiber sensing is that for fiber core diameters of ~ 85 μm or greater, a large number of modes are sustained which overlap to give a more uniform intensity distribution compared with fig. 1.5. Also for a given numerical aperture and fiber radius half as many modes are sustained with a parabolic index profile than with a step index profile. In addition, equations (1.14), (1.15), (1.16) also imply that different optical wavelengths are preferentially, although not exclusively, associated with different propagation modes which have wavelength-dependent cut-offs [2]. Axial modes (transverse electric or magnetic) which are associated with meridional rays have more a preponderance of shorter wavelengths than peripheral modes which are associated with skew rays. This increases the complexity of the system behavior, as is apparent from the separate wavelength integration demanded in equation (1.6) for each mode and which gives a nonuniform spectral distribution across the fiber cross-section.

Tolerances for detector, connectors and modulator alignment thus require more careful consideration than dictated by simple geometric optics considerations.

It also needs to be appreciated that external influences such as excessive fiber bending or the addition of extra fiber lengths may affect not only thc optical attenuation (section 1.3.2) but also the wavelength redistribution amongst the various modes. In general it is the longer wavelengths which are propagated preferentially via the peripheral modes and which are most susceptible to such influences (fig. 1.5).

Power exchange between the various modes occurs over a finite length of optical fiber (the mixing length) before an equilibrium mode distribution is reached (fig. 1.6) [2]. Quantification of the mixing length is difficult and depends upon the type of fiber involved. Nonetheless the implication is that following and abrupt perturbation of the propagation (due for instance to cladding refractive index change, the presence of connectors or fiber bends), a finite length of fiber is needed for an equilibrium mode distribution to be reached. Coupled with the preferential wavelength selectivity of the different modes and the wavelength dependent attenuation (section 1.3.2) the spectral profice of the optical signal may therefore vary with the length of fiber. The use of mode scrambling devices can be advantageous for producing equilibrium mode distribution over a shorter length of fiber and so reduce such effects.

The question of the number of propagation modes is important when considering the design of homodyne interferometers for vibration studies [3]. Major advantages of multimode fibers include their efficient delivery of sufficient optical power to the sensor without difficult source/detector/sensor alignment problems. When used with highly polychromatic sources phase noise is not a problem because of the incoherent nature of the emission from such sources.

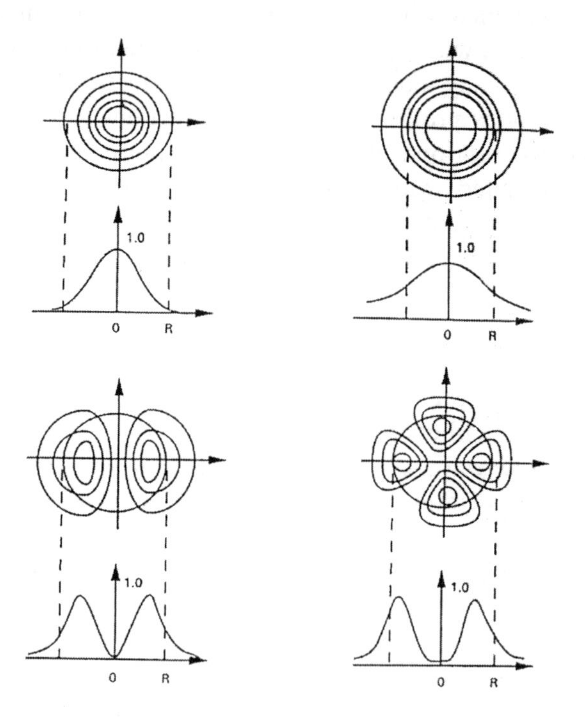

Figure 1.5. Radial distribution of optical intensity for various propagation modes. (a) LP_{01} modes confined to core; (b) LP_{01} modes overlapping cladding; (c) LP_{11} modes overlapping cladding; (d) LP_{21} modes overlapping cladding

1.4 SOME IMPORTANT MODULATION MECHANISMS

Sensors used in multimode systems fundamentally affect the amplitude of the optical signal via the factor $M_1(\lambda)$ in equation (1.6). Tolerances for detector, connectors and modulator alignment thus require more careful consideration than dictated by simple geometric optics considerations.

It also needs to be appreciated that external influences such as excessive fiber bending or the addition of extra fiber lengths may affect not only the optical attenuation (section 1.3.2) but also the wavelength redistribution amongst the various modes. In general, it is the longer wavelengths which are propagated preferentially via the peripheral modes which are most susceptible to such influences (fig. 1.5).

Power exchange between the various modes occurs over a finite length of optical fiber (the mixing length) before an equilibrium mode distribution is reached (fig. 1.6) [2]. Quantification of the mixing length is difficult and depends upon the type of fiber involved. Nonetheless the implication is that following and abrupt perturbation of the propagation (due for instance to cladding refractive index change, the presence of connectors or fiber bends), a finite length of fiber is needed for an equilibrium mode distribution to be reached. Coupled with the preferential wavelength selectivity of the different modes and the wavelength dependent attenuation (section 1.3.2) the spectral profice of the optical signal may therefore vary with the length of fiber. The use of mode scrambling devices can be advantageous for producing equilibrium mode distribution over a shorter length of fiber and so reduce such effects.

The question of the number of propagation modes is important when considering the design of homodyne interferometers for vibration studies [3]. Major advantages of multimode fibers include their efficient delivery of sufficient power to the sensor without difficult source / detector / sensor alignment problems. When used with highly polychromatic sources phase noise is not a problem because of the incoherent nature of the emission from such sources.

The primary types of physical processes upon which such modulators rely are:

- **radiation absorption:** the sensors rely upon the measurand to affect directly the attenuation of light in a medium by changing the absorption coefficient α (equation (1.13)) of the medium;
- **transmission or reflection:** the sensors change the transverse cross-section of the light propagating channel;
- **refractive index changes:** the sensors change the refractivity of a medium or modulator arrangement to induce amplitude modulation;
- **inherent light generation:** this involves the production or regeneration of light; and
- **optical path length changes which may produce interference.**

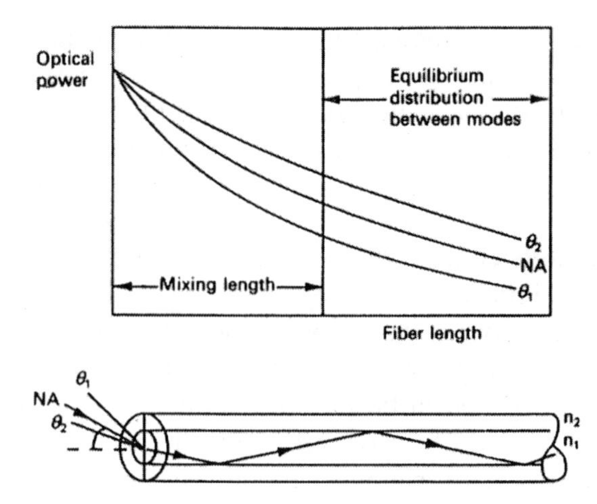

Figure 1.6. Mixing length for reaching equilibrium mode distribution. NA: numerical aperture of fiber. θ_2<NA< θ_1: launch angle of light

1.4.1 Extrinsic transmission-reflection modulation

1.4.1.1 Fundamental principles

These forms of modulators are used for measuring lateral or longitudinal displacement in a variety of ways. The generic form of such modulators is represented by the planar hutter moving transversely across the optical channel to reduce the channel crossection and hence the transmitted light intensity. The resulting intensity modulation of light captured by the receiving fiber is governed by the modulation factor (equation (1.11)) [1].

$$\delta M_1(\lambda) = \left[\frac{a}{d \tan \theta_m}\right]^2 \left[\frac{\delta A}{A}\right] \tag{1.17}$$

where a is the fiber core radius, θ_m the fiber numerical aperture, d the separation of the light delivering and receiving fibers and $\delta A/A$ the fractional area transmitting light into the receiving fiber [1].

$$\frac{\delta A}{A} = \frac{1}{\Pi}\left[\cos^{-1}\left(1-\frac{\delta x}{a}\right) - \left(1-\frac{\delta x}{a}\right)\sin\left(\cos^{-1}\left(1-\frac{\delta x}{a}\right)\right)\right] \tag{1.18}$$

with δx the shutter edge displacement.

Similar analyses may be applied to a modulator in which the light beam from the delivery fiber is collimated by a lens before being focussed onto the receiving fiber [4] and to two overlapping gratings one moving transversely with respect to the other within such a collimated beam.

A further variant of the latter is when the fringes on the moving and stationary gratings are inclined to each other to form Moiré fringes [4]. The opaque elements of such modulators may be replaced by transparent refracting elements (wedges or spherical lenses) to direct rather than block the light rays which would otherwise be captured by the receiving fiber [5].

Longitudinal displacement may be measured using a reflective form of such a modulator, the light intensity captured by the receiving fiber being determined by the displacement of a reflecting element [1][6]

$$M_1(\lambda) = \left(\frac{\delta x}{a}\right)\left(\frac{a}{2d\tan\theta_m}\right)\left(\frac{\delta A}{A}\right) \tag{1.19}$$

with d the displacement of the reflector from the fiber tip. A more precise mathematical description of such a modulator needs to take account of the radial intensity profiles of both incident and reflected beams [6].

More sophisticated forms of reflective sensors utilize multiple fibers with a number N_i illuminating the reflector and a number N_r receiving the reflected light (fig. 1.7). The behavior of such modulators may be calculated computationally using an extension of the two-fiber model which integrates the total overlap of profiles for all N_i illuminating and N_r receiving fibers. Various geometric distributions of illuminating and receiving fibers (e.g. random mix, semicircular, concentric annuli) give modulator response curves of similar form (and which are similar to the simple shutter response but which give different modulation range and resolution (fig. 1.7)).

1.4.1.2 Modulation performance

The modulation range of transmissive and reflective sensors is mainly governed by the displacement geometry. The range of the basic shutter modulators is determined by the cross-section of the optical fiber (equation (1.17)) which is typically up to 1mm maximum. The parallel beam and refractive shutter type modulators relax this constraint to some extent but at the expense of elongating the sensor element and so compromising the transducer size. Fiber bundle reflective sensors allow an additional control of range through a choice of distribution of illuminating and receiving fibers (fig. 1.7(b)) but the upper limit is typically 300 μm. The overlapping fringe modulator allows, in principle, an infinite range to be achieved at the expense of a loss of absolute position measurement under static conditions.

The fundamental limit to the resolution of fiber systems is governed by optoelectronic noise considerations which requires an optical power of typically 1 μW to be incident on the detector (section 1.2.2). This means that the resolution of a particular modulator type depends upon its optical efficiency.

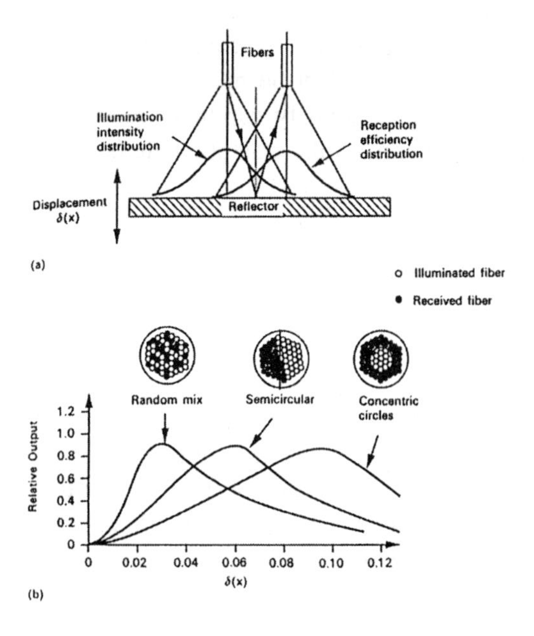

Figure 1.7. Moving reflector monitor: (a) principle; (b) multiple fiber forms and their characteristics

In the case of a two-fiber reflective sensor (fig. 1.7(a)) optical coupling efficiencies of the order of 5% are ideally possible with 200 μm fibers having a numerical aperture of 0.5 (e.g. [1]). With the 1 μW stipulation for received power this leads to a resolution of approximately 1 nm in 200 μm (1 in 2 x 10^4). Because of their greater optical efficiencies the multiple fiber modulators have the potential of higher resolutions.

For the parallel beam, planar shutter sensor an ideal resolution of 1 in 10^4 is possible, whilst the simple shutter without the lenses and with an interfiber gap of 1mm (200 μm core, 0.5 NA) has a reduced resolution of 1 in 10 on account of the reduced power collected by the receiving fiber due to numerical aperture effects. The resolution of the overlapping fringe modulator is of the same order as the other modulators but the resolution is with respect to the fringe widths rather than fiber core radius so that smaller displacement may, in principle, be resolved.

Although the reflective and transmissive modulators primarily detect displacement they may be used in combination with other primary transduction means for monitoring other parameters such as pressure (via membrane displacement) and electric current (via electromagnetic displacement) (e.g. [7]).

1.4.2 Quasi-intrinsic modulation

1.4.2.1 Fundamental principles

Intrinsic attenuation modulation is due to changes in the light intensity propagating through a fiber when the normal process of total internal reflection, which assists the transmission, is perturbed. In terms of geometric optics, the condition for modulation is

$$\theta_m < \sin^{-1}(n_2 / n_1) \tag{1.20}$$

This may be achieved by varying the refractive indices of the core or cladding (n_1, n_2 respectively), or the geometry of the fiber core so that the ray propagation angle (θ_m) is affected.

With the alternative wave propagation description of fiber transmission the number of modes propagated in the fiber core is bounded by the inequality (15). The permissible values of the propagation constant are limited be the values of $n_1$and n_2. The power associated with the lost core modes is transferred to energize modes in the cladding. The implications are that it is the higher order modes which are eliminated and that because of the different wavelength predominance in the various modes there is an accompanying change in the spectral distribution of the overall signal. Core boundary conditions also affect the wave equation solution (equation (1.14)).

Such perturbations of the optical fiber lead to optical power reductions with length along the fiber as shown in fig. 1.6 on account of changes in the numerical aperture, leading to a different equilibrium mode distribution. The intensity reduction depends not only upon the magnitude of the refractive index change but also upon the length of fiber over which it is maintained. The effect is incorporated in equation (1.6) via the factor F(λ) which has a different dependence upon length along the fiber for each propagation mode, *l, m*.

1.4.2.1.1 Geometric effects

Sensors which rely upon losses caused by fiber bending (section 1.3.2) are available for measuring displacement, pressure and force [4]. Such bend-induced losses are enhanced in the microbend modulator. This consists of stressing the fiber at regular intervals along its length (fig. 1.8) so that the stress has a spatial periodicity which enhances the coupling of the propagation modes in the fiber core to those in the cladding. For enhancing such coupling the periodicity Ω needs to satisfy (e.g. [1])

$$\Omega = 2\pi\,(\beta_{core} - \beta_{cladding}) \tag{1.21}$$

For adjacent modes m, m+1

$$\Omega = \left(\frac{S_c}{S_c + 2}\right)^{1/2} \left(\frac{2\sqrt{n_2 + n_1}}{a}\right) \left|\frac{m}{M}\right|^{(2-S_c)/(2+S_c)} \tag{1.22}$$

where S_c is the exponent defining the refractive index profile in the core, a the core radius, M the total number of modes (equation (1.16)) and m the mode designation.

The periodicity required according to equation (1.22) is typically of the order of a millimeter.

The attenuation (equation (1.13)) suffered by the light signal as a result of microbending varies with fiber properties according to

$$\alpha_{\mu} = K\left(\frac{a}{a_c}\right)^2 (NA)^{-4} \tag{1.23}$$

where a, a_c are the core and cladding radii, NA is the numerical aperture, and K a constant of proportionality.

1.4.2.1.2 Core refractive index effects

The refractive index of a fiber core may also be affected by induced birefringence effects from the compressions and rarefactions of acoustical waves propagating through the fiber. As a result the optical propagation modes are perturbed leading to coherent optical signals in the fiber interfering with each other and producing a ‘speckle pattern’ at the fiber output. This forms the basis of the homodyne interferometer which has been successfully used for monitoring acoustic waves (e.g. [3]). Such an ‘interferometer’ has the advantage of not requiring a distinctly separate reference arm but suffers from the disadvantage of less specific control over the referencing. Clearly multimode propagation is necessary for homodyne interference to be manifest.

1.4.2.1.3 Cladding refractive index effects

Changes in the refractive index of the cladding may be induced in a number of ways, each corresponding to the potential for sensing a different parameter. The cladding may be made from electro-optic, magneto-optic or photoelastic material (e.g. [7]) to form electric field, magnetic field or pressure sensitive modulators. Liquid level detectors (e.g.[4]) have been demonstrated whereby the presence of the liquid adjacent to the fiber causes the necessary refractive index change to reduce the signal intensity in the fiber core. Sensors for temperature alarm monitoring are available which rely on the different temperature dependence of the core and cladding refractive indices [1].

1.4.2.1.4 Evanescent wave effects

Refractive index changes can also be induced simply by the microdisplacement of two fibers so that a thin layer of the surrounding medium with its different refractive index is introduced. The phenomenon of total internal reflection involves the penetration of the electromagnetic wave into the adjoining medium to a depth dp (evanescent field) (e.g. [7]) which is of the order of the optical wavelength. Thus displacement of the two fibers relative to each other (fig. 1.8) over distances of this order changes the condition for internal reflection by perturbing the evanescent field associated with the wave and so modulates the light flux in the fiber.

Several forms of modulators based upon such evanescent wave coupling have been proposed (e.g. [4]) but quantification of the effect is most conveniently demonstrated with the split fiber sensor shown in fig. 1.8. The transmission coefficient (synonymous with $M_1(\lambda)$ in equation (1.19)) of two fibers cleaved at an angle θ to their axes and separated by a microgap d is given by [8]

$$M_1(\lambda) = 1 - \left| \left(z^2 + \delta^2\right) \left[\left(z^2 - \delta^2\right) + 4\delta^2 z^2 \cot\left(\frac{h^2\beta}{2}\right) \right]^{-1} \right| \qquad (1.24)$$

where $\beta = \left(\frac{4\pi d}{\lambda}\right)\left(n_1^2 \sin^2\theta - 1\right)^{1/2}$

$$z_{\perp} = (n_1 \cos\theta)^{-1} \qquad z_{\parallel} = \cos n_1 \theta$$

$$\delta_{\perp} = (n_1^2 \sin^2\theta - 1)^{1/2} \qquad \delta_{\parallel} = (n_2^3 \sin^2\theta - 1)^{1/2}$$

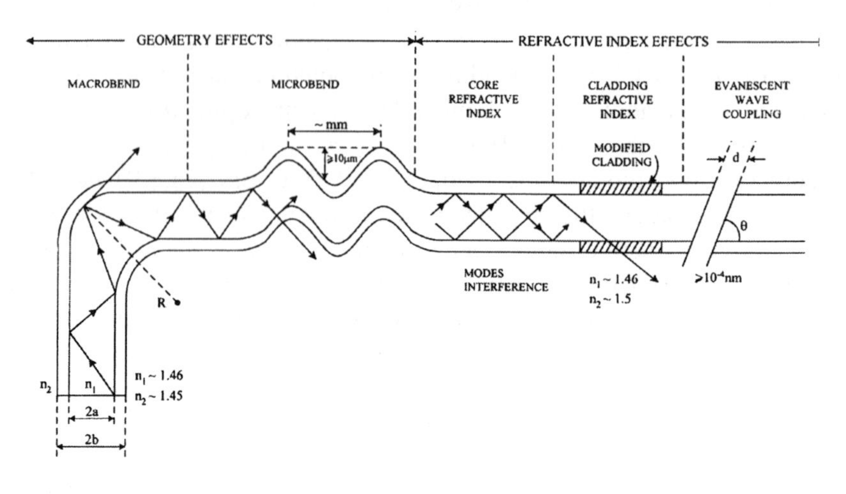

Figure 1.8. Modulation mechanisms in intrinsic fiber sensors based upon multimode fibers (a) macrobending (b) microbending (c) core refractive index changes (d) cladding refractive index changes (e) evanescent wave coupling

An implication of equation (1.24) is that the transmission is dependent upon the direction of polarization of the wave through the subscripts $\perp$ and $\parallel$ perpendicular and parallel to the plane of the diagram and also upon the optical wavelength via λ, n_1 and n_2.

Evanescent wave effects may also be utilized in conjunction with a modified fiber cladding whose spectral absorption or fluorescence properties are made dependent upon the amount of chemical change occurring in the surrounding medium. Such sensing is discussed further in section 1.4.3.

1.4.2.2 Modulation performance

For microbend modulators, displacements of the order of micrometers are sufficient to cause a few percent change in the optical attenuation.

The range can be tailored by extending the length of the fiber over which microbending is applied. The fiber may be repeatedly folded for minimizing the area of the bend inducing mechanism (serrated plate, helically wound wire) and so produce a small (few square centimeters) sensing element. Alternatively, an extended distributed sensor element may be configured using a wire helically wound around the fiber. Typically a modulation depth of at least 60% is achievable [9] without incurring serious penalties from fiber fatigue effects.

In the case of microbend sensors resolution estimates based upon electronic noise limitations suggest that displacements of the order of 10 μm should be detectable. For force sensing, changes in attenuation coefficient ($\Delta\alpha_{\mu}$ equation (1.23)) of 0.2 N^{-1} have been reported for a 2 m long fiber, leading to a pressure sensitivity of 1.3 x 10^{-4} Pa [7].

Homodyne interferometers are not easily quantified in general terms. However their performance appears to depend upon the type of fiber and cladding used, the length of fiber involved in the sensing and the spatial filtering employed for speckle pattern discrimination. Typically acoustic amplitudes of magnitude 10^{-12} m appear to be detectable [3] with acoustical frequencies up to approximately 20 kHz.

The modulation range of the cladding refractive index modulators is governed by the magnitude of the refractive index change and length of the optical fiber along which the changed refractivity extends. In the case of electro-optic, magneto-optic and photoelastic effects the refractive index changes are small and so extended lengths of fiber are needed to produce measurable effects. However, where a change of medium is involved (e.g. liquid level detectors) significant refractive index changes are possible ($\Delta n > 0.1$), only one or two internal reflections being required to provide a significant effect. Consequently separate modulation elements (e.g. prisms) attached to the fiber tips are feasible for, for instance, liquid level detection.

The resolution of cladding sensors is difficult to establish universally. A major difficulty with such sensors for liquid level monitoring is that liquid drops easily remain attached to the fiber after the liquid level has dropped, leading to erroneous level indications because of the highly localized nature of the sensing action.

With evanescent wave modulators the range of displacement which can be used is of the same order as the wavelength of the light wave, leading to a displacement range of typically 5 μm, displacement resolutions of typically 5×10^{-4} nm have been reported [8].

The pressure resolutions achievable with both the microbend and evanescent wave sensors are similar to those achievable with interferometric sensors which make them suitable for hydrophone applications.

The refractive index difference between core and cladding, which is necessary for microbend induced effects, can be minimized with respect to temperature variations. This is a distinguishing advantage over interferometric methods for which a 1°C temperature change can cause a 10^6 times greater change than a pressure of 1 Pa [6]. However, this potential advantage is offset by frequency range limitations (kHz) and concern about aging effects induced by microbend strains in the fiber.

It should be noted that all the intrinsic effects described above can conversely act in a deleterious manner to produce unacceptable fiber sensor performance unless they are carefully and properly controlled.

1.4.3 Wavelength dependent modulation

Equation (1.6) shows that the output signal from a fiber sensing system is, in general, dependent upon wavelength distribution as well as intensity. Modulation may therefore be induced by taking a broadband source and arranging for the measurand to affect a restricted band of wavelengths before detection. In this manner the spectral emission of the source is intensity modulated by a wavelength-dependent function.*

1.4.3.1 Fundamental principles

Fundamental physical processes which can be utilized for wavelength dependent modulation are wavelength dependent absorption, luminescence, dispersion, interference and scattering.

1.4.3.1.1 Absorption modulation

Absorptive modulation is based upon a multicomponent form of the Beer-Lambert law

$$M_1(\lambda) = \exp\left(-\sum_h \alpha_h(\lambda) C_h l_h\right) \tag{1.25}$$

where $\alpha_h(\lambda)$ is the extinction coefficient, l_h the path length, C_h the molar concentration of the absorbing species and $M_1(\lambda)$ is the modulation factor of equation (1.6). The wavelength selective nature of the modulation relies upon the wavelength dependence of the extinction coefficient. The subscript h distinguishes the number of absorbing components in the system.

The modulation may be induced by changes in either the path length, concentration or extinction coefficient. The latter may for instance be affected by temperature (fig. 1.9 (a)) or nuclear radiation.

The optical length dependence l_h may be used for monitoring volume changes as in medical pulsimetry or displacements via refractive shutters (section 1.4.1) having appropriate extinction coefficients [5]. The

* It should be noted that the term 'wavelength modulation' which has been used hitherto in the literature is inaccurate because it is not possible to change the wavelength associated with a photon in an optical fiber by any significant amount.

concentration dependence C_h is utilized for instance for measuring blood oxygen saturation in medicine (fig. 1.9 (b)).

Fernando et al [10] have described the use of such a spectral absorption technique for monitoring the curing of an epoxy/amine resin system and the ingress of moisture into such a system; Smith et al [11] have produced a system for detecting the thickness of a Kerosene film several microns thick on water using similar principles.*

This class of sensors also includes those in which the fiber cladding has been modified to a matrix supporting chemicals which modify the spectral signature of light coupled into the cladding from the fiber core before returning to the fiber core for onward transmission and detection. Small changes in concentration in the associated chemistry (e.g. at ppm levels) are detectable. Major difficulties with such approaches relate to trapping the optically reactive chemical within a support matrix which can form the base of the fiber cladding without prejudicing the sensitivity of the chemical indicator and yet provide sufficiently rapid time response. By way of examples Solvachromatic dyes for detecting small quantities of water in oil have been reported by Russell et al [12]; Khavaz and Jones [13] have reported the use of a cobalt chloride based colorimetric reagent immobilized on the core of a multimode fiber with a gelatine film for humidity sensing.

1.4.3.1.2 Luminescence-fluorescence modulation

Luminescent and fluorescent modulation involve the emission of light by the modulation element itself. The emission may be inherent (as in the case of monitoring a luminous event such as an electrical discharge [14]) or induced via, for instance, photoluminescence (which involves stimulation with light of a different wavelength [4] (fig. 1.9).

In the former case the modulation occurs via the source term $P(\lambda)$ and with $M_1(\lambda)=1$ in equation (1.6).

For the case of photoluminescence $M_1(\lambda)=1$ also whilst

$$P(\lambda)=\frac{\eta Nhc}{\lambda_s} \tag{1.26a}$$

where

$$N=\sum_{1.m}\int_{\Delta\lambda_p}\left(\int P_p(\lambda)dl\right)d\lambda \tag{1.26b}$$

Luminescence occurring at a different wavelength λ_{LT_1} to the excitation wavelength depends upon temperature shifting λ_{LT_2} for higher temperatures. The luminescence decay time also depends on temperature. h is Plancks' constant, c the velocity of light in free space, η the efficiency of luminescence, $\Delta\lambda_p$ the range of exciting wavelengths produced by the source of power $P_p(\lambda)$ and λ_s is the wavelength of the light produced by luminescence. The detector response $R(\lambda)$ is sensitive to λ_s only so that equation (1.6) simplifies to equation (1.8) and the same fiber may be used for pumping and delivering the signal. Since λ_s varies with temperature, the method may be used for temperature monitoring (fig. 1.10).

This class of sensor also includes sensors based on fluorescing materials incorporated into a supporting matrix replacing the normal fiber cladding (section 1.4.2). As an example MacCraith et al [15] have described an oxygen sensor based upon a sol - gel cladding having a response time of a few seconds.

1.4.3.1.3 Dispersion based Modulation

Modulation based upon chromatic dispersion utilizes wavelength dispersing elements (prisms, diffraction gratings, achromatic lenses and zone plates* [16]) to produce spatial distribution of wavelengths for position encoding (fig. 1.11). Thus linear or angular displacement may be determined by calibration using equation (1.6) with

$$M_1(\lambda)\begin{cases} =1, \lambda=\lambda_S \\ =0, \lambda \lessgtr \lambda_S \end{cases} \tag{1.27a}$$

and

$$\lambda_s = f(x) \tag{1.27b}$$

1.4.3.1.4 Interference based Modulation

Interference modulation may be produced with polychromatic light as well as the more conventional monochromatic light. The modulation factor in equation (1.28) takes the general form

* The zone plate may be regarded as a spherical concave mirror or lens with a wavelength-dependent focal length.

$$M_1(\lambda) = P_A(\lambda) + P_B(\lambda) - 2\sqrt{P_A(\lambda)P_B(\lambda)} \quad \cos^2\delta \tag{1.28}$$

where P_A, P_B are the optical powers of the interfering beams and δ is the phase difference between the beams. Thus, as in the monochromatic case, the power varies as cosine squared with the maximum modulation sensitivity on the steepest part of the curve and lowest sensitivity as $\delta \rightarrow \pi/2$.

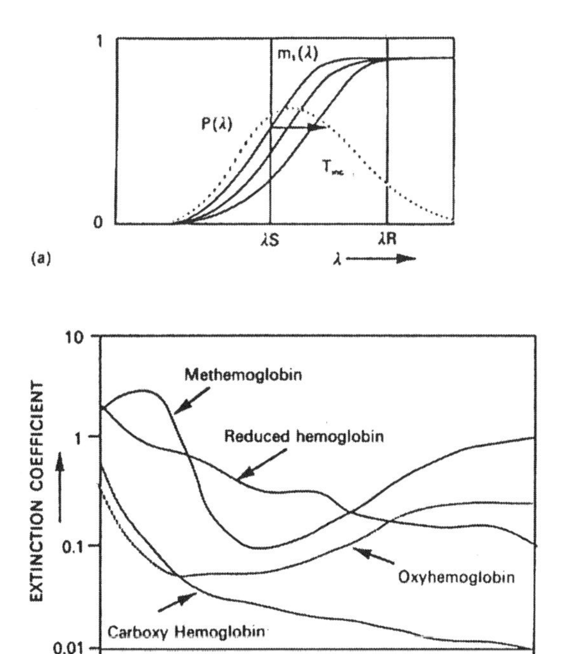

Figure 1.9. Wavelength selective absorption: (a) temperature dependence of spectral transmission, (b) spectral transmission of hemoglobin and various derivatives

In the polychromatic case, the interference modulation is effectively wavelength integrated (fig. 1.12) since $M_1(\lambda)$ is embedded within the integral of equation 1.11. An implication of this factor is that there will always be for a continuous spectral source one wavelength range operating close to the maximum sensitivity part of the $\cos^2\delta$ function. Thus with an appropriate detection and signal processing methodology a high sensitivity may be sustainable over a greater range than with a monochromatic system and fading may be largely averted.

A number of polychromatic interference cases may be identified leading to different physical realizations of the phase difference δ. These include

- interference cavities for measuring thickness [17], strain [10], temperature [18], [19], [20], pressure [21] for which

$$\delta = \Pi n \,{}^{\delta x}\!/_{\lambda} \tag{1.29}$$

n = refractive index of cavity filling, δx = path difference between the two interfering beams, λ = wavelength. Such multiwavelength based interference sensors are discussed further in section 1.5.2.

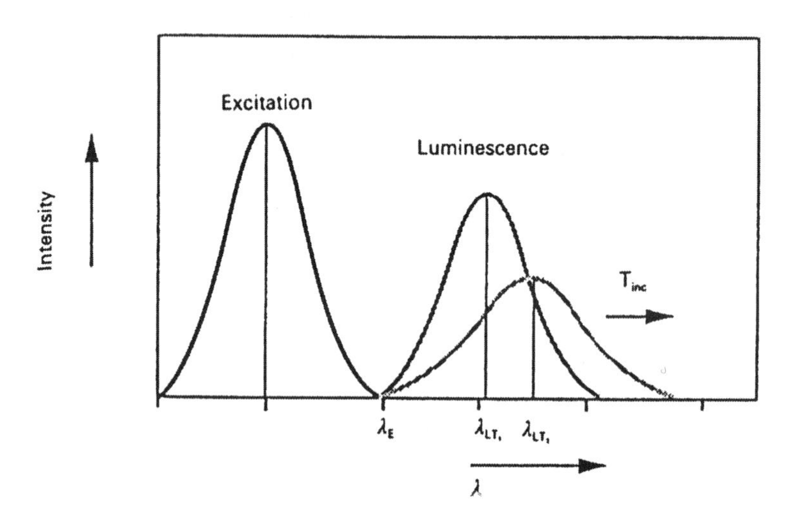

Figure 1.10. Temperature dependence of luminescence

- Optically active materials such as birefringent materials wherein the refractive index depends upon the orientation of the wave polarization and is defined in tensor form. Thus waves polarized along these different orientations have different propagation constants and phase velocities and consequently develop phase differences. On recombining, the waves interfere to give a modulation factor

$$\delta = \left\lfloor \Pi (n_o - n_c)\, {}^{\ell}\!/_{\lambda} \right\rfloor \tag{1.30}$$

where n_o, n_c are the refractive indices for the differently polarized waves, ℓ is the length of the birefringent material, λ the light wavelength. In photoelastic materials $n_o - n_c$ is sensitive to stress

$$n_o - n_c = {}^{1}\!/_{4}\left(\eta_0 p_{11} s\right) \tag{1.31}$$

(p_{11} = photoelastic coefficient, s = strain)
leading to the manifestation of an optical fiber stress sensor, polychromatically addressed (e.g. [22], [23]). Similarly for electro optic materials, $(n_o - n_c)$ is sensitive to the electric field to which the material is subjected

$$n_o - n_c = n_0^3 r_{ij} V \tag{1.32}$$

(r_{ij} = electrooptic coefficient; V= voltage)
this forms the basis of an optical fiber voltage sensing element

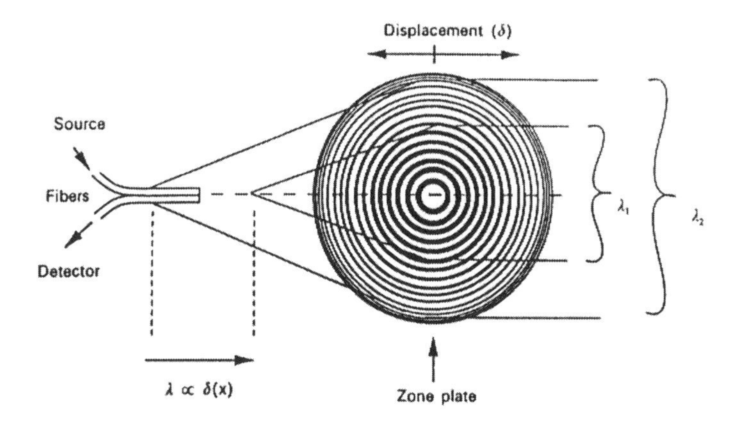

Figure 1.11. Zone plate used as a chromatic modulator

– magneto optic materials in which the Faraday effect occurs are also governed by the $\cos^2 \delta$ type variation. In this case the effect arises because of the rotation of the plane of polarization of the lightwave in proportion to the magnetic field (B) applied to the material i.e.

$$\delta = V(\lambda)\ell B \tag{1.33}$$

$V(\lambda)$ = Verdet constant which is inherently dependent upon the optical wavelength (λ), ℓ is the optical path aligned with B. It should be noted that optical polarizers used with such materials have a wavelength dependent extinction coefficient.

In all cases the wavelength dependencies of δ opens the possibility of polychromatic fiber sensing. This has been realized for thin film [17], pressure, temperature [18], voltage [24] [25] [26] [27] and current sensors.

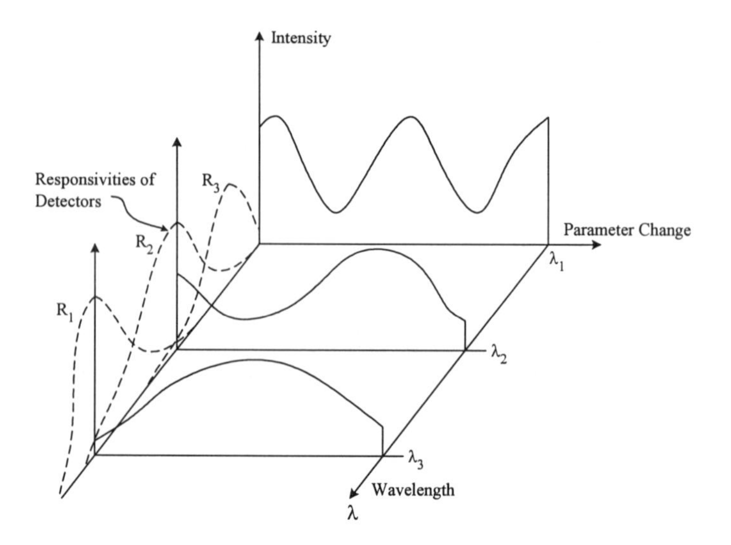

Figure 1.12. Interferometric ($\cos^2\delta$) variation of signals at different wavelengths with changes in value of measurand. Also shown are typical wavelength responsibilities (R_1, R_2, R_3) of photo detectors which might be used to monitor each wavelength

1.4.3.1.5 Scattering based modulation

Light scattering is produced when a light beam is incident upon a conglomeration of particles, the nature and amount of scattering being dependent upon the optical wavelength, the direction of observation, the size and concentration of the particles. For particles which are smaller than optical wavelengths ($<10^{-1} R_p$, where R_p is the particle radius shown if fig. 1.13) the scattering varies as λ^{-4} (Rayleigh scattering).

$$M_1(\lambda) = \frac{q_1}{\lambda^4} \tag{1.34a}$$

For particles having dimensions similar to the light wavelength the extinction coefficient of the forward scattered light (Mie scattering) varies with the parameter $(2\pi R_p / \lambda)$ as shown in fig. 1.13, i.e.

$$M_1(\lambda) = 1 - C\left[1 - \mathbf{f}\left(\frac{2\pi R_p}{\lambda}\right)\right] \tag{1.34b}$$

where C is the concentration of the scattering particles. For particles moving with a velocity v, there is also a shift in the wavelength λ_o of the scattered light due to the Doppler effect [3]. For forward scattered light

$$M_1(\lambda) = \lambda_o(1 - (v/c)) \tag{1.35}$$

where c is the velocity of light in free space. Since in practice $v/c \approx 10^{-6}$, the wavelength shift is small. This is a case of rigorously true wavelength modulation (*footnote, section 1.4.3).

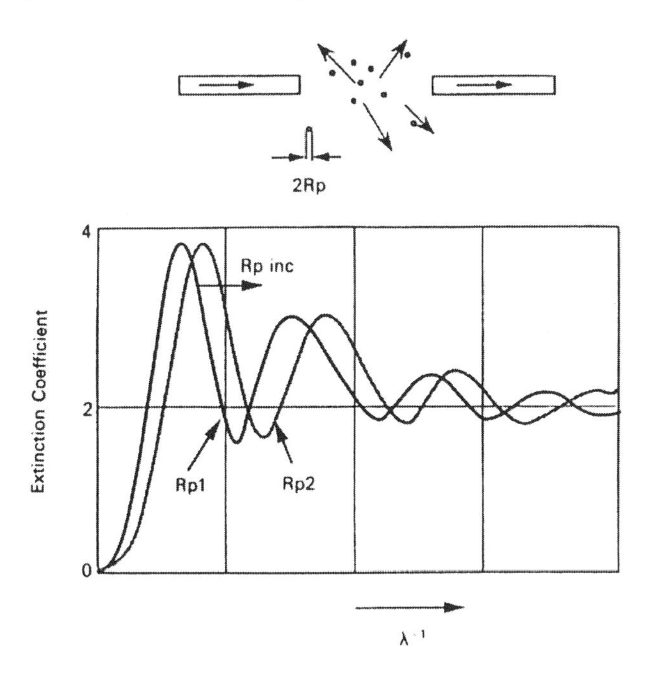

Figure 1.13. Extinction coefficient for Mie scattering as a function of wavelength and with particle radius, R_p, as parameter

Scattering-based modulation may therefore be induced by changes in particle size, concentration, velocity or orientation of the scattered light. Such methods have been used for monitoring particles in power switchgear [28], oil drops in water (e.g. [4]) and for temperature monitoring using the temperature dependent scattering from Christiansen filters. Fiber-based laser Doppler velocimeters are also available.

1.4.3.2 Modulation performance

The performance of wavelength dependent sensors depends upon the nature of the spectral changes produced by the measurands. Often these changes are not simple (e.g. a distinctive change in the wavelength of a single, isolated spectral line) but may be complicated variations in a wideband spectral pattern. Such complex changes may be regarded as combinations of three fundamental changes in a spectral distribution (figure 1.14):

- shift in mean wavelength [figure 1.14(a)]
- change in wavelength bandwidth [figure 1.14(b)]
- change in signal intensity [figure 1.14(c)]

(A change in signal intensity alone of course degenerates into intensity modulation already discussed in section 1.4.1).

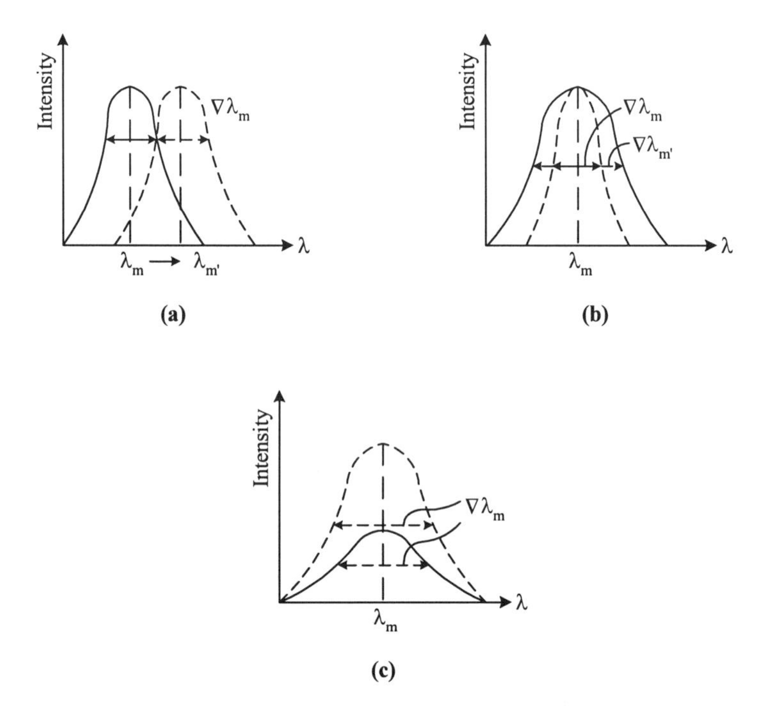

Figure 1.14. Three fundamental spectral distribution changes (a) change in mean wavelength only (b) change in wavelength bandwidth only (c) change in intensity only

Other spectral changes may be regarded as combinations of these three basic distribution changes, by way of examples

- simple wavelength shifting (figure 1.14(a)) is typical of sensors based upon photo luminescence, Christiansen filters and interference filters (governed by wavelength varying with inclination angle θ)

$$\lambda_\theta = \lambda_o \left(n^2 - \sin\theta\right)^{1/2} n^{-1} \rightarrow M_1(\lambda) \tag{1.36}$$

Sensors based upon dispersion modulators, thermochromic materials (spectral transmission varying with temperature) and a rotating polarizer - stationary analyzer combination (which only polarizes at wavelengths below the infrared part of the addressing spectrum (e.g. Faraday current sensors [24]) are typically governed by a combination of wavelength shift [figure 1.14(a)] and bandwidth change [figure 1.14(b)].
- Hemoglobin saturation sensors (oximeters) [29], [30], cobalt chloride type temperature sensors [31] are based upon a combination of wavelength shift [figure 1.14(a)] and relative intensity changes [figure 1.14(c)].
- Electrical plasma emission monitoring involves complicated superpositions of wavelength shift [figure 1.14(a)], wavelength bandwidth [figure 1.14(b)] and intensity changes [figure 1.14(c)] [32].

The significance of such considerations is that *a priori* knowledge of the type of spectral signature change produced by a given sensor type enables the optimum signal detection and processing means to be determined, so providing optimum performance and system efficiency. (The three fundamental changes shown on figures 1.14 (a), (b), (c) correspond to the Hue, Saturation, Lightness (H,S,L) chromatic attributes discussed further in section 1.5.1.4)

The measurement range of modulators based upon mean wavelength shifts (fig. 1.14 (a)) is governed by the wavelength range over which spectral changes are produced and in general modulators tailored to particular applications can be selected. In practice, the fundamental range limitation is governed more by the wavelength dependent attenuation of the fiber ($F(\lambda)$ in equation (1.6)) and the spectral distribution of the source $(P(\lambda))$. Typically a wavelength range of 450 nm to 1.6 μm is available with conventional fibers. This may be extended through the use of special fibers but is reduced to about 1 μm if Si-based detectors are used. Broadband light sources are available for taking advantage of the full wavelength range over which the sensor modulation occurs; the choice is typically between

- tungsten halogen lamp (450 nm - 1.6 μm)
- white LED (450 - 600 nm)c
- conventional LED ($\Delta\lambda \sim$ 10 - 50 nm, λ variable)

Detection of spectrally modulated signals may be achieved at various level of detail ranging from the use of spectroscopic detection at one extreme to the ratio of intensities at two different wavelengths at the other, whilst chromatic detection techniques offers major advantages for a range of applications. The spectral approach compromises signal sensitivity and cost against the acquisition of unnecessary detail; the two wavelength approach sacrifices spectral information for over simplicity; the chromatic approach constitutes a versatile, optically efficient method for monitoring spectral signature changes as small as 0.04 nm cost effectively without capturing unnecessary data [33]. The extraction of meaningful sensor output information from spectral signature detectors is strongly coupled to signal processing so that these techniques are better discussed in section 1.5 below.

By suitable choice of modulating elements the range and resolution of the primary measurand can be adjusted appropriately. For instance, the temperature range covered by monitoring the shift of the spectral absorption edge of ruby glass can in principle be extended to several hundred degrees Celsius. Displacement monitoring with a zone plate [16] is feasible over typically a range of a few centimeters. Conversely through the choice of other modulating elements a temperature resolution of 10^{-3}°C is possible; displacement resolutions with a zone plate of 1 μm and less are conveniently achievable.

1.5 SIGNAL PROCESSING AND SYSTEM ARCHITECTURE

Although intensity is the only variable available for modulation with multimode fibers the above discussions indicate that it is not well conserved in an optical fiber. For instance, the attenuation per unit length can be variable, the introduction of splices, extra connectors etc. changes the intensity transmitted by a fiber dramatically. This means that the intensity change produced by a modulator is multiplied by an unknown factor along the fiber length. In addition, to the random multiplicative noise or fading, there is also additive noise which arises in active and passive electrical components of the system (section 1.2.2).

The bulk of the publications and patents relating to intensity modulation sensors concern techniques for overcoming this one fundamental problem. The general approach is to obtain more information about the condition of the system and the modulator by taking more measurements. These additional measurements may either be made using a separate reference channel, using another part of the optical spectrum (thereby increasing the information content of the signal by multiplexing signals at two or more different wavelengths onto the same fiber), or by using the time domain to send signal and reference information alternately (thereby increasing the bandwidth of the information transmitted). These constitute the range of options open to the designer to improve the performance of analog sensors.

The trend for improving performance of electronic systems has been towards digital methods. Similarly a digital solution can offer advantages with optical fiber systems and there are examples of digitally encoded rotary and linear displacement sensors with 11- or 12-bit resolution. However, it is not always possible to devise such an elegant optical analog-to-digital converter for all measurands, and other approaches need to be taken. One such approach involves the so-called frequency-out transducers which use digital pulse detection techniques, but actually rely upon the use of time as an analog variable, The sensor head produces a pulse train whose frequency is measured in order to determine the value of the measurand.

Another approach is to utilize a hybrid system/whereby digital electronic sensing is combined with digital optical transmission, energization being achieved optically via the power advantages of multimode fibers.

There are situation where a purely digital sensor is not acceptable (e.g. power switchgear monitoring in which the transducer supply may be temporarily disabled leading to problems of data identification with digital systems). In such cases a combination of digital and analogue approaches is advantageous (e.g. linear travel recorder reported by Isaac et al ([34])).

These represent the different system architectures and signal processing options which are available.

1.5.1 Analog techniques

1.5.1.1 Intensity referencing

The simple expediency of using as a reference channel a separate, parallel fiber system which bypasses the modulating element is fraught with difficulties which are highlighted by equation (1.6). For instance the components of the reference channel $P(\lambda)$, $R(\lambda)$, $F(\lambda)$, $M_2(\lambda)$ tend to behave differently from the corresponding components of the signal channel and, in practice, such effects can be pronounced.

An elegant approach to referencing which overcomes some but not all of the above difficulties has been developed for incorporation into the reflection modulator employing fiber bundles (fig. 1.7). The method utilizes a composite fiber bundle consisting of an inner group of randomly mixed illuminating and receiving fibers surrounded by an outer annulus of receiving fibers only (fig. 1.15).

The modulation characteristic for each set of receiving fibers differs, as shown in fig. 1.15. Referencing against nonmodulated intensity changes, sensor element aging etc. is possible by forming the ratio of the two outputs.

A rigorous examination of this referencing approach using equation (1.6) to define the output from each measurement channel leads to

$$\frac{V_s}{V_R} = \frac{q_s}{q_R}\left[\frac{\sum_{l,m}\int_{\lambda}\left(\int_{l} P(\lambda)F(\lambda)M_{2s}(\lambda)dl\right)M_{1s}(\lambda)R_s(\lambda)d\lambda}{\sum_{l,m}\int_{\lambda}\left(\int_{l} P(\lambda)F(\lambda)M_{2s}(\lambda)dl\right)M_{1R}(\lambda)R_R(\lambda)d\lambda}\right]^p \tag{1.37}$$

where V_s, V_R are the voltage outputs of the signal and reference channels respectively.

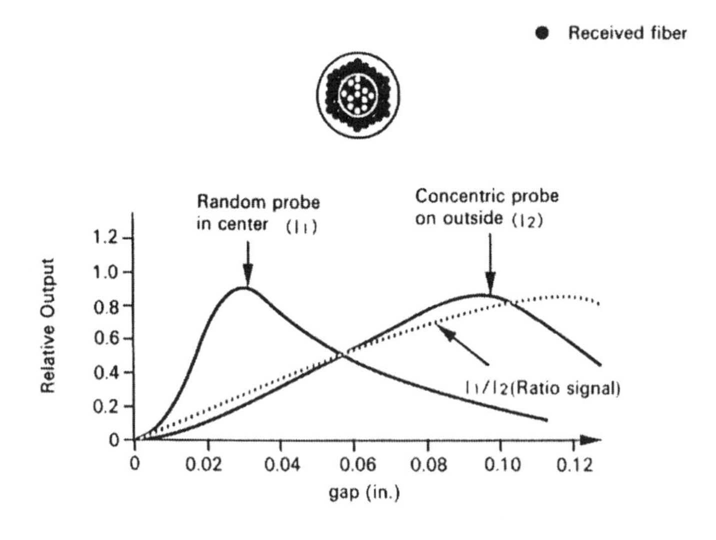

Figure 1.15. Self-referencing reflective modulator

Equation (1.37) shows that the referencing is rigorous if the amplitude changes due to modulator, source or fiber-bending effects are identical for both received signals. The referencing would, however, be susceptible to error if different amplitude changes occurred on each channel which could be the case for instance for the fracture of fiber elements forming the bundle.

Equation (1.37) also leads to the less obvious observation that referencing errors can arise from changes in the wavelength dependence of any of the previously defined system factors $P(\lambda)$, $F(\lambda)$, $M_1(\lambda)$, $M_2(\lambda)$, $R_s(\lambda)$,

$R_R(\lambda)$. Such errors would, however, be reduced the more monochromatic the source.

Thus, although such an approach is capable of providing reasonable referencing the method is restricted to displacement monitoring alone and is not easily applied for monitoring other parameters except via a secondary transduction process.

1.5.1.2 Two wavelength monitoring

Two wavelength monitoring involves detection at two separate wavelengths λ_s, λ_R using separate detectors and suitable narrowband optical filters. The trend has been to make one wavelength λ_R independent of the modulation whilst the other wavelength suffers both modulation and system effects. The ratio of the signals at the two separate wavelengths is used for referencing against intensity variations in a similar manner to the method used with the fiber bundle displacement sensor (section 1.4.1(a)). However, the output signal in this case is given by

$$\frac{V_s}{V_R} = \frac{q_s}{q_R}\left[\frac{\sum_{l,m} P(\lambda_s)F_a(\lambda_s)M_{2a}(\lambda_s)M_1(\lambda_s)R_a(\lambda_s)}{\sum_{l,m} P(\lambda_R)F_b(\lambda_R)M_{2b}(\lambda_R)M_1(\lambda_R)R_b(\lambda_R)}\right]^p \tag{1.38}$$

where the subscripts S, R refer to signal and reference and the subscripts a, b distinguish system components which are different for the reference and signal channels. Equation (1.38) indicates that error-free referencing is feasible against nonmodulated intensity effects provided these effects are identical for both signal (λ_s) and reference (λ_R) wavelengths. In practice such an assurance is difficult to obtain because of the wavelength dependence of each of the system factors in equation (1.38). For instance source drifts, fiber-bending effects, detector responses and propagation mode changes are all susceptible to produce spectral variations to accompany intensity changes. Nonetheless some degree of successful referencing is feasible and has been realized in practice.

1.5.1.3 Spectrometric techniques

The most detailed spectral information is obtained using spectrometric methods which rely on the dispersion of the optical signal using prisms or diffraction gratings. The resulting spectrum is focused onto an array of photodetectors (fig. 1.16) which could take the form of a charge-coupled device for facilitating subsequent signal processing. Although such a system provides most detailed spectral information it is susceptible to differential aging and responses of the array elements. Furthermore, such systems are cumbersome for optical fiber instrumentation design and are overgenerous in the amount of data acquired at the expense of optical efficiency. For instance, assuming the fundamental optical power requirement of 1 μW (section 1.2.2) by a single detector of alternative detection systems delivered for processing by a 1mm diameter optical fiber, this needs to be dispersed onto the detector array which is typically 2cm in length and contains 10^3 pixels of dimension 20 μm. Assuming only a linear dilution of optical power to fill the whole array, the implication is that each pixel receives only 10^{-6} μW of optical power whilst 10^3 data points are provided which require subsequent processing.

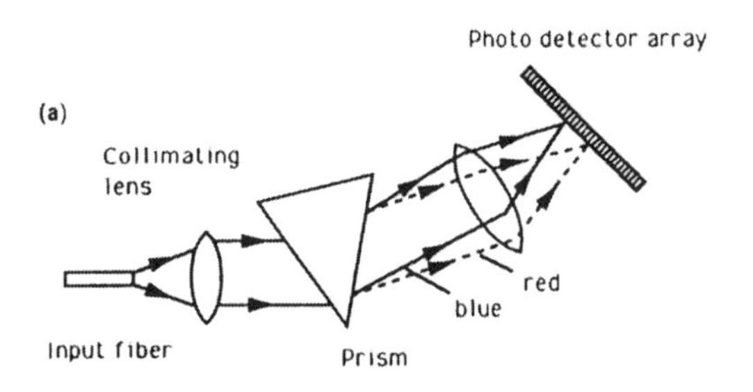

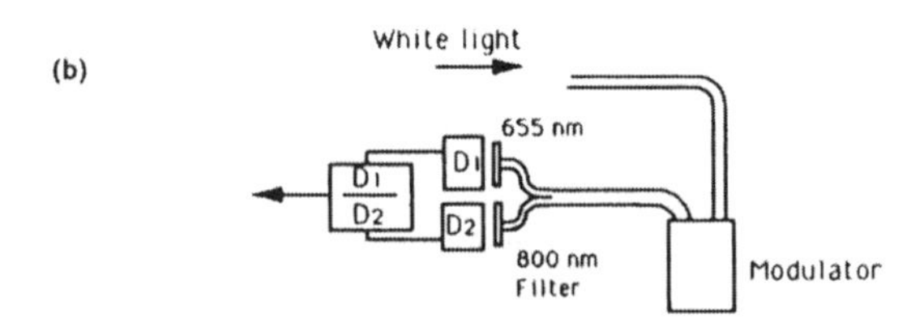

Figure 1.16. Detection system for wavelength dependent modulation: (a) spectrometric detection; (b) two-detector system

The information obtained from a single pixel at the limit of wavelength resolution is from equation (1.6)

$$V = q\left[\sum_{l,m} P(\lambda_1)F(\lambda_1)M_1(\lambda_1)M_2(\lambda_1)R(\lambda_1)\right]^p \tag{1.39}$$

For N pixels, N such data points need to be stored and handled to provide a single output representative of the degree of modulation of the sensor. The implication is that such an approach leads to poor system optimization. Little of the captured information is ultimately utilized to provide a measurand value whilst extra cost is involved in processing unnecessary data.

1.5.1.4 Chromatic monitoring

The complexity and weaknesses (including economics, signal power limitations etc.) of full spectral monitoring and the oversimplicity of the two wavelength approaches may be overcome using chromatic sensing and processing techniques (e.g. [33]).

Primitive chromatic monitoring methods have been described by Jones and Russell [33] and have since been extended for the processing of a wide range of optical sensor signals for electric plasma monitoring (e.g. [35], [36]) and more recently for other applications.

The method has been shown to be a particular form of wavelet transformation based upon a truncated Gabor expansion [37] to provide a non orthogonal means of sensor fusion. As such it has been employed for categorizing optoacoustic signals [38], [39], and two dimensional multi-wavelength interferometric monitoring (e.g. [40]). More recently the technique has been used to reduce the dimensionality and compress the information emerging from a multitude of different sensors measuring a variety of parameters.

Generically, chromatic monitoring may be regarded as the realization of a particular type of wavelet transform (e.g. [41]) combined with one of a number of sensor fusion algorithms for information extraction. With specific regard to spectral information characterized by optical wavelength (λ), a signal $P(\lambda)$ is detected with n detectors each of responsivity $R_n(\lambda, a, b)$ (a, b are parameters which characterize the detector response). The output from each x of the detectors is given by sets of equations of the form (1.6). For detectors for which R_x (λ, a, b) are of a Gaussian form (fig. 1.17)

$$\mathrm{R}(\lambda, a, b) = \mathrm{a}\exp\left[-\left(\frac{\lambda-\lambda_o}{b}\right)^2\right] \tag{1.40}$$

$a =$ amplitude, $b =$ half power width, $\lambda_o =$ wavelength of peak. Equation 1.6 becomes approximately a Gabor type transform [42]. Typical detector responses for a tristimulus ($n = 3$) system are shown on figure 1.18.

The use of n detectors leads to n equations of the form 6 which may be regarded as representing n terms of a Gabor expansion. $\mathrm{R}_n(\lambda, a, b)$ is essentially a wavelet associated with a Gaussian mother wavelet and may in practice be a priori specified. Limiting the number of terms *n* to a small finite number leads to a truncation of the Gabor series. Furthermore the *n* terms may be organized to be non orthogonal (fig.1.18) so optimizing the completeness of the information with respect to its acquisition.

The fusion of the outputs from *n* nonorthogonal sensors may be achieved algorithmically to discriminate signals in chromatic space. For example, tristimulus systems ($n=3$) provide a satisfactory means for discrimination in many situations and appropriate mapping algorithms may be chosen dictated by purpose. Thus if the three responsivities correspond to those of the human color vision [43], the appropriate map becomes the CIE diagram of color science [44] which may be used for subjective color mixing applications* Such mapping has also been used for optical fiber sensing applications (e.g. [33]). Alternatively, algorithms yielding Hue (H), Lightness (L), Saturation (S), (e.g. [45]) are better for defining signal properties objectively for process control (e.g. [32], [46]). (Other algorithms such as Lab [47] are available as are transforms between the various algorithm systems).

The H, L, S algorithms for n detectors are [45]

$$\mathrm{H} = 120(\mathrm{i}\text{-}2) = 120\left[\upsilon_{i-1} / \left(\upsilon_{i-1} + \upsilon_{i+1}\right)\right] \tag{1.41a}$$

$$\mathrm{S} = \left(V_{MAX} - V_{MIN}\right) / \left(V_{MAX} + V_{MIN}\right) \tag{1.41b}$$

$$\mathrm{L} = \left(V_{MAX} + V_{MIN}\right) /_{2} \tag{1.41c}$$

$$\upsilon_i = V_i - V_{MIN} \tag{1.41d}$$

* This corresponds to the algorithms used initially with chromatic modulation sensors and quoted in the first edition of this text. The algorithms take the form

$$\upsilon_x = \frac{V_x}{\sum\limits_{x=3} V_x}$$

leading to a chromatic map in the cartesian coordinates $\upsilon_1 : \upsilon_2$

$$V_{MIN} = \min\left(V_1, V_2 - - - - - V_{n-1}\right) \tag{1.41e}$$

$$V_{MAX} = \max\left(V_1, V_2 - - - - - - V_{n-1}\right) \tag{1.41f}$$

i = integer modulus

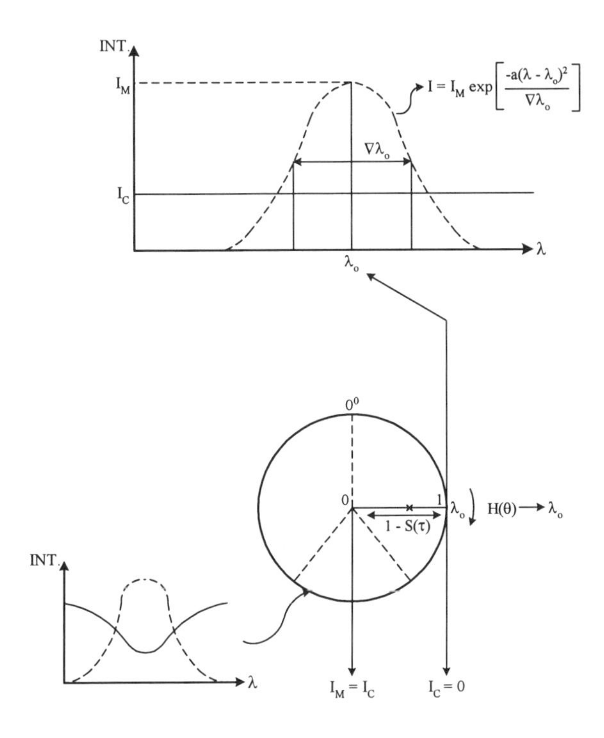

Figure 1.17. Interpretation of a signal in H, S space (showing the equivalent Gaussian)

H, L, S mapping * (figure 1.17) provides an approximate way of defining an optical signal from partially complete information in terms of a dominant wavelength (H) effective energy (L) and nominal bandwidth (S). Particular values of these parameters define completely a specific Gaussian distribution signal (figure 1.17) which represents a family of equivalent non Gaussian signals as detected by the three detectors.

* The H, L, S map is based upon the cylindrical coordinates θ (H), τ (S), z (L) with a monochromatic boundary at τ = 1 but with non circular loci of equal saturation.

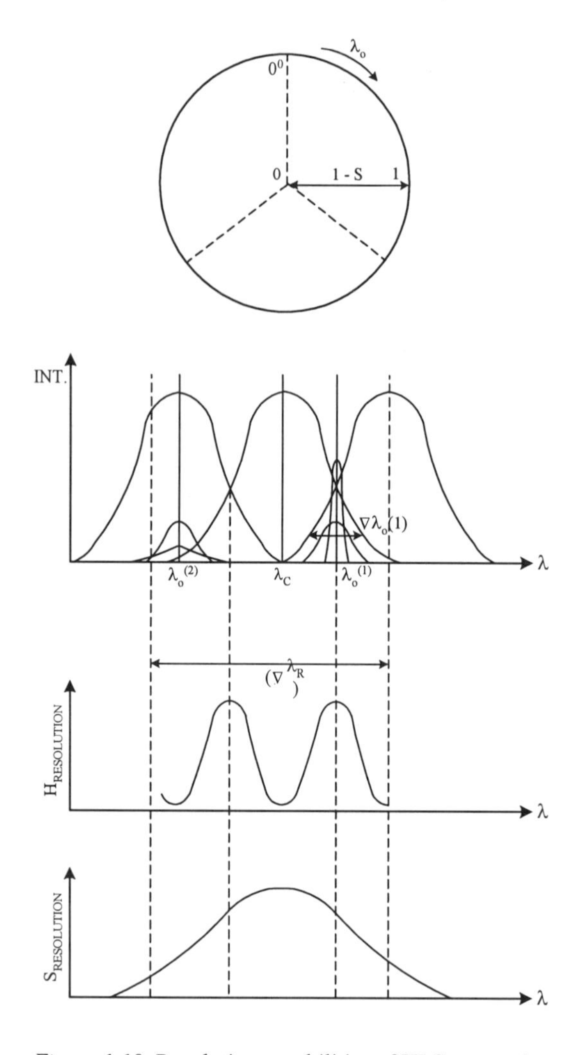

Figure 1.18. Resolution capabilities of HLS processing

Whereas a single detector system (n = 1) only provides optical intensity information, a two detector system (n = 2) with overlapping responsivities (distimulus) enables the dominant wavelength to be determined with automatic intensity compensation.

The measurement resolution of a tristimulus system (n = 3) may be assessed as the combination of two distimulus (n = 2) systems coupled via the nonorthogonality of the central detector responsivity. This leads to a wavelength dependent resolution (figure 1.18) which is a maximum at the steepest gradient point of the detector responsivity and a minimum when the second derivative is zero.

Computer based simulations suggest that tristimulus monitoring is optimum for most sensing applications. The use of additional detectors provides a greater degree of discrimination - for instance the use of six detectors (n=6) enables two equivalent Gaussian envelopes to be completely defined. Computer-based calculations [37] indicates a 95% degree of signal reproducibility leading to the important conclusion that significant truncation of the infinite series of Gaussian terms as proposed by Gabor [42] is sufficient for practical monitoring applications.

The optical efficiency of chromatic monitoring is inherently greater than both spectrometric or two wavelength approaches since the total optical power is utilized by the n detectors with overlapping wavelength responses.

The range and resolution may be chosen for particular applications by appropriately modifying the relative spectral responses of the detectors.

Resolutions as high as 0.04 nm in spectral wavelength changes have been conveniently achieved [33] and further extensions are possible

The chromatic monitoring approach may be used with optical fiber sensing in a variety of different ways and with a range of different modulation principles. Three main categories of chromatically based optical fiber sensors may be identified as shown in Table 1.1(a)

– 'physical' parameters such as temperature, pressure, electric current etc. (Table 1.1a)
– 'chemical' parameters which are not addressable interferometrically, such as electrical plasma and wet chemistry species, contamination, pH values etc. (Table 1.1b)
– 'non specific' parameters such as coloration, vibration, environmental condition etc. (Table 1.1c)

Table 1.1a. Chromatic optical fiber sensors for physical parameters

Sensor Type	Status	Reference
Pressure Photoelastic	Lab. Tested	[48]
μ Si interfer	Switchgear Tested	[21]
Temperature μ Si interfer	Switchgear Tested	[49], [18]
Linear Movement	Routine Switchgear Research	[34]
Rotary Encoder	Industrial Evaluation	[50]
Torque	Lab. Tested	[51]
Electric Current Faraday	Industrial Evaluation	[24], [50]
Hybrid	Industrial Evaluation	[26], [27]
Thin film thickness	Lab. Tests	[17]

Table 1.1b. Chromatic optical fiber sensors for chemical parameters

Sensor Type	Status	Reference
pH (indicator)	Lab. Tested	[52]
Plasma Processing	Chemical Instrument available	[32] [53]
Particle Detection	Routine use, Switchgear research	[28]
Oil Condition Monitor	Industrial Evaluation	[12], [54]
Oil layer on water	Industrial Instal	[11]
Water contamination of oil	Lab. Tested	[12]
Liquid petroleum products	Lab. Tests	[55]
Medical-blood oxygen	Chemical evaluation	[30], [29]

Table 1.1c. Chromatic optical fiber sensors - non specific parameters

Sensor Type	Status	Reference
Coloration change	Industrial Evaluation	[56]
Acoustic	Being developed for Various applications	[3], [57], [39]
Nuclear radiation susceptibility	Industrial tests	[11]

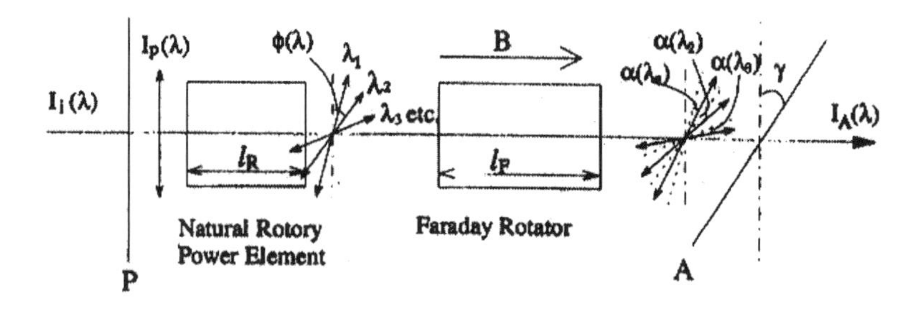

Figure 1.19. A chromatic Faraday current sensor arrangement incorporating a rotary power element

With regard to 'physical' parameter sensing, chromatic detection of polychromatic interference has been demonstrated for measuring temperature and pressure with miscrosilicon devices [21], [18], [49] and for measuring electric current with magneto optic devices [24]. In the latter case a BSO or quartz disc is used to wavelength encode the polarized light which is transmitted through a Faraday rotation element exposed to the B field produced by the current (figure 1.19). The approach has the advantage of maintaining a high sensitivity, independent of the B field magnitude (whereas intensity-based interferometry is susceptible to reduced sensitivity due to the cosine squared nature of the signal). The transducer output is the form of the dominant wavelength v current characteristic which is linear in nature.

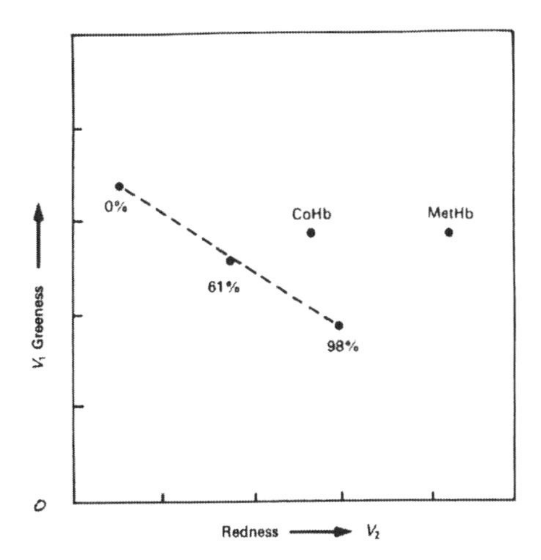

Figure 1.20. Chromatic distribution between various hemoglobin derivatives (also fig. 1.9(b)). ---: percentage oxyhemoglobin

An example of the 'chemical' parameter monitoring applications is the determination of proportions of propane and butane in liquid petroleum products. Such compounds have similar but not identical near infrared absorption spectra, which when superimposed are difficult to segregate by single wavelength, spectroscopic discrimination. The use of chromatic sensing with its integrated wavelength facility is capable of resolving the composition of such mixtures [55]. The approach has also been successfully used for monitoring the degradation of vacuum pump oils employed in the semiconductor processing industries [54] and for detecting the contamination of oil by water [12]. The discriminating capabilities of the tistimulus approach has been demonstrated in relation to pulse oximetry whereby not only the percentage of oxyhaemoglobin could be determined but also a distinction made between oxyhaemoglobin, methaemoglobin and carboxyhaemoglobin [29], [30] (fig. 1.20). The method is currently being extended to monitoring bilirubin levels in neonatals.

Non-parameter specific monitoring has been successfully applied to tracking changes in the integrated spectral emission from a radio frequency plasma utilized for the processing of semiconductor materials [46]. Such measurements enable a number of process affecting events to be sensitively identified from a single output (figure 1.21) including changes in the supply of various gases to the plasma chamber, changes in the radio frequency power driving the processing and the end point of the process. The approach has been successfully demonstrated for monitoring a wide range of semiconductor processes. The three chromaticity parameters can be utilized, in conjunction with an appropriate neural network, to provide reliable open loop control of the plasma process directly on line. A comparison of 'conventional' and 'neural network' models utilizing the chromatic parameters [58] shows that because of the non-linear relationships between the chromatic outputs and process control parameters, the neural approach gives a better and highly accurate agreement with the real situation.

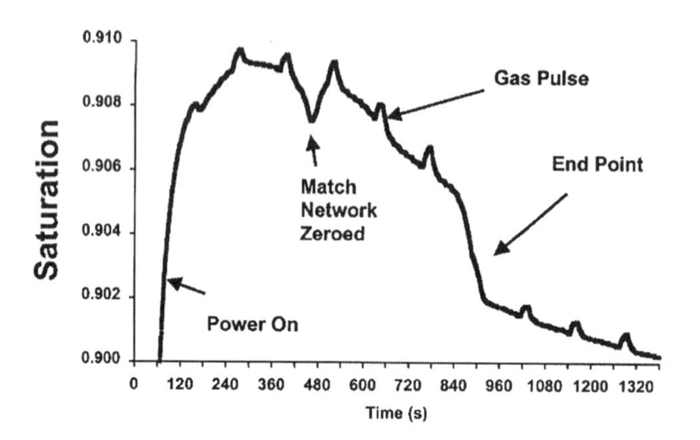

Figure 1.21. Semiconductor materials processing - the use of nominal bandwidth (saturation) for tracking process events

The possibility of self-calibration and compensation for system drift effects in chromatic modulation systems has been demonstrated using a photoelastic strain gauge by Humphries et al [23]. Although chromatic systems can be designed to be highly stable, the strain gauge demonstrator has shown that considerable spectral changes can be compensated, leading to accuracies under extreme change conditions of a fraction of a percent.

The chromatic approach is also extendible from the original domain of monitoring optical spectra to other domains such as the opto-acoustic monitoring of vibration [46]. In the latter case speckle pattern fluctuations have been mapped in an acoustical manifestation of a H,L,S map similar to the optical version of figures 1.17 and 1.18. As a result trends in the variation of acoustical signals from high voltage circuit breakers with different operating conditions (e.g. fault current being interrupted) have been derived in terms of acoustical versions of H,L,S.

The multiplexing of chromatic sensors has been demonstrated with a suite of photoelastic optical fiber strain gauges addressed via a spatial modulator [59]. Up to sixteen such sensors have been multiplexed.

More recent developments have led to the combination in chromatic space of the output signals from a number of optical fiber sensors measuring different parameters. For example a total of six signals – monitoring gas pressure, contact travel, acoustical emission, plasma optical emission, current and voltage – from a high voltage circuit breaker have been tracked in chromatic space for condition monitoring purposes (figure 1.22).

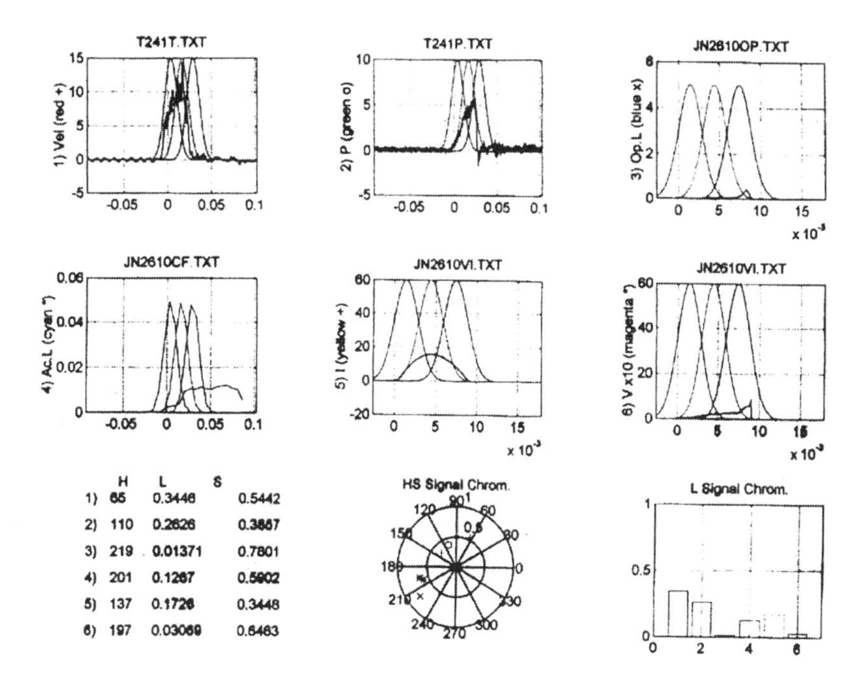

Figure 1.22. Fusion of outputs of six sensors using chromatic compression methodologies

1.5.1.5 Time-multiplexed referencing

In the case of the analog systems already described, sequential rather than parallel referencing is feasible by time multiplexing the modulated/reference signals along the same fiber and detecting with a single detector. Such a procedure removes the uncertainty highlighted by the situation in equation (1.38) of using two separate channels and detectors (subscripts a and b) which might have differently variable characteristics ($F(\lambda)$, $M_2(\lambda)$, $R(\lambda)$). This is achieved at the expense of introducing a multiplexing element into the system.

Formally, the term $M_1(\lambda)$ in equation (1.6) takes the form

$$M_1(\lambda) = \begin{cases} M_R(\lambda), & t_1 < t < t_2 \\ M_s(\lambda), & t_2 < t < t_3 \end{cases} \tag{1.42}$$

where $M_R(\lambda)$, $M_S(\lambda)$ are the referencing and signal monitoring functions confined respectively to the time intervals $(t_2 - t_1)$ and $(t_3 - t_2)$.

This analysis may, of course, be extended to several signals to form the basis of time multiplexing signals from several sensors along a single fiber link.

1.5.2 Broadband interferometric techniques

1.5.2.1 Introduction

Polychromatic or broadband interference, which was described in simple terms in section 1.4.3, (fig. 1.12), also provides a route for overcoming the deficiencies of absolute intensity modulation. An understanding of the signal processing, sensing mechanisms and practical system configurations used for broadband interferometry requires a more detailed description of the interference process. For this purpose, the two-beam interference equation is analyzed in the temporal and spectral domains.

1.5.2.2 Basic principles

The elements of a representative system are shown in fig. 1.23. Light from a spectrally broadband source L is transmitted to the interferometric sensor S by means of a 3 dB Y coupler and fiber optic link. At the sensor, the amplitude of the incident light is divided into two components and an optical path difference introduced between them in response to the action of the measurand. The sensor is designed such that a defined relationship exists between the path difference and the measurand. (Mechanisms for impressing the required phase modulations are summarized in section 1.5.2.3.3). The output of the sensor is then coupled via the common input fiber back to an optical processor, P, where P consists of a second optical system, the output of which is a function of the path difference generated at the sensor. The final element of the system is a photoelectric detector placed at the output of the optical processor. This generates an analog electrical output from which the value of the path difference, and hence the measurand, is derived.

In practice, the optical processor is either a second interferometer or a spectrometer. Thus broadband interferometric sensing systems may be divided into interferometer/interferometer and interferometer/spectrometer configurations. To understand and to analyze fully these systems it is necessary to consider their operation in both spectral and temporal domains. Although both treatments are fundamentally equivalent (through Fourier transform relationships) each lends its own insights and one is better suited to the explanation of a particular function than the other.

The analysis which follows is confined to two-beam interferometers such as a Michelson or low-finesse Fabry-Perot, although the basic principles are not significantly affected if multiple beam interferometers are used.

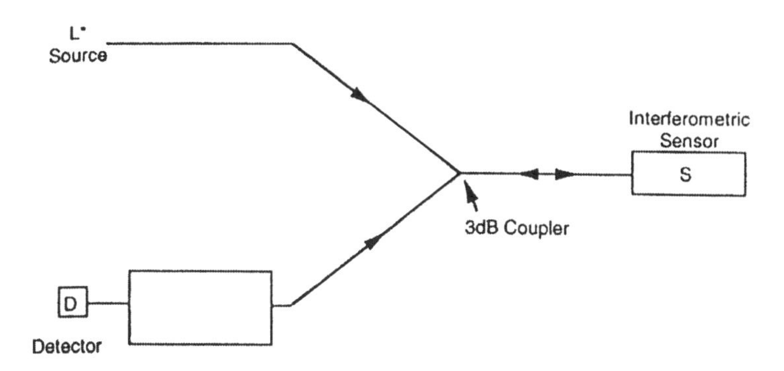

Figure 1.23. Basic elements of the system

1.5.2.2.1 Temporal domain analysis

Consider the operation of the two-beam Michelson interferometer illustrated in fig. 1.24. The path length of the reference arm, a, as defined by the position of the reference mirror M_r is assumed to be fixed, and that of the measurement arm, b, to vary in response to the value of the measurand. The optical path difference D is then given by

$$\Delta = 2n(a-b) \tag{1.43}$$

For simplicity it has been assumed that the refractive index of both arms is equal to *n*.

In general the magnitude of the measurand, and hence Δ, will vary with respect to time, t. It then follows from the two-beam interference equation that the time-dependent intensity transmitted by the interferometer, I_t (t), is given by:

$$I_t(t) = 2I\left[1+\gamma(\tau)\cos\frac{2\pi}{\lambda}\Delta(t)\right] \tag{1.44}$$

In equation (1.44), Δ(t) is the signal the interferometer is required to recover, and γ(τ) is the degree of coherence of the source as a function of the temporal delay caused by the path difference of the interferometer, where

$$\tau = \frac{\Delta}{c} \tag{1.45}$$

c = velocity of light.

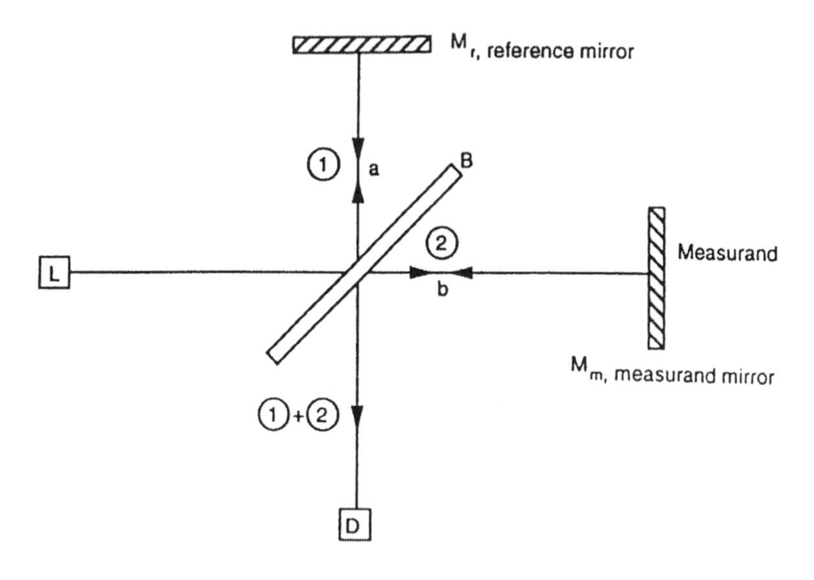

Figure 1.24. Two-beam Michelson interferometer

$\gamma(\tau)$ is equal to the correlation function of the interfering beams, which have a relative time delay, τ [60], where:

$$\gamma(\tau) = \frac{\langle f(t) f^*(t+\tau) \rangle}{\langle |f(t)|^2 \rangle} \tag{1.46}$$

and $\langle\rangle$ indicates a time average.

When the interfering beams have equal power, $\gamma(\tau)$ may be interpreted as the visibility of the fringes observed in the interferometer. That is

$$\gamma(\tau) = \frac{I_{max} - I_{min}}{I_{max} + I_{min}} \tag{1.47}$$

where

$$\gamma(0) = 1 \tag{1.48}$$

$\gamma(\tau)$ therefore defines the reduction of contrast of the interference fringes that occurs as the path difference departs from zero and is slow compared with the cosine variation of equation (1.44). $I(t)$ hence consists of a cosine function multiplied by the coherence envelope. The shape and width of this envelope depends on the characteristics of the source and is often defined in terms of the source coherence length. This is the path difference, l_c, for which the contrast of the fringes is l/e^2 of the contrast at zero path difference ($\tau = 0$), and for a source of central wavelength λ_c, and spectral width $\Delta\lambda$, is approximately equal to $\lambda_c^2/\Delta\lambda$. (A near infrared LED of the form described in section 1.5.2.3.1 with λ_c=700 nm and $\Delta\lambda$=100 nm could be used for spectrally broad interferometry and has a coherence length of about 5 μm. Narrowband sources for which $\Delta\lambda$ may be considerably less than a nanometer will have coherence lengths of the order of meters).

When spectrally broad instead of narrow bandwidth sources are used in interferometry, the unambiguous (i.e. absolute) measurement range is extended from a single to many wavelengths, where the precise range will depend upon the method of fringe processing used (section 1.5.2.3.1). This is a major advantage because it brings absolute measurement with interferometric resolution into a range suitable for a large number of sensing applications. This important feature derives from the coherence function fringe modulation described above, which enables the fringe order number to be determined relative to the maximum visibility zero-order fringe for which the path difference is zero. This is not possible in the case of a spectrally narrowband source for which $\gamma(\tau)$ is effectively unity over many fringes (i.e. $l_c >> \lambda_c$) and the unambiguous range is therefore limited to a single wavelength. The resolution is limited ultimately by the coherent source noise, which for a broadband source has been determined to be of the order of 10^{-7}rad rms/Hz$^{1/2}$ [61]. Spectrally narrowband sources have lower phase noise, but this is difficult to exploit due to the mechanical noise floor of practical systems.

A second advantage of spectrally broadband interferometry is that it enables the technique of interferometer linkage to be used. This transfers the interference occurring at the remote passive sensor to an active processing interferometer within which a number of fringe processing techniques may be used advantageously, whilst still maintaining sensor simplicity (section 1.5.2.3.1).

1.5.2.2.2 Spectral domain analysis

The behavior of the interferometer, as illustrated in fig. 1.24, can alternatively be analyzed in the spectral domain. The spectral transmission of a two-beam interferometer is given by

$$I_t(\sigma,t) = 2I(\sigma)[I + \cos 2\pi\sigma\Delta(t)] \tag{1.49}$$

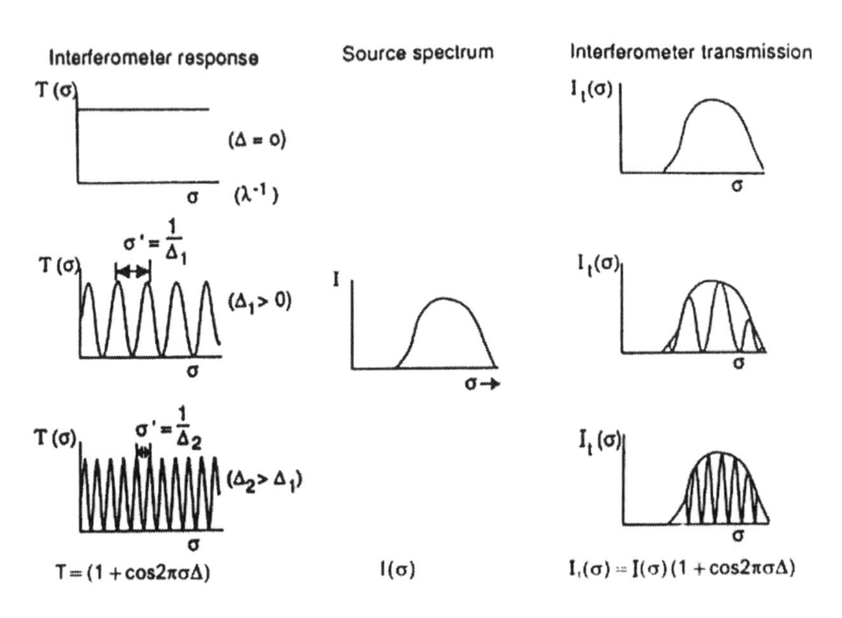

Figure 1.25. Spectral modulation by a two-beam interferometer

where $I_t(\sigma)$ is the transmitted intensity, $I(\sigma)$ is the input intensity, $\sigma(1/\lambda)$ the wavenumber and Δ the optical path difference. The spectral transmission therefore generally varies cosinusoidally with wavenumber. The frequency of this spectral modulation is controlled by Δ. The larger this value the greater the frequency of spectral modulation. This is illustrated schematically in fig. 1.25 for three values of Δ. It should be noted that for $\Delta=0$ there is no spectral modulation, i.e. the interferometer is transparent, otherwise peaks occur for $\sigma\Delta = n$ (where n is an integer) such that the spacing of adjacent transmission peaks is given by $\sigma' = 1/\Delta$. The output spectrum of the interferometer is therefore the product of the input spectrum and the path difference dependent spectral modulation. The number of cycles of modulation will depend on both σ' (and hence on Δ) and the width of the input spectrum. It is notable that once σ' becomes significantly less than the input spectral width the transmission of the interferometer will be 50% of its peak value (at $\Delta = 0$). Beyond this point (i.e. as Δ increases and σ' reduces) the transmission will remain at 50%, but the transmitted spectrum will be increasingly rapidly modulated. The value of Δ at which the transmission first drops to 50% is inversely related to the width of the input spectrum. Thus, once the OPD exceeds a certain value (which depends on the input spectrum), the value of the measurand is encoded solely via the frequency of the spectral modulation.

1.5.2.3 Relationship between phase and spectral domain parameters

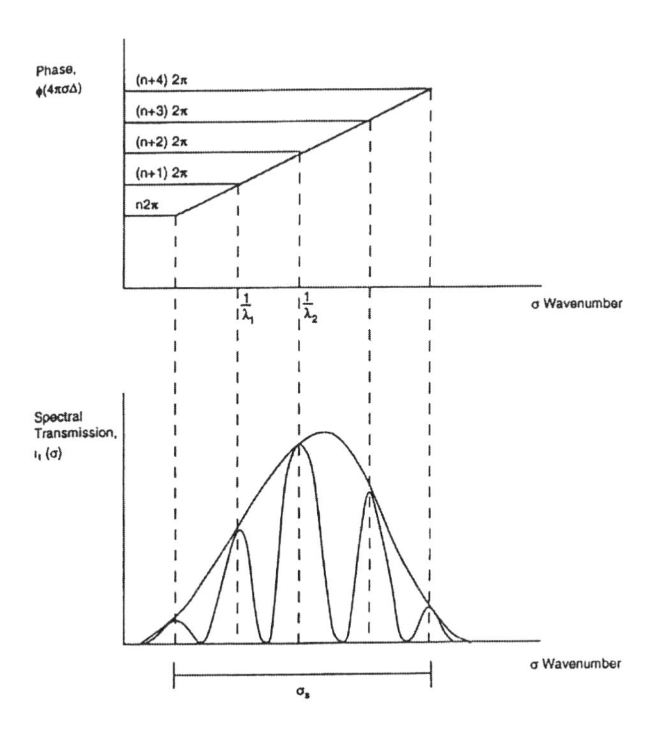

Figure 1.26. The relationship between phase and spectral domain parameters for a given path difference Δ (>$1/\sigma_s$)

Consider now the operation of the two-beam Michelson interferometer with a spectrally broadband source of $1/e^2$ spectral width, δλ, corresponding to wavelengths λ_{min}, where λ_{min} is considered about the central wavelength λ_c. The corresponding wavenumber width, σ_s, is hence $(\lambda_{max} - \lambda_{min}) / \lambda_{max} \cdot \lambda_{min}$. The parameters that define the resultant phase and corresponding spectral modulation for a path difference greater than $1/\sigma_s$, are shown in fig. 1.26. The phase terms (4πσΔ) in equations (1.44) and (1.49) can be seen to be equal to the phase difference between the reference and the measurement arm integrated over the constituent wavelengths of the interfering beams.

Note also that the path difference in wavelength terms between adjacent transmission peaks of the scanned spectrum of the interfering beams is given by

$$\Delta = \frac{\lambda_1 \lambda_2}{\lambda_1 - \lambda_2} \tag{1.50}$$

1.5.2.4 System design

It has been noted in section 1.5.2.2 that the spectrally broadband interferometric sensing system may be based on interferometric sensors used in combination with a processor consisting of either a second interferometer or a spectrometer. In most practical arrangements one interferometer is sited remotely and is linked to the receiver – either the spectrometer or interferometer – by an optical fiber link. The fact that the information is encoded purely spectrally is of great value in such a system since it is particularly rugged and largely resilient to the inevitable variations in the attenuation of the link. The spectrally-encoded information will also be unaffected by fiber dispersion and multimode connection can hence be used. In contrast, dispersion within the interferometer(s) is a potential source of error (unless identical in each interferometer) since the spectral modulation period will become wavenumber dependent.

In this section the design of sensing systems based on the above configurations is discussed, together with the associated signal processing schemes. The section concludes with a summary of sensor phase modulation mechanisms.

1.5.2.4.1 Interferometer/interferometer configuration

In this arrangement, proposed initially by Ulrich [62], the second, processing interferometer is interposed between the sensing interferometer (assumed remote) and the detector (fig. 1.27). At the sensing interferometer light component 1 is reflected from the reference mirror M_1 and component 2 from the sensing mirror M_2, where the latter is subject to the external stimulus. The temporal delay between these fields is τ_s where:

$$\tau_s = \frac{b-a}{c} \tag{1.51}$$

If we now make the assumption that the path difference in the sensor is adjusted such that $\tau_s > l_c/c$ for all values of the external stimulus, it is clear that there will be no interference between beams 1 and 2 and consequently no interferometric intensity modulation. At the processing interferometer beams 1 and 2 can propagate simultaneously along paths 3 and 4. The temporal delays, $\tau_{s,p}$, for these possible paths are given by the following expressions:

Case (i): Paths $1 \rightarrow 3$, $2 \rightarrow 4$

$$\left[\tau_{s,p}\right]' = \left|\frac{(a+c)-(b+d)}{c}\right| \tag{1.52a}$$

Case (ii): Paths $1 \rightarrow 4$, $2 \rightarrow 3$

$$\left[\tau_{s,p}\right]'' = \left|\frac{(a+d)-(b+c)}{c}\right| \tag{1.52b}$$

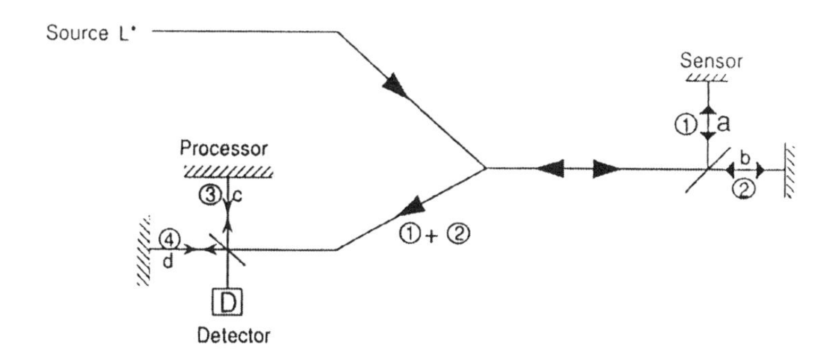

Figure 1.27. Twin interferometer system

It is clear from equations (1.52a) and (1.52b) that for a given value of Δ, there will exist values of d, d' and d" for which $\tau_{s,p}'$ and $\tau_{s,p}''$ are zero, i.e. where

Case (i)

$$d' < c, \qquad \frac{c-d'}{c} = \tau_s \tag{1.53a}$$

Case (ii)

$$d'' < c, \qquad \frac{c - d''}{c} = \tau_s \tag{1.53b}$$

In addition the difference in path lengths of the beams reflected from the mirrors in the processing interferometer only will be zero when

$$c - d = 0 \tag{1.53c}$$

Hence cross correlation terms symmetrically positioned at $\pm\tau_s$ about a central correlation component at τ=0 will be observed as d is varied. The general form of this output is shown in fig. 1.27. The position of the cross correlation peaks relative to the central, fixed component depends on Δ, and the modulation envelope of the finer interferometric fringe structure is defined by the source coherence function (equation (1.46)).

The operation of the interferometer/interferometer configuration may also be analyzed in the spectral domain. It may be assumed initially that the second interferometer is identical to the first. Under these conditions its spectral transmission will vary with its own path difference, also identically. Generally the transmission characteristics of the two interferometers will not match and it can be shown that in this case the second interferometer will typically transmit 50% of the received light. However, when the path differences of the two interferometers are equal, their spectral transmissions will also match. In this case the second interferometer will become effectively completely transmitting to the received light. Thus if the path difference of the second interferometer is scanned away from zero, there will be a peak transmission when its own path difference matches the unknown path difference of the first interferometer. If the path difference of the second interferometer is measured at this point, the unknown value is recovered and the spectral information has effectively been decoded. (There is always an additional transmission peak when the path difference of the second interferometer is zero since it is then transparent.) This means of decoding can be understood as one of matched spectral filtering or spectral correlation, and the narrowness of the correlation peak is inversely related to the spectral width. There is fine structure in the correlation which is more readily understood in the time domain as individual interference fringes and the output of the interferometer /interferometer configuration is necessarily identical to that previously illustrated in fig. 1.28.

It may now be assumed that the combination of path length within the sensor and processing interferometer are adjusted such that at zero the output of the processor is at the center of either one of the $\pm\tau_s$ correlation peaks.

Under these conditions, the output of the processor will vary in accordance with equation (1.44) with the fringe contrast decreasing as the measurand departs from zero. In order to measure the absolute phase (and hence path difference) it is necessary that the fringe be either temporally or spatially modulated in phase in one arm of the processing interferometer. Coherence modulation may then be used to identify the zero order fringe from an output that will have the general form

$$I(t) = I\left\{1 + \gamma(t)\cos\frac{2\pi}{\lambda}(p(t) - \Delta_m)\right\} \tag{1.54a}$$

where

$$\gamma(t) = \gamma\left\{\frac{\Delta_m \pm p(t)}{c}\right\} \tag{1.54b}$$

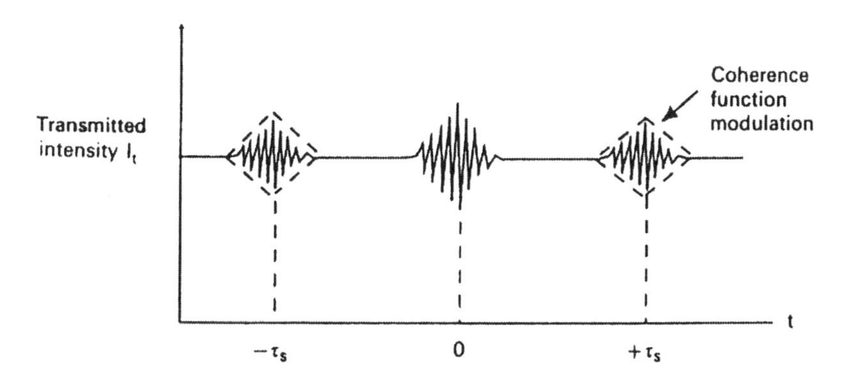

Figure 1.28. Output of the interferometer/interferometer system

Note that in equation (1.54a) the maximum intensity is half that in equation (1.44) due to the halving of the power between the correlation peaks at $\pm\,\tau_S$. p(t) is the cyclic time-dependent phase modulation which has been assumed to have a frequency considerably greater than that of the measurand. The amplitude of p(t) will be governed by the range of measurement required and the method of fringe identification used. The latter may be based either on central peak discrimination, CPD[62-65], or differential fringe visibility measurement, DFV[66]. These modes are illustrated in fig. 1.29. In the CPD mode the complete coherence envelope is scanned and the zero order fringe identified by determining the position of peak visibility. This is achieved either by the mechanical translation of a single mirror in the processing interferometer and the correlation of the peak with the encoded mirror position, or by projection of the complete coherence envelope onto a CCD array by means of a wedge angle in one arm of the interferometer. In the former case the measurement range will be equal to the amplitude of p(t), which in practice is limited by the bandwidth and mechanical constraints of the translation mechanism. (For piezo devices this will be of the order of 10^2 μm, but if mechanical stages are used p(t) may be extended to typically 10^2 mm). When a CCD array is used, limits imposed by the pixel size will reduce the range to typically 10^2 μm. DFV eliminates the need to measure the complete coherence envelope by sampling the fringe contrast symmetrically over a few fringes about the coherence peak. A processing interferometer of the form shown in fig. 1.30 may be used for this purpose. The normalized differential fringe visibility output V_{12} is given by,

$$V_{12} = \frac{V_1 - V_2}{V_1 + V_2} \tag{1.55}$$

where V_1 is fringe visibility measured in interferometer channel 1, and V_2 is fringe visibility measured in interferometer channel 2.

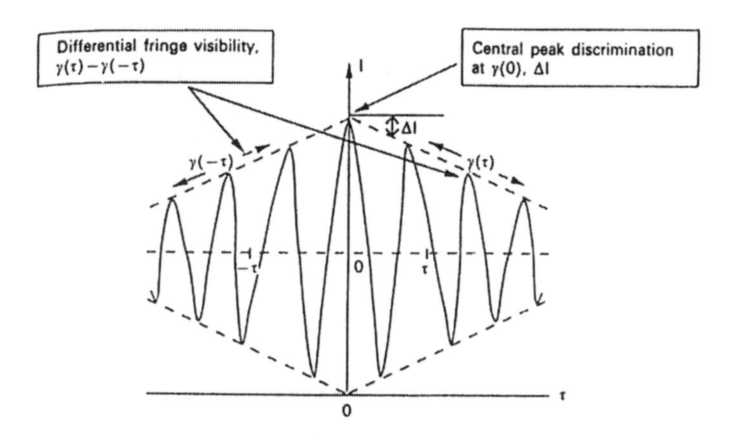

Figure 1.29. Zero-order fringe identification techniques

V_{12} is a linear function over a range approximately equal to the source coherence length (fig. 1.31) and has a fringe contrast ratio that is typically 30 dB greater than the equivalent CPD output because it operates over the region of maximum variation of γ with respect to τ.

The above techniques may be used to measure the absolute fringe order to sub-wavelength resolution. Maximum resolution may be obtained by analyzing the individual fringes within the coherence envelope. Fringe processing of this type is conventionally carried out using the sub-fringe resolution counting of quadrature fringe fields [67]. Fringe fields suitable for this application may be generated by introducing a differential path length of λ/8 between the stepped mirrors in the processing interferometer shown in fig. 1.30 or by the use of polarizing optics.

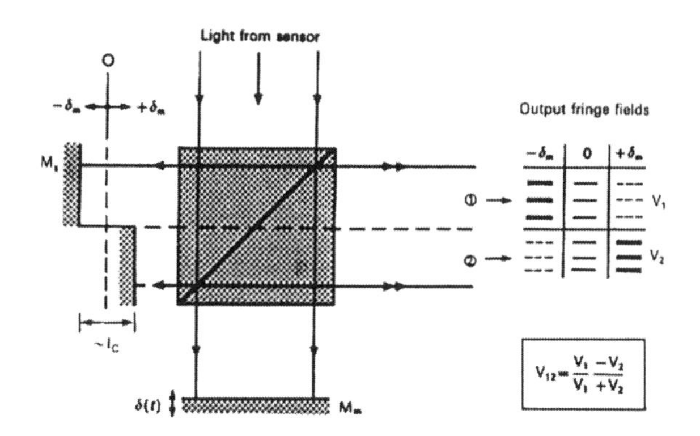

Figure 1.30. Differential fringe visibility measurement interferometer

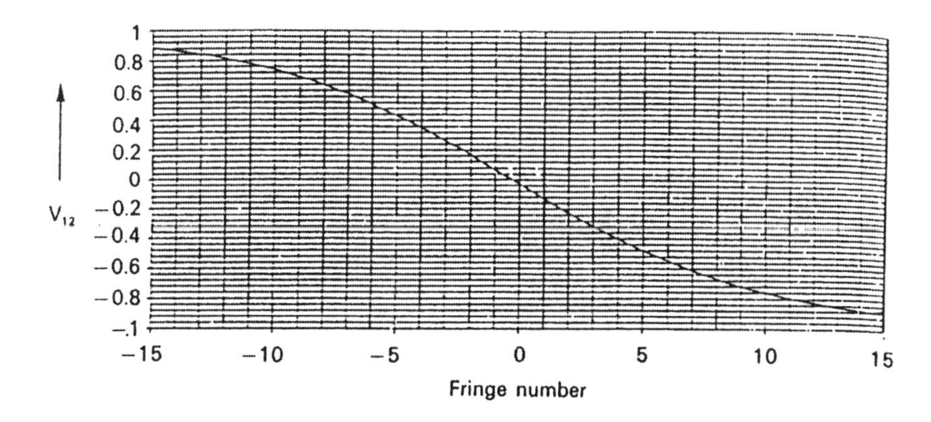

Figure 1.31. Experimental linear calibration function derived from DFV processing [20]

Quadrature fringe fields are also essential when it is required to measure high frequency sub-fringe path differences superimposed upon a DC drift of magnitude greater than a wavelength. Fringe signal fade will occur under these conditions when the fringe passes through fringe maximum and minimum in a single channel. The use of quadrature channels means that this situation will not occur simultaneously and the required path difference can always be measured. This technique is preferred for the measurement of the resonator frequency in fiber optic linked silicon resonant sensors.

1.5.2.4.2 Interferometric/spectrometer configuration

In the arrangement, the output of the interferometric sensor is linked by an optical fiber to the spectrometer which then produces as the primary output the spectral characteristics of the input field generated at the sensor. There is no scope for path length compensation in this scheme (as is the case for the interferometer/interferometer configuration) and interference must therefore be arranged to occur at the sensor by ensuring that the path length difference between the reference and measurement arm is always less than the source coherence length. Recovery of the path difference simply requires that the wavenumber separation between transmission peaks be measured (figs 1.25 and 1.26). The path difference is then given by

$$\Delta = 1/\sigma' \tag{1.56}$$

This method for the recovery of the path length information is the basis for the operation of all interferometer/spectrometer configuration. The performance of such a sensor will depend on the properties of the spectrometer and in particular its resolution. The latter will limit both the range of operation since the spectral modulation frequency increases with OPD and the measurement sensitivity since this depends on the smallest change in $\sigma\Delta$ that can be detected.

1.5.2.4.3 Phase modulation mechanisms

All of the sensors described in the previous sections rely for their operation on the phase change in the measurement arm induced by the action of the external stimulus. It may be assumed that under ambient conditions the measurement arm has a length l and refractive index n. The optical path is hence nl and the change in phase with respect to a general measurand, s, is given by $\delta\phi/\delta s$ where

$$\frac{\delta\phi}{\delta s} = \frac{2\pi}{\lambda}\left\{\frac{l\delta n}{\delta s} + \frac{n\delta l}{\delta s}\right\} \tag{1.57}$$

The first term within the brackets in equation (1.57) is due to the refractive index change within the sensing arm and the second is a result of its mechanical deformation. The index, n, and length, l, will be affected independently by different external stimuli and a key requirement is to design the sensor so that it is predominantly sensitive to a single measurand. This problem may be approached within the framework of a number of basic phase modulation mechanisms. These are described below.

Mechanical displacement A simple means by which a mechanical displacement may be created in response to an external stimulus is to couple the external sensing element to the sensing mirror of the interferometer. This is the principle of a number of pressure transducers based on the Fabry-Perot geometry. Alternatively, mechanical displacement may be coupled directly to the sensing medium through which the light is propagating. For this purpose, birefringent materials are used commonly since they enable the range of unambiguous measurement to be extended. This is clear from the following equation that describes the polarimetric phase ϕ_{pq} of the two polarization components that results from their mutual interference, i.e.

$$\frac{\delta\phi_{pq}}{\delta s} = \frac{2\pi}{\lambda}(n_p - n_q)\frac{\delta l}{\delta s} \tag{1.58}$$

where, in practice,

$$n_p - n_q << n_{p,q}$$

The interference of either the p or q mode with an independent reference component will result in conventional interferometric sensitivity.

Refractive index modulation The refractive index of various materials is functionally dependent on a number of stimuli (e.g. temperature and pressure) and it may be used as a sensing element which need not necessarily be subject to mechanical load or displacement. The equations describing its operation are equivalent to (1.57) and (1.58) above. Compact sensing elements from which both of the abovc types of phase modulation may be detected are based on high birefringent fiber in which equal powers of light are coupled into the orthogonal eigen modes within the fiber. It is interesting to note that the resultant ϕ_{pq} phase modulation component, conventionally measured polarimetrically by a 45° analyzer at the fiber output, may also be measured using a remote spectrometer, as described in section 1.5.2.3.2.

1.5.2.5 Application to in-process surface finish measurement

Changes in surface finish characteristics during mechanical manufacturing are often precursors of process error. If surface finish can be monitored, in-process these changes may be identified and the process halted before waste is generated and production time lost.

An optical fiber-linked, broad band interferometric sensor is suitable for this application in that the measurement probe can he made compact and robust allowing access to complex work pieces whilst the processor may be mounted and be remote from the measurement environment. We describe below the experimental evaluation of this technique with particular reference to the simultaneous measurement of the surface finish and form of engineering surfaces.

1.5.2.5.1 The experimental system

The test system is shown in figure 1.32 and consists of a sensing interferometer linked by a multimode optical fiber F_2 to the processing interferometer. Both interferometers have a Michelson geometry. The input illumination is derived from a monomode fiber pigtailed SLED (S, F_1). Microscope objectives L_1 and L_2 are incorporated in the respective arms of the Sensing Interferometer. The plane of M_1 and the object at the center of the measurement range correspond to the mutually conjugate focal planes of L_1 and L_2. The path lengths in the two arms are off-set by a distance greater than the source coherence length by translating the L_1, M_1 combination parallel to their optical axis. This off-set is compensated in the dual channel processing interferometer in which fringe signals with differential envelopes of coherence modulation (see figures 1.29, 1.30) are detected and amplified at the photo diodes D_1, D_2. The signals from the latter V_1, V_2 (equation 1.55) are fed directly to the PC analog to digital converter and processed as described below.

The object positioning mechanism is coupled mechanically to the sensing interferometer and consists of a stepper motor driven X scan unit (range 100 mm, step resolution 0.1 μm) upon which is mounted a piezo actuated Z scan translator (range: 50 μm, resolution: 0.01 μm). The stepper stage is programmed to scan the specimen over a 5mm sample length with the X co-ordinate derived from the stepper control unit. The piezo maintains the object Z co-ordinate within the linear range of measurement and can be used to measure non-plane specimens. In the latter application the Z object co-ordinate is measured by the capacitative probe C (figure 1.32). Alternatively the object may be translated over short X scan lengths by a single piezo stage mounted on a static support.

1.5.2.5.2 Continuously scanned surface finish measurement and signal processing

In this mode of operation, the object is scanned continuously in the X direction and the surface finish measured directly from the filtered and calibrated V_{12} output of the interferometer. This requires that the variations in height of the object across the scan length be less than the 10 μm linear measurement range of the sensor (fig. 1.33).

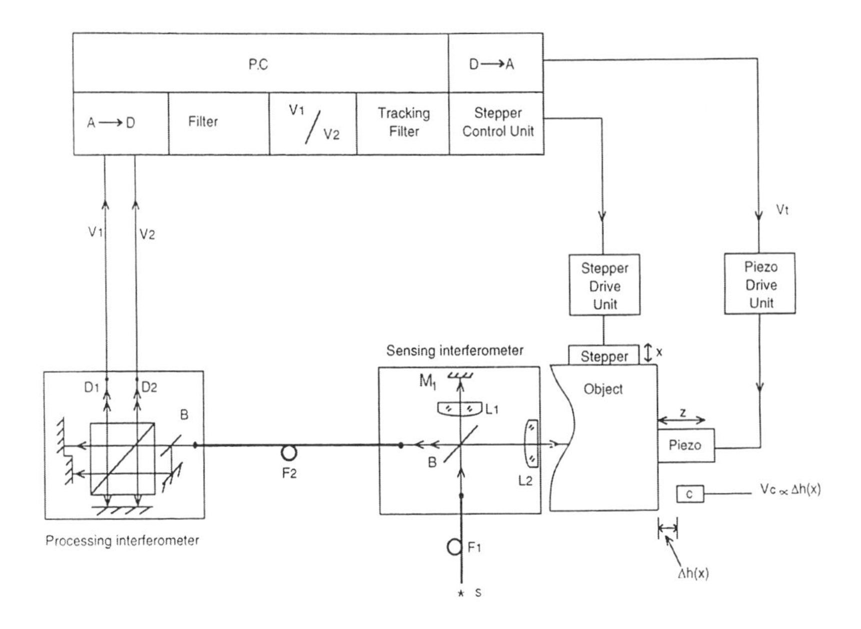

Figure 1.32. The experimental system

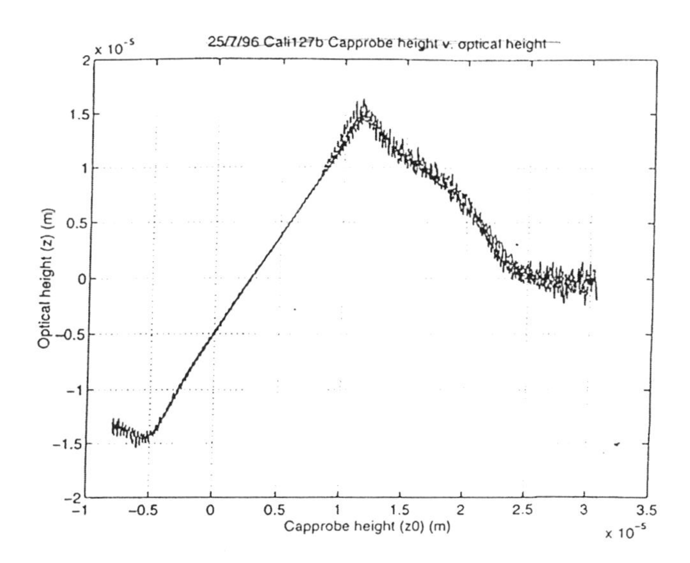

Figure 1.33. Capacitive probe height versus optical height

The calculation of V_{12} is in practice difficult when measuring surface finish, not only does each fringe amplitude vary with surface reflectivity, so also does the accompanying unmodulated light varying with surface reflectivity. This near DC signal has to be almost completely removed before fringe amplitude can be estimated.

In the following it is presumed that the fringes are being scanned out by a mirror oscillating with frequency f_{mirror}, and that both surface reflectivity and height contain no frequency components higher than $f_{mirror}/2$. The specific figures below relate to one particular low-speed implementation.

The filtering required falls into 3 stages: initial removal of the low-frequency band from $-f_{mirror}/2$ to $+f_{mirror}/2$ ("DC blocking filter"), a bandpass filter over the same range after rectification ("bandpass filter"), and notch filters for removal of specific harmonics of f_{mirror} that have not been adequately removed by the bandpass filter. In each case it is critical to preserve the entire shape of the waveform generated by the frequencies passed, as the final division operation causes any phase distortion to cause wide fluctuations in the output signal. This demands a linear phase-delay, not just a constant group-delay in the filtering. Practicable analog filters with this performance are hard to realize, and the signal processing therefore needs to be digital.

The sampling rate required is determined by the requirement not to alias the frequencies present in significant quantity in the input signal. The sharpness of cut-off required for each filter is determined by the maximum of the time bandwidth of the reflectivity signal and the height; let us call this maximum F (both these bandwidths are of course related to the corresponding spatial bandwidths by the scanning speed). The DC blocking filter needs to pass frequencies outside $\pm(2f_{mirror} - F)$ with negligible distortion, while the bandpass filter needs to pass frequencies of up to $\pm F$ with negligible distortion. Stopband attenuation and passband distortion limits of 60 dB have proven adequate; this can readily be achieved for example using f_{mirror} of 30 Hz and F of 10 Hz. The residual phase distortion introduced by the subsequent use of single-pole narrow notch filters to remove {1, 2, 3} f_{mirror} has been negligible as these filters are very nearly linear-phase throughout the vast majority of the passband.

In practice scanning at 5 μm/s with the above parameters has involved sampling at 2 kHz, the combination of a number of different FIR digital filters to form the DC blocking filter, the use of multiple iterations of the FIR bandpass filter, and several applications of each notch filter. However, since only the DC blocking filter and the first part of the bandpass filter run at the maximum sample rate, it is these that determine the processing power required, which for the current implementation with no quadrature channels amounts to 200 multiply-adds/ms. This figure is expected to scale linearly with scanning speed and to slightly more than double if quadrature channels are used.

1.5.2.5.3 Continuously scanned, simultaneous measurement of surface finish and form measurement

This mode of operation is used when it is necessary to measure the surface finish of objects with form height variations greater than the linear interferometric measurement range. Figure 1.34 shows the elements of the system that are used specially for this purpose. The output from the tracking loop as derived from the differential interferometer signals (V_{12}) drives the piezo translator to maintain the distance, h, between the interferometer reference plane and object constant. In practice h will correspond to the path length in the L_2 arm of the interferometer at the center of the measurement range. The time constant of this loop is significantly greater than that of the surface finish measurement.

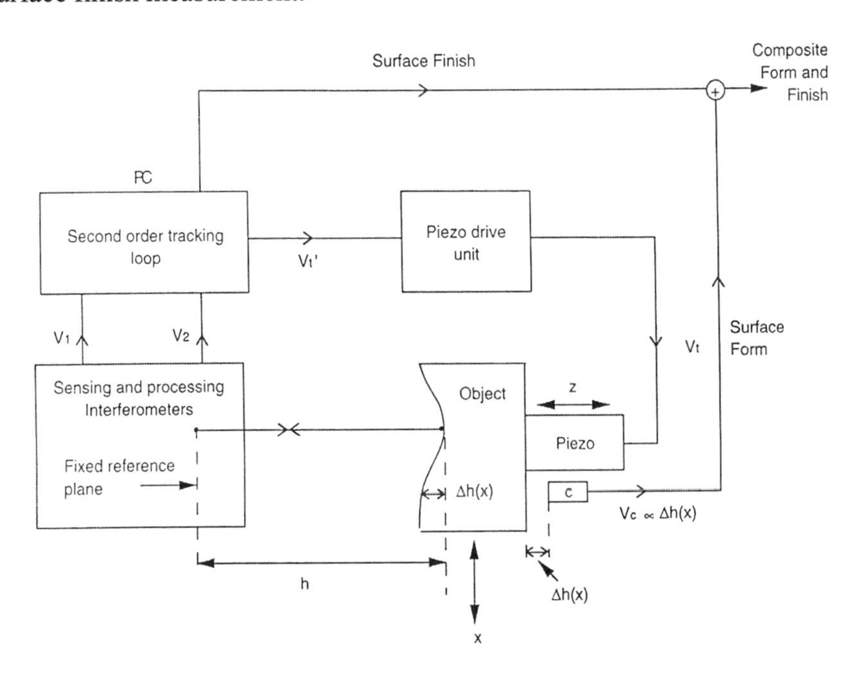

Figure 1.34. Elements of the experimental system used for simultaneous finish and surface form measurement. V_t is the tracking applied to the piezo translator that maintains h constant as an object with varying height, $\Delta h(x)$, is scanned in the x direction. $\Delta h(x)$ is measured by the capacitative transducer C

Under the above conditions, the low frequency from height variations in the object, $\Delta h(x)$, observed as it is scanned in the X direction are compensated by corresponding displacements of the piezo. $\Delta h(x)$ is measured by the capacitative probe C and defines the surface form. The higher frequency surface finish measured directly from the ratiometric interferometer output is simultaneously superimposed upon the surface form to generate the composite finish and form output.

1.5.2.5.4 Experimental results

The system was tested using independently calibrated transducers, surfaces and machined components.

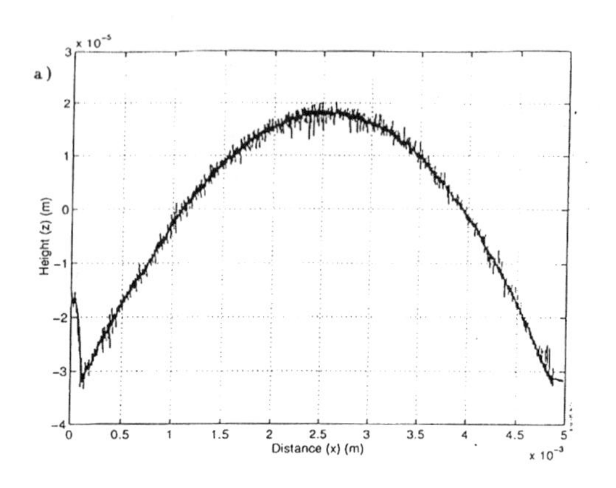

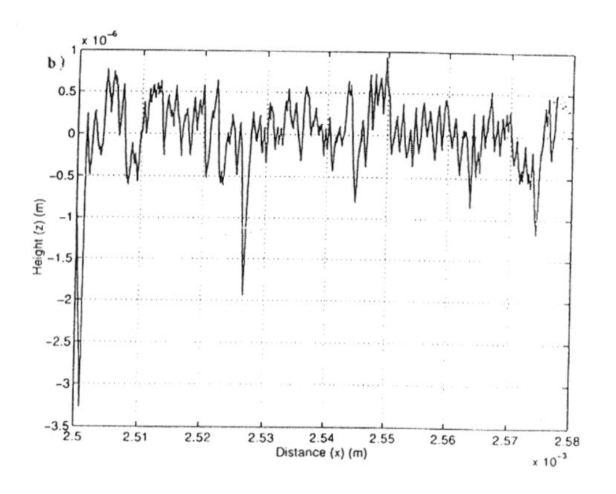

Figure 1.35. Surface form (a) and finish traces (b) for the inner surface of a machined component

Figure 1.35 shows the experimental calibration of the sensor derived from the interferometer output plotted against the displacement measured using the calibrated capacitative probe. This indicates a linear height measurement range of 10 μm with an estimated noise limited measurement resolution of a few nm in an operating bandwidth of 100 Hz.

Table 1.2 shows the good correlation obtained between the R_A (Surface Roughness Average) values of the Rubert calibration surfaces and those measured interferometrically.

Figure 1.35 illustrates the combination of surface finish and form measurement. Figure 1.35(a) is the geometrical form of the surface upon which the surface finish is superimposed. In figure 1.35(b) a section of this scan is magnified and the low frequency surface form component subtracted to reveal the detailed surface finish trace.

The general conclusion of this work is that optical fiber linked, broadband interferometric sensors provide a powerful technique for the high precision, in-process measurement of engineering surfaces.

Table 1.2. Comparison between the Rubert calibrated R_A value and those measured interferometrically (WL1 R_A values)

Specimen	Type	Rubert R_A value (μm)	WL1 RA value (μm)
A3	Flat lapped, reamed	.2	.290
B1	Ground	.05	.060
B3	Ground	0.2	.233
B6	Ground	1.6	1.65
C1	Horizontally milled	.4	.35

1.5.2.6 Discussion

Spectrally broadband interferometry covers a surprisingly wide range of sensing techniques and systems. These are not all directed towards the same goals and so cannot simply be judged against a single overriding requirement. There are, however, a number of attributes which are characteristic of all such systems and these are summarized below.

Rugged transmission: an important feature of the different systems is that the spectrally encoded output, common to all configurations, enables absolute measurand data to be transmitted in a mode that is immune to loss variations in the transmission media. The technology is therefore better suited for use with optical fiber linked sensors than most others available.

Compact sensor design: the interferometric sensors can have simple geometries such as the Fabry-Perot which, provides the basis for the design of robust and compact sensing heads. This attribute is of particular relevance when comparisons are made with non-interferometric alternatives and in combination with rugged transmission is responsible for much of the attraction of the systems. The Fabry-Perot geometry is particularly well suited to pressure sensors where a diaphragm, very similar to conventional devices, can be used.

Absolute output: the short coherence length enables an absolute output to be obtained over a range which is limited only by practical scanning techniques. This attribute is of specific relevance in comparison to conventional interferometric systems which have a non-ambiguous range which extends only over one fringe.

Simple fringe processing: the relatively simple implementation of fringe processing schemes made possible by the use of a second processing interferometer (for example, the generation of phase quadrature fringes, discussed in section 1.5.2.3.1 is an important feature of the technique).

The technical advantages that accrue from the above features account for the significant level of work that has been undertaken in this area over the past few years and the emerging systems based on the technology.

1.5.3 Digital and time domain techniques

The difficulties associated with the analog transmission of data have been encountered previously in telecommunications and have led to the use of pulse code or digital modulation. This technique may also be adapted for use in optical fiber sensor systems. The basic concepts are apparent from a consideration of a status indicator of which the overlapping grating sensor (section 1.4.1) is a specific example. This has only one of two possible outputs - light on or off.

The formal incorporation of such a modulation principle into equation (1.6) is relatively simple and involves setting

$$M_1(\lambda) = M_0(\lambda) f(t) \tag{1.59a}$$

with

$$f(t) = \begin{Bmatrix} 1 & t_1 < t < t_2 \\ 0 & t < t_1 < t_2 \end{Bmatrix} \tag{1.59b}$$

The advantage of such an approach with regard to noise limitations (section 1.2.1) is apparent from fig. 1.36. A threshold level is set, above which the signal is taken to be on and below which it is taken to be off. This means that the resolution (section 1.2.1.3) needs only to be half of the full scale for satisfactory operation, and any excess dynamic range (section 1.2.1.4) is a margin which can be allowed for subsequent degradation. Transmission of more information (e.g. discrimination between several rather than two states) is achievable using several binary pulses. Each pulse is generally referred to as a bit and any number of bits may be used to describe the measurand to the required precision. Such a system has a dynamic range which is much less dependent upon the transmission medium, and is simply determined by the number of bits chosen. In electrical measurements, this function of digitization is performed by an analog-to-digital converter (ADC). However, the optical equivalents of these ADCs necessary for building such systems do not yet exist (nor are imminently likely) and hence the realization of true binary digitally modulated fiber optic sensors is limited to few cases (section 1.5.3.1). Nonetheless, a compromise solution exists, which is quasi-digital in nature and with which the information is carried by pulses, being in an analog manner using time as the proportional variable (section 1.5.3.2).

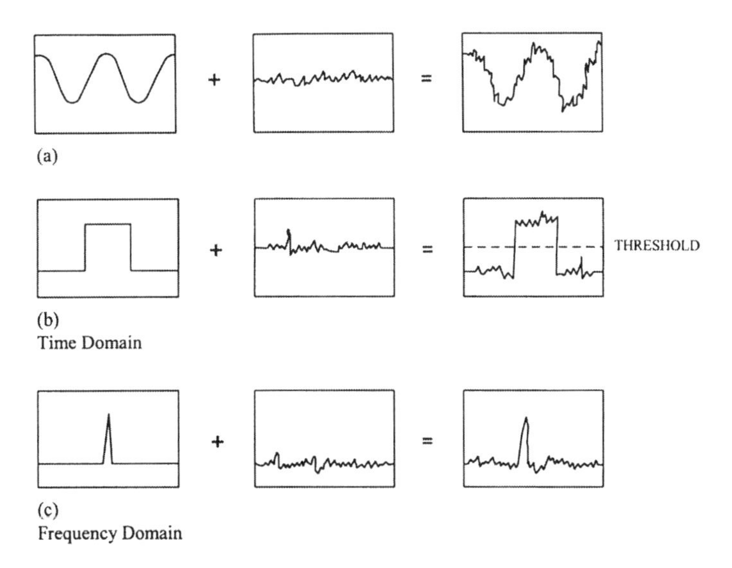

Figure 1.36. Comparison of analog (a) and digital (b) signals and noise in the time domain and a representation in the frequency domain (c)

1.5.3.1 Binary digital modulation

An ideal binary digital system transmits from the sensor to the receiver a number of bits in the form of a 'word' at a sufficient rate to give the sensor the desired frequency response or rise time. The realization of such sensors is limited to a few special cases, the most notable of which is the encoder. This is a sensor of position, either linear or rotary. Optical digital encoders are based on binary-coded reflectors (fig. 1.37). White regions in the figure are reflective, while the black regions are absorbing. Each track is interrogated by a simple status sensor. The output is in the form of a parallel binary word which represents the measurand to the required precision. The disadvantages of this system are that one complete optical sensor system (defined by equation (1.6)) is required per bit of resolution and that the size of the coded strip or disc increases with the number of bits. Nonetheless the system provides an absolute measurement despite its digital nature.

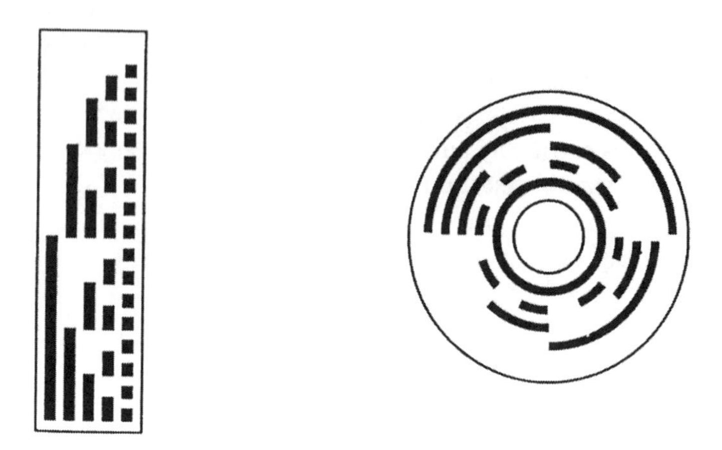

Figure 1.37. Linear and rotary encoders

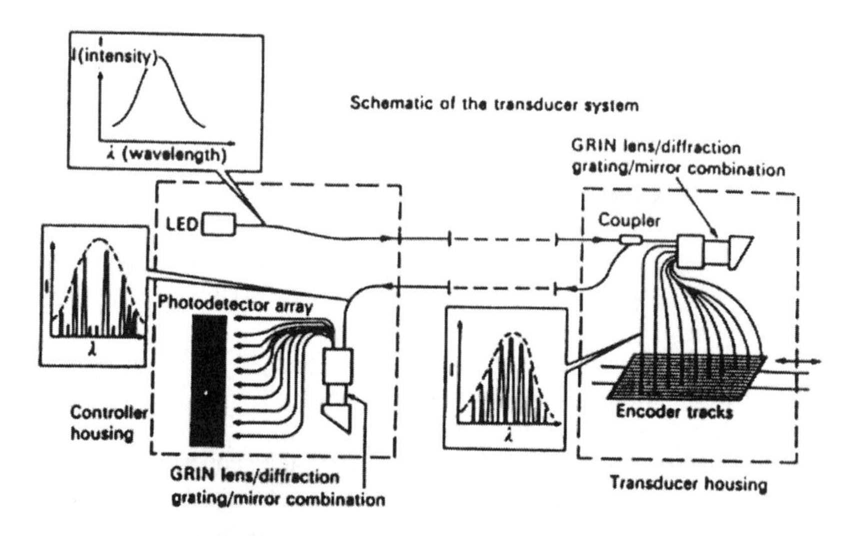

Figure 1.38. Wavelength division multiplexed encoder (Schlumberger)

Some system economy can be achieved through the use of wavelength division multiplexing whereby a single broadband LED in conjunction with a wavelength discriminating element at the receiver allows several digital channels to be communicated via a single fiber link (fig. 1.38). An example of such a device with a stroke of 38mm and a resolution of 38 μm is under development by the Transducer Division of Schlumberger Industries. Rather than a simple binary code this device makes use of a Gray code in which only one bit changes at each transition. This allows error detection to be performed.

1.5.3.2 Quasi-digital modulation

Quasi-digital monitoring allows many of the advantages of digital modulation to be obtained for measurands where an optical digitizer is not available. The technique uses digital pulses to transmit the information from the sensor head, but in the form of a stream, rather than as a series of words. The information is carried as the rate of pulse transmission, the interval between successive pulses or the width of each pulse. In all cases it is necessary to provide an accurate clock or timing circuit at the receiver in order to demodulate the signal.

An example of such a sensor system is a rotating vane flowmeter arranged to output one or more optical pulse per revolution so that the flow rate becomes proportional to the frequency of occurrence of the optical pulses.

Another example is the medical monitoring of patient pulse rate by measuring the periodicity of spectral fluctuations suffered by an optical signal transmitted through pulsating blood in the patient's artery. This case corresponds to

$$f(t) = 1 + \sum_M A_M \sin(\omega_M t) \tag{1.60}$$

where ω_M is the angular frequency of the blood flow and A_M is an amplitude term.

Such methods are being extended to sensors for a range of measurands whose output is not intrinsically periodic. These include electromagnetically excited vibrating wires, piezoelectrically excited quartz crystals, and optothermally excited micromachined silicon resonators. The common feature of these devices is the incorporation of a mechanically resonant element whose motion modulates the optical signal by a shutter action (section 1.4.1) and whose resonant frequency is proportional to the measurand. The transducer is operated at resonance to minimize the drive power requirement which may be supplied by a long-life lithium battery within the sensor head, by a photovoltaic device which is illuminated by radiation transmitted down the fiber, or by direct optothermal means where the optical energy is absorbed, converted into heat and hence into mechanical motion.

The major difference between the frequency out and conventional analog systems relates to the signal demodulation process.

Two basic measuring techniques are available – the first involves counting the number of cycles (N_c) in a given time interval whilst the second involves measuring the time for a fixed number of cycles (t_{Nc}). In the first case the measurement resolution is to one cycle of the modulation frequency, i.e

$$R_s = \pm (N_c)^{-1} \tag{1.61}$$

In the second case the resolution is governed by the frequency of a clock signal (f_c) used for timing purposes, i.e.

$$R_s = (f_c t_{Nc})^{-1} \tag{1.62}$$

Equations (1.61) and (1.62) imply that the resolution is inversely proportional to the sampling time (and hence transient response). However, in the second case (equation 1.62), the resolution is also dependent upon the clock frequency which may be independently varied to preserve resolution at high transient responses. Typically sampling a 1 kHz modulation signal every 100 ms with a 10 kHz clock yields a resolution of 0.1%. Such time scales are long compared with light propagation time scales (3×10^8 ms^{-1}) along optical fibers, so signal dispersion in multimode fibers associated with intermodal and chromatic effects is not a limiting factor as in the case of high data rate transmission systems.

1.5.4 Full hybrid techniques

The 'frequency-out' methods described in section 1.5.3.2 require the optomechanical transducer to be energized through the provision of auxiliary power (battery, photovoltaic or optothermal) before it is capable of performing its sensing function. These transducers are therefore active rather than passive in nature and the sensing systems may be regarded as semihybrid on account of the need for subsidiary activation of the optomechanical function (fig. 1.39(a)).

The above concept may be extended to realize systems which are fully hybrid. These rely upon the use of conventional electronic transducers for performing the sensing function. Auxiliary power is needed for energizing not only the transducer but also an optoelectronic source and modulating circuitry to provide the optical signal for transmission from the transducer unit by the optical fiber (fig. 1.39(b)). This power may be provided from a photovoltaic device energized remotely via a fiber link or in the case of power system monitoring by electromagnetic induction from the electric power line being monitored.

The performance of such hybrid systems is formally described by equation (1.6) with

$$M_1(\lambda) = 1 \quad (1.63a)$$

$$P(\lambda) = P_s(\lambda) f(t) \quad (1.63b)$$

where $P_s(\lambda)$ $f(t)$ is the spectral power distribution of the modulated optical source (fig. 1.39(b)). In the case of the remotely energized photovoltaic power source.

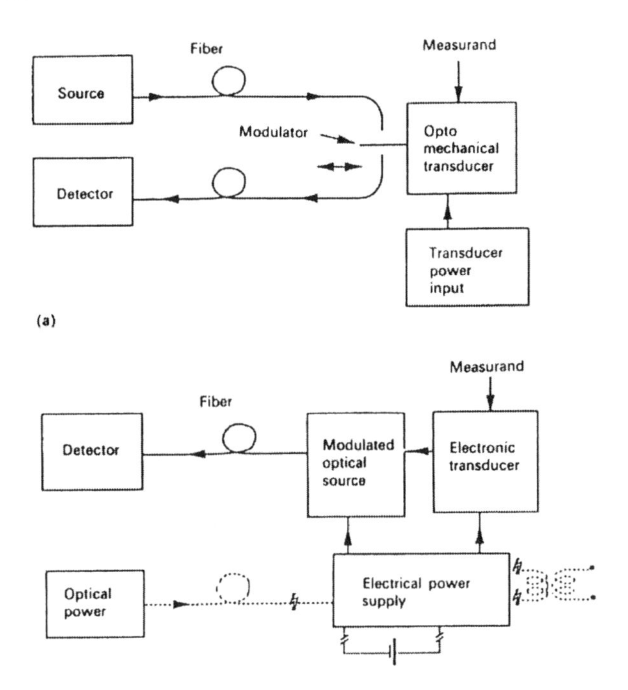

Figure 1.39. Distinction between (a) semi- and (b) full-hybrid systems

$$P_s(\lambda) = \eta(P_0 - P_E) \quad (1.64a)$$

$$P_0 = \sum_{l,m} \int_{\lambda} \left(\int_{l} P_p(\lambda) F_p(\lambda) M_{p2}(\lambda) dl \right) R_p(\lambda) d\lambda \tag{1.64b}$$

where η is the quantum efficiency of the optical source driving the fiber forming the signal channel, and $P_p(\lambda)$ is the optical output power of the source activating the remote electronic system, $F_p(\lambda)$, $M_{p2}(\lambda)$ are fiber and component losses in the fiber delivering power to activate the remote electronics, $R_p(\lambda)$ is the response of the optoelectronic converter used to drive the electronic circuit, and P_E is the power required to drive the electronic circuit. Optical power budget difficulties are paramount since sufficient optical power needs to be propagated via the optical fiber for energizing the electronic sensor (P_E) and the remotely located light emitting diode $P_p(\lambda)$.

Although such sensing systems fail to take the full advantage of complete optical monitoring, their use is probably more attractive for power system monitoring [22] when the electrical isolation provided by fiber transmission is advantageous.

Such fiber sensing systems are attractive for monitoring electric current on high voltage transmission and distribution systems on account of the inherent electrical insulation potential of the optical fibers and the possible availability of fiber optic telemetry on the transmission systems. The possibility of energizing an electromagnetic current transformer based system has been demonstrated by Pilling et al [25] and more recently such a hybrid system has been demonstrated which meets the demanding metering requirements of the electric power industry [27]. The system has been shown to meet the performance requirements over the temperature range – 25 °C to +70°C. In order to meet the demands of longevity required by such applications and the need for fail safe operation, the system may be energized directly from the power line on which the current is monitored or from a remote, fiber linked laser source. Automatic switching between the two alternative power sources has been implemented to conserve laser life on the one hand and to measure near zero line currents on the other.

1.6 CONCLUSIONS

The general behavior of multimode fiber sensors is incorporated within equation (1.6) from which limitations, performance and design aspects all can be determined, in principle, for a variety of applications.

Much progress continues to be made in producing practical sensors which meet the rigorous demands of many industries and the robustness of multimode fiber sensors is proving to be advantageous. Broadband approaches to interferometry (section 1.5.2) continue to provide methods for overcoming fading and other limitations of monochromatic interferometric sensors. Microsilicon cavity pressure and temperature sensors [18] are providing good applications feedback.

Digital and time domain techniques (section 1.5.3) linked to spectral encoding [50] have been implemented and are showing good reliability in service.

Progress has been maintained in realizing hybrid sensors and such systems for monitoring electric current on high voltage transmission systems have been developed [27] and are undergoing industrial evaluation.

The chromatic modulation methodology has been deployed with a number of different sensing elements and this has been extended for processing opto acoustical signals [3].

It is likely that future trends will demand systems that will be capable of intelligent monitoring for widespread applications. Generic needs for such intelligent monitoring systems are summarized on figure 1.40 [68]. It is likely that multimode optical fiber sensing will have a major role to play in such systems on account of their versatility for monitoring physical, chemical and non parameter specific situations, their robustness and their cost effectiveness. Already evidence of the realization of such potential of multimode fiber sensors is emerging. For instance Fernando et al [10] has described the design and operation of one manifestation of multiwavelength sensing based upon Fabry-Perot interferometry for simultaneously monitoring strain, curing chemistry, moisture ingression, temperature and vibration in relation to epoxy/resin curing. A suite of multimode fiber sensors is available for measuring seven different parameters- gas pressure, contact temperature, contact travel, arc plasma emission, particulate contamination, electric current and acoustical vibration – within the extremely aggressive environment of SF_6 filled high voltage switchgear. Isaac [28] has described the use of a single multiwavelength sensor for the combined monitoring of arc plasma emission and particle contamination accumulation.

The output from a range of such multiwavelength, multimode fiber sensors has recently been demonstrated as being capable of combination and compression via information processing and representation in chromatic space (section 1.5.1.4). As an example the output from six different parameter sensors of SF_6 switchgear may be Gabor transformed via a family of Gaussian wavelets for representation on a single H,L,S chromatic map (figure 1.22). Further, the different sensor points on the H,L,S map may be combined using well established additive/subtractive laws of chromatic processing to reduce the dimensionality of the sensors output to an appropriate level for particular applications. Trends in system behavior within acceptable limits may then be tracked in H,L,S space. A simple form of such tracking has already been demonstrated by Russell et al [46] to yield commercially valuable results in the monitoring of radio frequency plasma for semiconductor materials processing.

Further developments of such intelligent monitoring are likely to be realized in the future and these will feed back on the type, nature and performance of multimode fiber systems which will be required to optimize the performance of such systems. Multimode fiber sensing has sufficient flexibility to allow its incorporation into such intelligent monitoring systems.

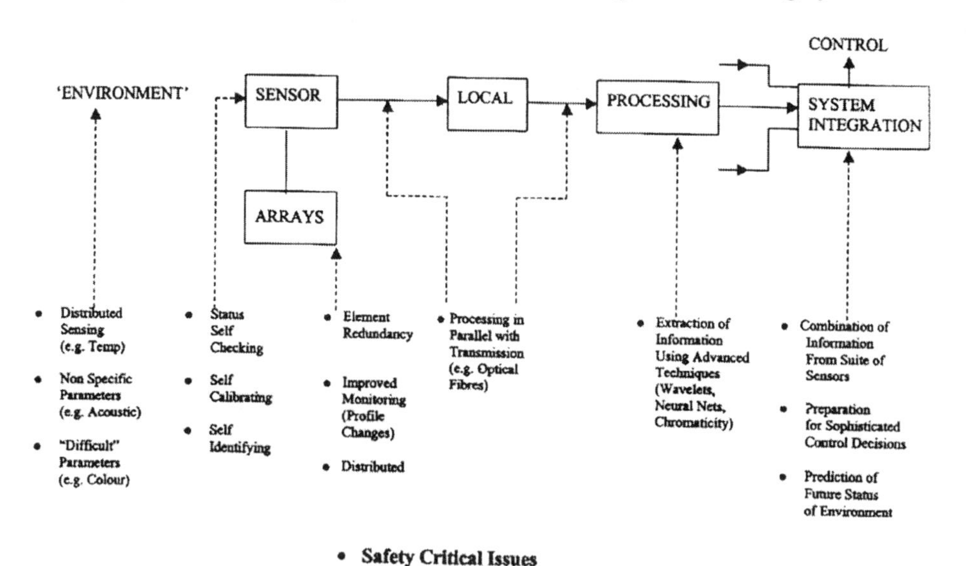

Figure 1.40. Elements of intelligent monitoring

1.7 REFERENCES

1. Culshaw, B. (1994) *Optical Fibre Sensing and Signal Processing*, Peter Peregrians.
2. Halley, P. (1995) *Fibre Optic Systems*, Wiley.
3. Cosgrave, J. A., Russel, P.C., Hall, W., Jones, G. R.(1997) Optoacoustic monitoring of electric avcs in high voltage circuit breakers. *Proc. 3rd Int Conf. On Switching Arcs*, Xian Jiatong University, P R China, 598-605
4. Medlock, R. S. (1986) Review of modulating techniques for fibre optic sensors. *J. Opt. Sensors,* **1**(1), 43-68.
5. Kwan, S., Beaven, C. M. and Jones, G. R. (1990) Displacement measurement using a focussing chromatic modulator. *Meas. Sci. Technol.*, **1**, 207-15.

6. Hoogenboom, L., Hull-Allen, G. and Wang, S. (1984) Theoretical and experimental analysis of a fiber optic proximity probe. *Proc. SPIE,* **478**, *Fiber Optic and Laser Sensors, 11,* 46-57.
7. Busurin, V. I., Semenov, A. S., and Udalov, N. P. (1985) Optical and fiber-optic sensors (review). *Sov. J. Quant. Electron.* **15**(5), 595-621.
8. Spillman, W. B. and McMahon, D. H. (1980) *Appl. Oprics,* **19**, 113.
9. Saad, M., Jones, G. R. and Stevens, A. (1990) Fiber optic vehicle sensing. *Report,* TRRL.
10. Fernando, G. F., Lin, T., Crosby, P., Doyle, C., Martin, A., Brooks, D., Ralph, B., Badcock, R. (1997) A multipurpose optical fiber sensor design for fiber reinforced composite materials. *Meas. Sci. Technol.* **8**, 1065-79.
11. Smith, R., Spencer, J. W., Jones, G. R., Lightfood, J. and Dean, E. (1994). Fiber Optic probe for determining the thickness of immiscible layers, *IEE Proc. Optoelectron.* **141**, 275-9.
12. Russell, P.C., Jones, G. R., Crowther, D. C., Jones, M. Ahmed, S. U., Huggett, P. (1996) A chromatic sensor for detecting water in organic solvents, *Proc. OPTO 96 2nd Congress for Optical Sensor Technology, Measuring Techniques and Electronics*, 145-50.
13. Khavaz, A., Jones, B. E. (1995) A distributed optical fiber sensing system for multi-point humidity measurement. *Sensors and Actuators A*, **46-47**, 491-3.
14. Jones, G. R., Lewis, E., Kwan, S. *et al* (1989) Optical fiber monitoring of power circuit breakers. *Proc. SPIE,* **1120**, *Fibre Optics '89,* 224-35.
15. MacCraith, B. D., O'Keeffee, G., McDonagh, C., McEvoy, A. K. (1994) LED-based fiber optic oxygen sensor using sol-gel coating. *Electronics Letters*, **30**, 888-9.
16. Hutley, M. C., Stevens, R. F. and Putland, D. E. (1986) Wavelength encoded optical fiber sensors. *J. Opt. Sensors,* **1**(2), 153-62.
17. Jones, G. R., Russell, P. C., Khandaker, I. (1994) Chromatic Interferometry for an Intellegent Plasma Processing System. *Meas. Sci, Technol.* **5**, 639-47.
18. Singh, P. T., Spencer, J. W., Li, G., Humphries, J. E., Jones, G. R., Doncaster, J. L., Pinnock, J. L., Dean, E., Simpson, H. (1996) A white light interferometric technique for monitoring temperatures. *Proc. XI Int. Conf. on Optical Fiber Sensors (Advanced Sensing Photonics)*, **1**, 344-7.
19. Ezbiri, A., Tatam, R. P. (1996) Interogation of low finesse optical fiber Fabry Perot interferometers using a four wavelength technique. *Meas. Sci. Technol.* **7**, 117-20.
20. Ezbiri, A., Tatam, R. P. (1997) Five wavelength interogation technique for miniature fiber optic Fabry - Perot sensors. *Optics Commun.*, **133**, 62-66.
21. Saac, L. T. (1997) Puffer circuit breaker diagnostics using novel optical fibre sensors. *Ph.D. Thesis*, University of Liverpool
22. Murphy, M. M., Jones, G. R. (1992) Polychromatic Birefringence Sensing for Optical Fiber Monitoring of Surface. Strain, *Sensors and Actuators A*, **32**(1-3), 691-5.
23. Humphries, J., Harrison, J. A., Spencer, J. W., Jones, G. R. (1996) On line compensation and calibration of a chromatically addressed, photoelastic strain sensor with optical fiber transmission, *IEE Proc - Sci. Meas. Technol.* **143**(3), 166-70.
24. Jones, G. R., Li, G., Spencer, J. W., Aspey, R. A., Kong, M. G. (1998) Faraday current sensing employing chromatic modulation, *Optics Commun.*, **145**, 203-12.
25. Pilling, N. A., Holmes, R. and Jones, G. R. (1993) optically powered hybrid current measurement system, *Electronics Letters*, **29**, 1049-51.
26. Pilling, N. A. (1992) Optical fiber measurements in power systems, *Ph.D. Thesis*, University of Liverpool.

27. Donaldson, E. (1998) *Private Communication*
28. Isaac, L., Spencer, J. W., Humphries, J., Jones, G. R. (1997) Particle formation by SF_6 circuit breaker arcs, *Proc. 12th Int. Conf. on Gas Discharges and their Application* (Greifswald, Germany), 123-6.
29. West, I. P., Holmes, R. and Jones, G. R. (1994) Pulse oximeter for accurate measurement of oxygen saturation in the presence of carboxy heamoglobin. *Proc. Appl. Optics and Optoelectronics Conf.* (10P) (York), 401-2.
30. Scully, P. J., Holmes, R. and Jones, G. R. (1994) Remote in - vivo monitoring of blood oxygen saturation of patients using optical fibers. *Proc. European Conf. on Lasers and Electro Optics* (Amsterdam), 265-6.
31. Scheggi, A. M., Bacci, M. and Brenci, M. (1985) Compact temperature measurement system for medical applications. *Proc. SPIE*, **586**, *Fiber Optics Sensors*, 110-3.
32. Russell, P. C., Alston, D., Smith, R. V., Jones, G. R.., Huggett, P. (1996) On line fiber optic based inspection using chromatic modulation for semiconductor materials processing. *Nondestr. Test. Eval.*, **12**, 379-389.
33. Jones, G. R. and Russell, P. C. (1993) Chromatic modulation based metrology, *Pure and Applied Optics*, **2**, 87-110.
34. Isaac, L. T., Spencer, J. W., Jones, G. R., Jones, C., Hall, W. B., Taylor, B. (1995) Live monitoring of contact travel on EHV circuit breakers using a novel optical fibre techniques. *Proc. XI Int. Conf. on Gas Discharges and their Application* (Tokyo, Japan) 288.
35. Jones, G. R. (1993) Plasma monitoring using chromatically processed optical signals, *Proc. XXI Int. Conf. on Phenomena in Ionized Gases, Proc III* (Invited Lectures), (Bochum, Germany), 24-33.
36. Jones, G. R. (1995) Electric arc monitoring utilising intelligent optical fiber systems. *Proc X Int. Conf. on Gas Discharges and their Applications, part II* (invited lecture), 504-512 (Tokyo, Japan) ISBN 4-88686-449-6C3054 P30000E
37. Stergoulas, L. (1997) Frequency : time mapping for signal extraction, *Ph.D Thesis*, University of Liverpool.
38. Jones, G. R., Russell, P. C., Cosgrave, J., Spencer, J. W., Vourdas, A., Hall, W., Wilson, A. (1996) Chromatic processing of optoacoustic signals for identifying incipient faults on electric power equipment. *IEE Colloquim on Intelligent Sensors* (Leicester). Digest **121**, 3/1-4.
39. Russell, P. C., Tomtsis, D., Cosgrave, J., Vourdas, A., Stergioulas, L. and Jones, G. R. (1998) Extraction of information from acoustic vibration signals using Gabor devices. *Meas. Sci. Technol.*, **9**, 1282-90
40. Ahmed, S. U., Russell, P. C., Lisboa, P. J. and Jones, G. R. (1997) Parameter monitoring using meural network processed chromaticity, *IEE Proc. Sci. Meas. Technol.*, **144**(6), 257-62.
41. Weiss, L. G. (1994) Wavelets and Wideband Correlation Processing, *IEE Signal Processing Magazine* (January), 13-32.
42. Gabor, D. (1946) Theory of communication, *J. IEE 93* (3), 429-457.
43. Wysecki, G. and Stiles, W. S. (1982) *Colour science : concepts and methods. quantitative data and fomulae.* second edition, Wiley, New York.
44. CIE Publications 15.2 '*Colorimetry*' (1986)
45. Levkovitz, H. and Herman, G. T. (1993) GLHS : a generalised Lightness, Hue and Saturation color model, *CVGIP: Lyraphic Models and Image Processing*, **55**(4), 271-85.
46. Russell, P. C., Spencer, J. W., Jones, G. R. (1998) Optical fiber sensing for intelligent monitoring using chromatic methodologies, *Sensors Review*, **18**, 44-8.

47. Schwarz, M. W., Cowan, W. B., Beatty, J. C. (1987) An experimental comparison of RGB, YIQ, LAB, HSV and opponent colour models. *ACM Trans. Graphics*, **6**(2) 123-58.
48. Murphy, M. M. (1991) Optical Fiber Structure Monitoring. *Ph.D. Thesis*, University of Liverpool.
49. Messent, D. N., Singh, P. T., Humphries, J. E., Spencer, J. W., Jones, G. R., Lewis, K. G., Hall, W. (1997) Optical fiber measurement of contact stalk temperature in an SF_6 circuit breaker following fault- current arcing. *Proc. 12th Int. Conf. on Gas Discharges and their Applications* (Greifswald, Germany) 543-6.
50. Lewis, G. L., Jones, R. E., Jones, G. R. (1995) A tap-changer monitoring system incorporating optical sensors, *Proc. IEE Conf. Reliability of Tramsmission and Distribution Equipment.*
51. Tranter, A. D. (1996) Non contact optical strain and Torque Measurement. *Ph.D. Thesis*, University of Liverpool.
52. Kershaw, D., Holmes, R., Henderson, P. and Jones, G. R. (1991) Chromatic pH measurements, *Proc. Sensors and their Applications, V. Conf.* (Edinburgh), 3-8.
53. Yu, R. J., Lisboa, P. J. G., Russell, P. C., Jones, G. R. (1997) Resolution Capabilities of Chromatic Sensing in the Monitoring of Semiconductor Plasma Processing Systems. *Non. Destr. Test. Eval.* **13**, 347-60.
54. Khandaker, I. I. (1993) Optical fiber sensors for the optimisation of plasma processing. *Ph.D. Thesis*, University of Liverpool.
55. Ryan, J. D., Russell, P. C., Tinture, E., Jones, G. R., Dwars, Strachan, D. (1997) Near infrared tichniques for LPG control. *Environmental Sens. 97*, 289-94 (Munich).
56. Beavan, C. (1989) Colour Measurement in Optical Metrology. *Ph.D. Thesis*, University of Liverpool.
57. Russell, P. C., O'Keefe, G., Cosgrave, J. and Jones, G. R. (1996) Vibration monitoring using 2-D speckle pattern images, *Proc. 22 Int. Conf. on High Speed Photography and Photonics* (USA), 33-4.
58. Meng, H., Lisboa, P. J. G., Russell, P. C., Jones, G. R. (1996) The modelling of plasma etching processes using neural networks and statistical techniques. *Proc. IEEE Int. Symp. on Intellegent Control* (Dearborn USA), 218-23.
59. Walker, J. C., Holmes, R., Jones, G. R. (1995) Code division multiplexing optical fiber sensors using a spatial light modulators, *Pure and Applied Optics*, 105-17
60. Lipson, S. G. and Lipson, H. (1981) *Optical Physics, 2nd Edn*, 221-32, Cambridge University Press.
61. Krakenas, K. and Blotekjaer, K. (1993), *J. Lightwave Technol.*, **11**(4), 643-53.
62. Ulrich, R. Patent Number 4,596,466, June 24, 1986 (filed November 20, 1981).
63. Chen, S., Palmer, A. W., Grattan, K. T. V., Meggitt, B. T. (1992) Digital signal-processing techniques for electronically scanned optical-fiber white-light interferometry. *Appl. Optics,* **31**(28), 6003-10.
64. Sandoz, P. and Tribillon, G. (1993) Profilometry by zero-order interference fringe identification. *J. Mod. Optics,* **10**(9), 1691-700.
65. Caber, P. J. (1993) Interferometric profiler for rough surfaces. *Appl. Optics,* **32**(19), 3438-41.
66. Jones, R. (1992) Micro system technologies, in *Proceedings 3rd International Conference on Micro Systems,* Berlin, Oct. 21-23, 1992, 147-55, VDE-Verlag GmbH.
67. Birch, K. P. (1990) *Prec. Eng.,* **12**(4), 195-8.
68. Jones, G. R. (1998) Optical fiber sensing for intelligent monitoring, *Sensors Review* (ISSN-0260-2288) **18**, 5-6.

2

Bragg Gratings in Optical Fibers: Fundamentals and Applications

A. Othonos

2.1 INTRODUCTION

The development of fiber optics has revolutionized the field of telecommunications making possible high-quality, high-capacity, long distance telephone links. Over the past three decades, the advancements in optical fiber has undoubtedly improved and reshaped fiber optic technology. Today optical fibers are synonymous with the word "telecommunication". In addition to applications in telecommunications, optical fibers are also utilized in the rapidly growing field of fiber sensors. Despite the improvements in optical fiber manufacturing and advancements in the field in general, basic optical components such as mirrors, wavelength filters, and partial reflectors have been a challenge to integrate with fiber optics. Recently, however, all these have changed with the ability to alter the core index of refraction in a single mode optical fiber by optical absorption of UV light. This photosensitivity of optical fibers allows the fabrication of phase structures in the core of fibers called the *fiber Bragg grating* (figure 2.1). Photosensitivity refers to a permanent change in the index of refraction of the fiber core when exposed to light with characteristic wavelength and intensity that depend on the core material. The fiber Bragg grating can perform many primary functions, such as reflection and filtering, in a highly efficient, low loss manner. This versatility has stimulated a number of significant innovations [1-3].

For a conventional fiber Bragg grating, the periodicity of the index modulation has a physical spacing that is one half of the wavelength of light propagating in the waveguide (it is phase matching between the grating planes and incident light that results in coherent back reflection). Reflectivities approaching 100% are possible, with the grating bandwidth

K.T.V. Grattan and B.T. Meggitt (eds.), Optical Fiber Sensor Technology, 79–187.

tailored from typically 0.1 nm to in excess of tens of nanometers. These characteristics make Bragg gratings suitable for telecommunications [1], where they are used to reflect, filter or disperse light. Fiber lasers, capable of producing light at telecommunications windows, utilize Bragg gratings in forming both the high-reflectivity end mirror and output coupler to the laser cavity, realizing an efficient and inherently stable source. Moreover, the ability of gratings with non-uniform periodicity to compress or expand pulses is particularly important to high-bit-rate, long-haul communication systems. For example, grating-based dispersion compensation of 10Gb/s transmission systems over ~270 km has been demonstrated. Furthermore, the Bragg grating meets the demands of dense wavelength division multiplexing, which requires narrowband, wavelength selective components, offering very high extinction between information channels. There are numerous applications that exist for low loss, fiber optic filters, examples of which are ASE noise suppression in amplified systems, pump recycling in fiber amplifiers and soliton pulse control.

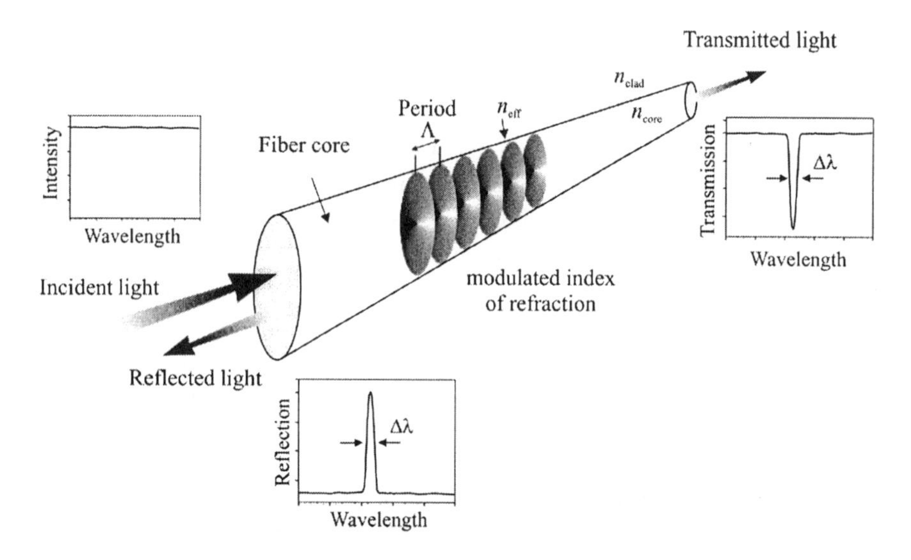

Figure 2.1. A schematic representation of a Bragg grating inscribed in the core of an optical fiber. The period of the index of refraction variation is represented by Λ. A broadband light is coupled into the core of the fiber. Part of the input light is reflected (at the Bragg condition) and the rest is transmitted. The bandwidth of the reflected and transmitted light depends on the characteristics of the Bragg grating, its length and modulation depth.

The grating planes are subject to temperature and strain perturbations, as is the host glass material, modifying the phase matching condition and leading to wavelength dependent reflectivity. Typically, at 1.5μm, the wavelength–strain (λ–ε) responsivity is ~1pm/nε, with a wavelength shift of

15pm/°C, for temperature excursions. Therefore by tracking the wavelength at which the Bragg reflection occurs the magnitude of an external perturbation may be obtained. This functionality approaches the ideal goal of optical fiber sensors; to have an intrinsic in-line, fiber-core structure that offers an absolute readout mechanism. The reliable detection of sensor signals is critical and spectrally encoded information is potentially the simplest approach, offering simple decoding that may even be facilitated by another grating. An alternative approach is to use the grating as a reflective marker, mapping out lengths of optical fiber. Optical time domain measurements allow for accurate length or strain monitoring.

The grating may be photoimprinted into the fiber core during the fiber manufacturing process, with no measurable loss to the mechanical strength of the host material. This makes it possible to place a large number of Bragg gratings at predetermined locations on the optical fiber to realize a quasi-distributed sensor network for structural monitoring, with relative ease and low cost. Importantly, the basic instrumentation applicable to conventional optical fiber sensor arrays may also incorporate grating sensors, permitting the combination of both sensor types. Bragg gratings are ideal candidates for sensors, measuring dynamic strain to a 1nε resolution in aerospace applications and as temperature sensors for medical applications. They also operate well in hostile environments such as high pressure, borehole drilling applications, principally as a result of the properties of host glass material.

Fiber optic photosensitivity has indeed opened a new era in the field of fiber optic based devices [1]. Innovating new Bragg grating structures find their way in telecommunication and sensor applications. Devices like fiber *Fabry-Perot Bragg gratings* for band-pass filters, *chirped gratings* for dispersion compensation and pulse shaping in ultrashort work, and *blazed gratings* for mode converters are becoming routine applications. Fiber optics sensing is an area that has embraced Bragg gratings since the early days of its discovery. Fiber Bragg gratings have become almost synonymous with the field itself. Most fiber optics sensor systems today make use of Bragg grating technology.

Within a few years from their initial development, fiber Bragg gratings have moved from laboratory interest and curiosity to the brink of implementation in optical communication and sensor systems. In a few years, it may be as difficult to think of fiber optic systems without fiber Bragg gratings as it is to think of bulk optics without the familiar laboratory mirror.

2.1.1 Historical prospective

Hill and co-workers [4,5] discovered photosensitivity in germanium-doped silica fiber at the Communication Research Center in Canada. During an experiment that was carried out to study the non-linear effects in a specially designed optical fiber, visible light from an argon ion laser was launched into the core of the fiber. Under prolonged exposure, an increase in the attenuation of the fiber was observed. Following that observation, it was determined that the intensity of the light back-reflected from the fiber increased significantly with time during the exposure. This increase in reflectivity was the result of a permanent refractive index grating being photoinduced in the fiber. This new non-linear photorefractive effect in optical fibers was called fiber photosensitivity. In their experiment the 488 nm laser light that was launched into the core of a specially designed fiber interfered with the Fresnel reflected beam (4% reflection from the cleaved end of the fiber) and initially formed a weak standing wave intensity pattern. The high intensity points altered the index of refraction in the photosensitive fiber core permanently. Thus, a refractive index perturbation that had the same spatial periodicity as the interference pattern was formed, with a length only limited by the coherence length of the writing radiation. This refractive index grating acted as a distributed reflector that coupled the forward propagating to the counter-propagating light beams. The coupling of the beams provided positive feedback, which enhanced the strength of the back-reflected light and thereby increased the intensity of the interference pattern, which in turn increased the index of refraction at the high intensity points. This process was continued until eventually the reflectivity of the grating reached a saturation level. These gratings were thus called *self-organized* or *self-induced* gratings since they formed spontaneously without human intervention. The specially designed fibers were supplied by Bell Northern Research, and had a small core diameter that was heavily doped with germanium.

For almost a decade after its discovery, research on fiber photosensitivity was pursued sporadically only in Canada using the special Bell Northern research fiber. During this time Lam and Garside [6] (1981) showed that the magnitude of the photoinduced refractive index change depended on the square of the writing power at the argon ion wavelength (488 nm). This suggested a two-photon process as the possible mechanism of refractive index change. The lack of international interest in fiber photosensitivity at the time was attributed to the effect being viewed as a phenomenon present only in this special fiber. Almost a decade later, it was proved otherwise [7]; Stone observed photosensitivity in many different fibers, all of which contained a relatively high concentration of germanium.

2.1.2 Externally inscribed Bragg gratings

Although the discovery of photosensitivity in the form of photoinduced index changes played a key role in the advancements of optical fiber technology, devices such as the *self-induced* gratings were not practical. This was largely because the Bragg resonance wavelength was limited to the argon ion writing wavelength (488 nm), with very small wavelength changes induced by straining the fiber. The key development that turned this phenomenon from a scientific curiosity to a mainstream tool was the side-writing technique first demonstrated at the United Technologies Research Center (this is sometimes called the transverse holographic technique). In 1989 Meltz et al [8], following the work by Lam and Garside [6], showed that a strong index of refraction change occurred when a germanium doped fiber was exposed to direct, single-photon, UV light close to 5eV. This coincides with the absorption peak of a germania-related defect at a wavelength range of 240-250 nm. Irradiating the side of the optical fiber with a periodic pattern, derived from the intersection of two, coherent 244 nm beams in an interferometer, resulted in a modulation of the core index of refraction, inducing a periodic grating. Changing the angle between the intersecting beams alters the spacing between the interference maxima, and this sets the periodicity of the gratings thus making possible reflectance at any wavelength. Even though the writing wavelength was at 244 nm, gratings could be fabricated to reflect at any wavelength permitting their use in modern telecommunication and sensor systems. A subtle, but important point is that this method relies on an increase in refractive index that is maintained in the long wavelength region of interest, i.e. 1300-1500 nm, even though the physical phenomenon is related to the absorption of light in the ultraviolet region. This technique was steadily refined such that by 1992 index changes as large as 2×10^{-3} were reported. By 1993 it was not uncommon for publications reporting values of Δn to be comparable with the core-cladding refractive index difference. The resulting competition between the guidance from the core-cladding refractive index difference and diffraction/mode mixing from Δn has made possible a wide variety of linear and nonlinear optical devices.

A basic comparison of the difference in efficiency between the two-photon (associated with a self-induced gratings) and single-photon writing process may be made by comparing the fluence levels required to induce comparable index changes. For the two-photon process the photoinduced refractive index saturates after exposures to fluence levels approaching 1GJ/cm^2: on the other hand, the single-photon process requires only 1kJ/cm^2 for the same index change, a factor of a million times less. The magnitude of the refractive index change has been shown to depend on many factors, the most important being the writing wavelength, the writing beam intensity and net dosage, the composition of the host material and any pre-processing that the fiber may have undergone. The most commonly used light sources are KrF and ArF excimer lasers, operating at 248 nm and 193 nm, respectively. These lasers typically generate pulses of duration 10-20 ns, at repetition rates of tens of pulses/s. A typical example indicates that exposure to laser irradiation lasting for several minutes, at intensities of 100-500 mJ/cm^2, will result in a Δn that is positive in Ge-doped, single-mode optical fiber, having a magnitude of 10^{-5} to 10^{-4}.

An issue common to internally and externally written gratings is a reflectivity dependent on the polarization of the probing light beam, that is to say, the refractive index change is birefringent. This fundamental property is relevant to understanding the photophysics of fiber Bragg gratings and has also proven to be useful to grating applications, such as the fabrication of polarization mode converting devices or rocking filters [9].

Further developments of consequence to be discussed in detail in the sections that follow are sensitization techniques, such as hydrogenation. A photosensitivity enhancement of an order of magnitude increase in grating reflectivity strength (Δn of 10^{-2}) has been realized for standard telecommunications fibers, through hydrogenation of fibers prior to UV exposure [10]. Additionally, the use of phase masks for grating fabrication has also made a tremendous impact to the field. Reliable mass-produced gratings - the reality of commercial grating based devices - may be realized through the use of phase masks [11]. This is a technique derived from conventional photolithography. The phase mask is a diffractive optical element that spatially modulates the UV writing beam, it is a surface relief structure made of silica glass. Interference between the diffracted plus and minus first orders results in a periodic, near field, high contrast intensity pattern, having half the phase mask grating pitch. For the correct UV wavelength, the interference pattern will photoinduce a Bragg grating into the fiber core. While the use of phase masks does not introduce any improvement in the strength of the index modulation, it does relax the laser source and stability requirements for the fabrication of Bragg gratings.

2.2 PHOTOSENSITIVITY OF OPTICAL FIBER

Hill et al [4,5] first discovered photosensitivity to light at 488 nm in germanosilicate optical fibers. The growth with laser power was associated to a two-photon process [6], and thus a connection with the well-known 248 nm absorption band was made. A transverse writing method was later used to photo-imprint Bragg gratings at a direct excitation wavelength of 240 nm

[8]. The absorption band centered on this excitation has been related with defect centers in germanosilicate glass [12,13]. Irradiation with a wavelength coincident with this band was shown to result in bleaching, while creating other absorption bands leading to a refractive index change that was described through the Kramers-Kronig relation [14]. In 1993 Lemaire et al [10] showed that significantly larger index changes could be achieved by "hydrogen-loading" the glass before exposure, and in some cases without variation of the 240 nm absorption band. The latest experimental findings indicate the formation of spectral changes below 240 nm, and 193 nm excitation of non-hydrogen loaded, low germanium content fiber can result in high index changes that are commensurate with the fiber core-cladding refractive index difference [15]. It appears that photosensitivity at 193 nm obeys one-photon dynamics in high-germanium content fiber, and two-photon dynamics in low-germanium content fiber. Recently, a two-photon process has also been observed in germanosilicate glass for various UV wavelengths [16]. The current consensus explains photosensitivity as being initiated through the formation of color-centers [17], that gives way to compaction of the UV irradiated glass [18,19]. The phenomenon of photosensitivity has resulted in arguably one of the most important in-fiber components called the fiber Bragg grating [1].

2.2.1 Photosensitivity in silicon-based optical fibers

Early studies on photosensitivity and grating growth pinpointed the essential requirement of germanium [7]. However, there are now numerous examples in the literature of photosensitivity in a wide range of fibers, many of which do not contain germanium as a dopant. Fibers doped with europium [20], cerium [21] and erbium:germanium [22] show varying degrees of sensitivity in a silica host optical fiber. One fiber doping producing large index modulations (of the order of 10^{-3}) is germanium-boron co-doping [23]. Photosensitivity has also been observed in a fluorozirconate fiber [24] doped with cerium:erbium where Bragg gratings were inscribed using 246 nm radiation. From a practical point of view, the most interesting photosensitive fibers are germanium core-doped, as they are used extensively in both the telecommunications industry and optical sensor applications.

Initially, when photosensitivity was thought to occur only in germanium-doped fiber it was believed that the germanium oxygen vacancy defects, such as a twofold coordinated neutral germanium atom (O-Ge-O or Ge_2^0 center) or the Ge-Si or Ge-Ge (the so called *wrong bonds*) were responsible for the photoinduced index changes. However, with the demonstration of photosensitivity in most types of fiber, it is apparent that photosensitivity is a function of various mechanisms (photochemical, photomechanical, thermochemical) and the relative contribution will be fiber dependent, in addition to intensity and wavelength. Several models, proposed to describe the photoinduced refractive index changes in germanium-doped fiber, share the common element of the germanium oxygen vacancy defects as precursors responsible for the photoinduced index changes. During the high-temperature gas-phase oxidation process of the modified chemical vapor deposition (MCVD) technique, GeO_2 dissociates to the GeO molecule (in other words the Ge^{2+} center) due to its higher stability at elevated temperatures. This species, when incorporated into the glass, can manifest itself in the form of oxygen vacancy Ge-Si and Ge-Ge wrong bonds. Regardless of which particular defect causes an oxygen deficient matrix in glass, it is linked to the 240-250 nm absorption band (peaking at 242 nm) and its centers are known as germanium oxygen-deficient centers (GODCs).

The growth dynamics of the Bragg gratings as they are exposed to the UV radiation give an important insight to the photosensitivity of fibers. One may distinguish three distinct dynamical regimes known as Type I, IIA and II [25]. The key differences are highlighted in Table 2.1. It is almost certain that the mechanisms responsible for Type I, Type IIA and II are different. The physical properties of these grating types may also be inferred through their growth dynamics and also by measurement of thermally induced decay. The accelerated decay is different for each grating type, with Type I being the least and Type II the most stable with temperature. Type IIA falls in between. This is not surprising given that Type I is related to local electronic defects, Type IIA to compaction and Type II to fusion of the glass matrix. It has been asserted that the growth dynamics of UV written Bragg gratings may be studied without making a distinction between gratings fabricated using pulsed or continuous wave (CW) lasers [26]. There is certainly evidence suggesting that the basic functional law for the growth of Type I gratings is the same under both writing conditions. It is also undoubtedly true that writing Type I gratings with pulsed laser light is more efficient than with CW radiation [27-29]. Similarly, Type IIA gratings have been written under pulsed and CW conditions [26], whereas Type II gratings have only been fabricated on exposure to high energy, pulsed radiation. Specific examples for each grating type are given and differences between them are highlighted.

2.2.1.1 Type I gratings

A comparison between Type I gratings written under CW and pulsed conditions is shown in figure 2.2(a) and (b). Figure 2.2(a) shows a high peak reflectivity Type I Bragg gratings which was inscribed in Ge-doped optical fiber (10 mol%) with a CW UV light beam at 244 nm. The fiber was exposed to 47 W/cm^2 for 15 minutes [28]. Near identical gratings, produced in the same Accutether fiber by use of a pulsed source, required only 5 W/cm^2 average intensity and a 200s exposure time (figure 2.2(b)) [29,30]. Assuming no errors in the reported result, it is possible that this difference in writing efficiency is related to transient heating, accelerating photothermal ionization of defects, or compaction of the glass matrix as a result of the high peak intensity of the pulsed light.

Table 2.1. Key differences between Type I, IIA and II Bragg gratings

Grating Type	Writing Conditions		Observations
	CW Fluence (cumulative)	Pulsed Typical pulse energies	
Type 1	Up to 500J/cm^2	100mJ/cm^2	$\Delta n > 0$ Associated with defects and density changes in glass matrix
Type IIA	>500J/cm^2	100mJ/cm^2	$\Delta n < 0$ Associated with compaction of the glass matrix
Type II	Not applicable	1000mJ/cm^2	$\Delta n > 0$ Associated with fusion of the glass matrix

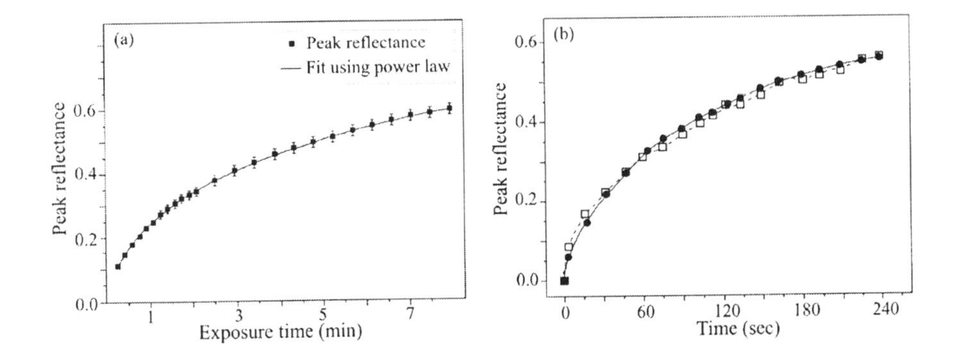

Figure 2.2. (a) Reflectance versus time for an exposure intensity of 46W/cm^2 and a final index modulation of ~8×10^{-5}. The solid curve is a fit to R that assumes that $\Delta n=Ct^b$ ($C = 4.3\times10^{-5}$, b=0.32, t is in minutes) (After [28]). (b) Growth rate of Bragg gratings written in AT&T Accutether fiber (10 mol% germania). The solid circles represent the growth rate of the initial Bragg grating, where as the squares correspond to the growth rate of the second writing after the first grating was thermally erased (after [29])

2.2.1.2 Type IIA gratings

Gratings referred to as Type IIA have most often been demonstrated in high Ge-content, small core fibers and have often been associated with the presence of high internal fiber stresses [32,33]. This is substantiated by the absence of Type IIA formation in low-Ge-doped fibers at 240 nm. The lack of Type IIA behavior for gratings in hydrogen loaded fibers implies that hydrogen treatment modifies a chemical or physical property of the fiber that changes the conditions for the initial photosensitivity mechanisms related to the color-center and compaction models. A typical result for the development of Type IIA from Type I gratings is shown in figure 2.3 [34]. A pulsed laser (12 ns at 10Hz) operating at 244 nm was used to record gratings in germanosilicate fiber (28 mol%). One observes a fast decrease in the transmission of the fiber corresponding to an increase of reflectivity of the Bragg grating of ~10%. The reflectivity eventually reaches 100%. Under continued irradiation the reflectivity decrease (transmission increases), indicating the disappearance of the index modulation. Beyond this threshold another index grating appears. This complex behavior has been seen in many different fibers and it can be concluded that it does not arise from instability in the writing process. Recent experiments have shown that the formation dynamics of Type I/Type IIA grating spectra are strongly affected when the gratings are written in strained fiber [32]. Straining a fiber during grating inscription limits the Type I index modulation while accelerating the formation of the Type IIA grating [33].

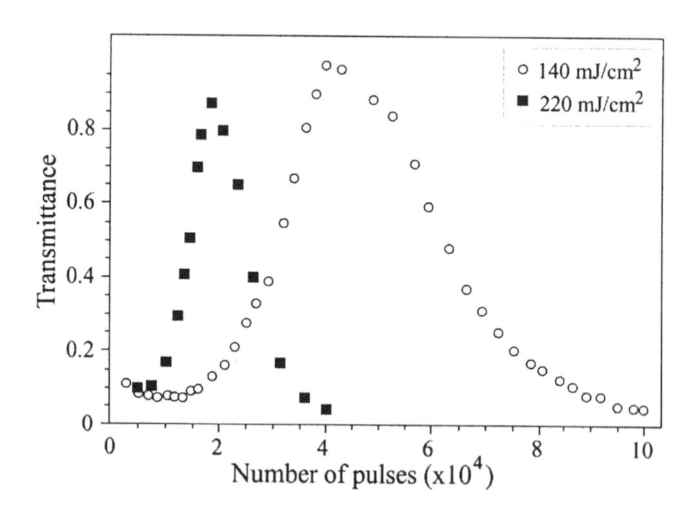

Figure 2.3. Transmission of a Bragg grating as a function of the number of pulses used for the inscription at two different energy densities (after [34])

2.2.1.3 Type II gratings

A single excimer light pulse can photoinduce large refractive index changes in small localized regions at the core/cladding boundary, resulting in the formation of the Type II grating [35]. This results from physical damage through localized fusion that is limited to the fiber core, and producing very large refractive index modulations estimated to be close to 10^{-2}. The growth dynamics of a Type II grating is shown in figure 2.4, produced with single high power KrF laser pulses. There is a sharp threshold corresponding to a pulse energy of ~0.65 J/cm^2. Doubling the pulse energy from 0.45 J/cm^2 to 0.9 J/cm^2 results in a photoinduced modulation index that increases by two orders of magnitude. Below the threshold, the index growth is linear.

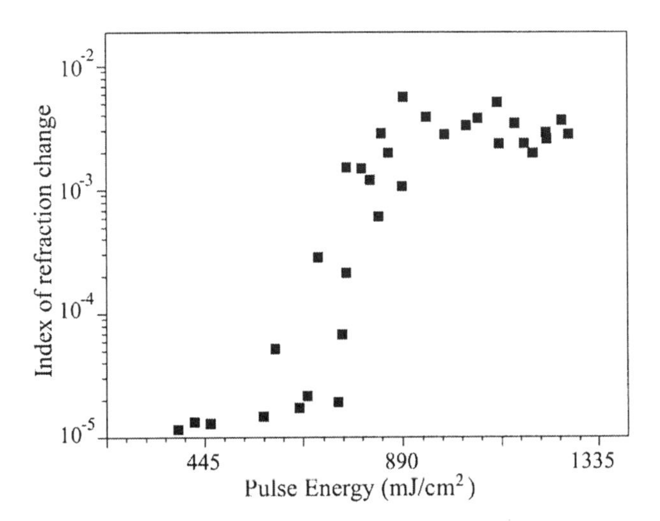

Figure 2.4. Index of refraction changes estimated for Bragg gratings inscribed in photosensitive fiber induced with a single excimer (KrF) laser pulse. The estimated refractive index change is plotted against the energy density of the laser pulse. Notice a sharp threshold at around 0.75J/cm^2 (after [35])

2.2.1.4 Temperature dependence

Bragg gratings are found to exhibit temperature-dependent decay of Δn with time after inscription. The resulting decay in reflectivity is characterized by a power law dependence with time which has a rapid initial decay followed by a decreasing decay rate, see figure 2.5 [36]. This behavior is consistent with the thermal depopulation of trapped states occupied by carriers that are photoexcited from their original band locations by UV irradiation. Thermally exciting carriers out of shallow traps causes the observed decay in the refractive index. Any residual carriers are related to the "stable" portion of the index change. Therefore, a grating may be preannealed to remove the portion of Δn that decays rapidly, leaving only the portion that has long-term stability. A noticeable difference between Type I and Type IIA gratings is their different thermal behavior [37]. Type I gratings were found to have reasonable short-term stability to 300°C, whereas Type IIA gratings demonstrated excellent stability at temperatures as high as 500°C. Type II gratings have been shown to be extremely stable up to 800°C.

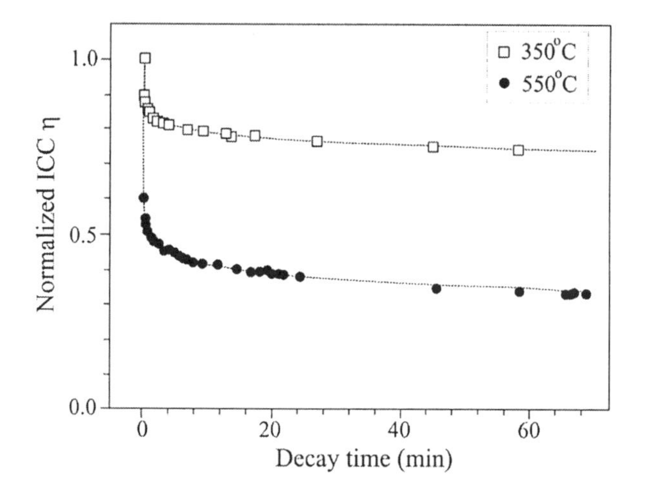

Figure 2.5. Measured integrated coupling constant normalized to starting value for two gratings heated to 350 and 550°C as a function of decay time. The lines are fit to the data (after [36])

2.2.1.5 Anisotropy in the photoinduced index change

Photoinduced birefringence in germanosilicate fibers has been studied in connection with self-organized gratings, which display a birefringence of the order of 10^{-6} [39], and externally written gratings for which the birefringence may be two orders of magnitude greater [8,40,41]. Similar anisotropy has been observed in the index change induced by the interference of visible light, again in germanium doped fiber [39,42-44]. The anisotropy of the UV induced index change has important implications for phase gratings, offering further insight into the mechanism of photosensitivity. Exploitation of the polarization-dependent reflectivity has resulted in demonstrations of significant telecommunication components, such as the single-mode operation of an erbium doped fiber grating laser [45], polarization mode converters and rocking filters [9] and in-fiber, in-line wave retarders [41].

2.2.2 Defects in germanosilicate fiber

The fact that the change of index of refraction in a germanosilica fiber is triggered by a single photon at energies well below the band gap (146 nm), implies that the point defects in the ideal glass tetrahedral network are responsible for this process. Defects in optical fibers first attracted attention because of the unwanted absorption band associated with them, which caused transmission losses. These defects are often called color centers due to their strong absorption. Normally these defects are caused by the fiber drawing process [46], and ionizing radiation [47]. A great deal of research has been directed in minimizing the formation of these defects. However, with their implication for fiber Bragg gratings, the role of the defects in optical fibers has changed dramatically. In the 1980s defects were implicated in the phenomenon of second harmonic generation [48-52], as well as the fabrication of phase gratings in optical fibers.

In 1956, Weeks [53] reported a narrow resonance in the ESR spectra of neutron irradiated crystalline quartz and silica from a species termed the E' center. This was the first study of point defects in amorphous silica and many other defects in silica and germanosilicate glass have since been characterized. Point defects in optical fibers are important to various phenomena, yet their origin, chemical structure and role are uncertain.

Germanosilicate fibers represent the most studied and important photosensitive fibers, therefore it is useful to look at the point defects in germanosilicate glass. Ge, unlike Si, has two moderately stable oxidation states, +2 and +4; thus germania can be expected as both GeO_2 and GeO in glass. From thermodynamic considerations it may be expected that the concentration of GeO_2 will be proportional to the GODC concentration in the glass. It is well known that the suboxide GeO becomes more stable than GeO_2 at high temperature [54,55]. This is important in fiber preform fabrication since the incorporation of GeO produced during the high temperature, gas phase oxidation process of MCVD results in an oxygen deficient matrix. Although GeO is sometimes considered to exist as discrete molecules in the germanosilicate matrix [56], it is very likely to manifest itself in the form of Ge-Si wrong bonds, suspected defect precursors. Indeed, it has been found by Jackson et al [55] that maintaining glass at an elevated temperature (~1600°C) followed by rapid cooling to room temperature results in the formation of a large number of defects. This parallels the drawing of optical fiber from a glass preform, thus it may be anticipated that the fiber will possess a defect distribution related to the drawing process, in addition to any variance in fiber or preform stress distributions.

Radiation studies of photoinduced, paramagnetic defect centers of germanosilicate fiber [57] have identified, via electron spin resonance (ESR), damage centers directly related to the Ge content in the core. The centers are labelled Ge(n) centers and denote diamagnetic Ge sites that can trap an electron to become $Ge(n)^-$ centers. The Ge(0)/Ge(3) centers are identified with GeE', the center having the deepest electron trap depth [58,59]. Further studies of radiation effects in Ge-doped silica have identified the paramagnetic GeE' defect center as a byproduct of the GODC bleaching, induced by exposure to visible light and radiation [60]. This behavior was accounted for by considering the GODC to be a Ge atom with a next-nearest-neighbor Si atom. The breakage of the bond between this pair formed the GeE' center and released an electron facilitating the formation of new defect centers, such as the paramagnetic $Ge(1)^-$ and $Ge(2)^-$, through charge retrapping elsewhere in the network [58].

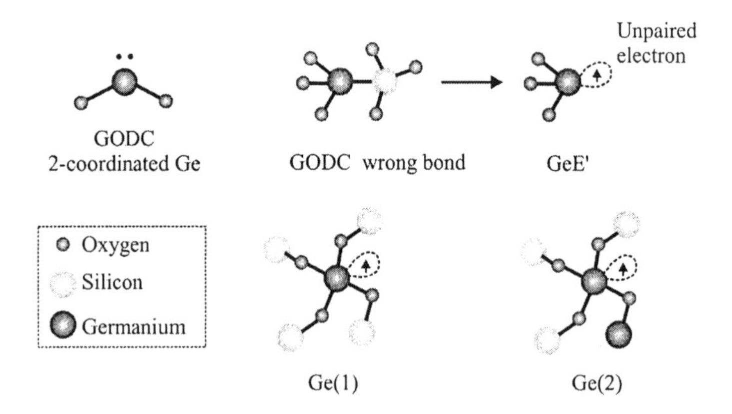

Figure 2.6. Possible GODC candidates. The GeE' center and the Ge(1) and Ge(2) electron trap centers

The Ge(1) center has a structure depicted by an electron trapped at a Ge atom coordinated to four O-Si next-nearest-neighbor atoms while the Ge(2) center is an electron trapped to a Ge atom coordinated to one O-Ge (i.e. ≡Ge-O-Ge≡) and three O-Si next-nearest-neighbor atoms, shown in figure 2.6, as proposed by Tsai et al [58,61]. $Ge(n)^-$ centers are likely to be sites of the tetrahedral network characterized by large distortions of the O-Ge-O angles from 109.4°, which mix in the Ge d-orbitals for the efficient capture of electrons into deep traps. Other ESR investigations indicate that the GeE' center is axially symmetric. The structures of the Ge-Si and GeE' centers are also shown in figure 2.6. Of significance is the large bond length increase between the Ge and Si atoms following ionization of the Ge-Si wrong bond. This is driven by the change of hybridization of the Si atom from sp^3 to sp^2, which draws the Si atoms towards the plane of the 3 O atoms to which it is bonded. There are implications to such large structural changes, both for stress relief and for compaction models.

2.2.3 Germanium-free optical fibers

Type I and Type IIA Bragg gratings have also been fabricated in Ge-free, N_2-doped silica-core fibers, with a change in refractive index $\sim 10^{-3}$ under 193 nm illumination without hydrogen loading [62]. The growth dynamics are similar to their Ge-doped counterparts, for both hydrogen loaded and unloaded cases. Whereas Type IIA behavior can be induced in unloaded fibers, with fluence levels of several kJ/cm^2, hydrogen loaded examples display a monotonic increase in the index modulation in complete agreement with hydrogen loaded, high Ge content fibers [33]. Temperature-induced changes in reflectivity also parallel the previous findings for Type I/Type IIA gratings, with the latter being more stable [37]. The mechanism for this fiber photosensitivity may be related to the two-photon absorption mechanism proposed by Albert et al for lightly Ge-doped fiber exposed to 193 nm irradiation [63]. As yet this remains unresolved.

2.2.4 Enhanced photosensitivity in silica optical fibers

Photosensitivity of optical fibers may be thought of as a measure of the amount of change that can be induced in the indcx of refraction in a fiber core following a specific exposure of UV light. Since the discovery of photosensitivity and the first demonstration of grating formation in germanosilica fibers, there has been considerable effort in understanding and increasing the photosensitivity in optical fibers. Initially, optical fibers that were fabricated with high germanium dopant levels or under reduced oxidizing conditions were proven to be highly photosensitive. Recently, *hydrogen loading (hydrogenation)*, *flame brushing*, and *boron codoping* have been used for enhancing the photosensitivity of germanosilica fibers.

2.2.4.1 Hydrogen loading (Hydrogenation)

A simple, but highly effective approach for achieving very high UV photosensitivity in optical fibers is the use of low temperature hydrogen treatment, prior to the UV exposure [10]. Fibers are soaked in hydrogen gas at temperatures ranging from 20-75°C and pressures of typically 150 atm, resulting in diffusion of hydrogen molecules into the fiber core. In excess of

95% equilibrium solubility at the fiber core can be achieved with room temperature treatment. Permanent changes in the fiber core refractive index of 0.01 are possible. One advantage of 'hydrogen loading' is the fabrication of Bragg gratings in any germanosilicate and germanium-free-fibers. Additionally, in unexposed fiber sections the hydrogen diffuses out, leaving negligible absorption losses at the important optical communication windows.

A comparison of the refractive index profile at the midpoint of a grating and compared with an untreated fiber is shown in figure 2.7. The average core index has increased by $\sim 3.4 \times 10^{-3}$. Similar results are obtained for MCVD and VAD-drawn fibers containing ~3% germania: the mechanism therefore is not dependent on fiber or preform processing, but rather on the interactions between germania and hydrogen molecules, coupled with the UV exposure conditions. Figure 2.8 shows the absorption spectrum changes in the infrared for a germanosilicate fiber exposed to 1 atm pressure of hydrogen gas at 100°C. The sharp absorption peak at 1.24μm, due to molecular hydrogen, is saturated after 10 hours. The absorption band due to OH formation is comprised of two closely spaced peaks at 1.39μm (Si-OH) and 1.41μm (Ge-OH) [64]. This indicates that hydrogen molecules react with germanosilicate glass and form OH absorbing species. On the other hand UV irradiation of untreated samples shows no OH formation [10,65]. Figure 2.8 (inset) shows the UV induced loss changes that occur in response to writing a strong grating in a 9% germania fiber loaded with 4.1% hydrogen. The OH ion concentration is consistently close to the germania content of the fiber. Optical absorption spectra of a germanosilicate preform rod heated in a hydrogen atmosphere at 500°C for different times are shown in figure 2.9 [66]. The growth of the broad 240 nm absorption band is clearly seen and indicates that the reaction of hydrogen molecules at Ge sites produces GODCs assigned to the broad 240 nm absorption band. Figures 2.8 and 9 suggest that the GODCs and OH species are formed from thermally driven reactions between hydrogen and Si-O-Ge glass sites. The inscription of Bragg gratings in hydrogen loaded fiber undoubtedly involves both thermal and photolytic mechanisms. Atkins and co-workers [65] investigated thermal effects by exposing germanosilicate glass (loaded 1 mol % hydrogen) to a CO_2 laser (CW mode) for 10 seconds, resulting in a glass temperature of 600°C. The UV absorption spectrum shows growth of the GODC band near 240 nm, from 20 dB/mm before heating to 380 dB/mm after heating. IR spectra of the sample indicate the formation of ~980 ppm OH. The tail of the OH broadband absorption peak at 1.39 and 1.41 μm introduces losses that are often unacceptable to telecommunication network systems designers. However, by loading fiber with deuterium instead of hydrogen the UV induced absorption peak is shifted to longer wavelengths, out of the erbium amplifier band of 1.55μm [67].

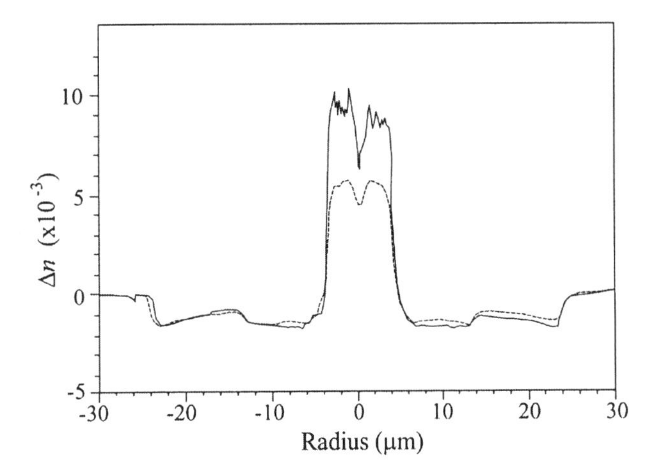

Figure 2.7. Refractive index profile for a standard single-mode fiber with 3% GeO_2, and for a grating that was UV written in the same fiber after loading with 3.3% hydrogen (solid curve) (after [10])

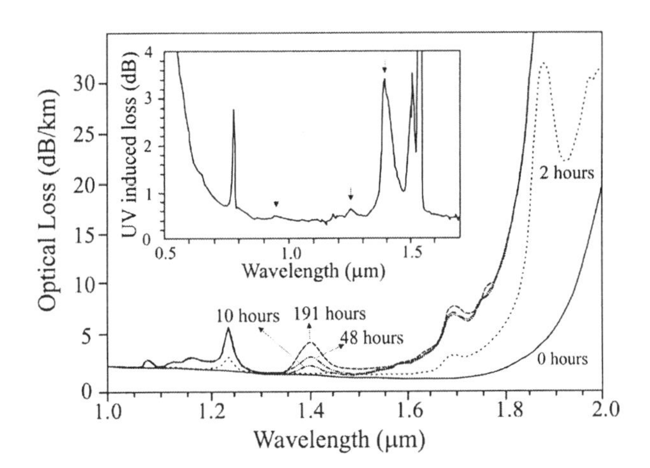

Figure 2.8. Absorption spectrum changes in the IR for a germanosilicate fiber exposed to 1 atmospheric pressure of hydrogen gas at 100°C (after [64]). Inset: UV-induced losses in ~5 mm long grating in fiber with 9% GeO_2. Features at 770 and 1500 nm are due to the gratings: marked peaks at 0.95, 1.24 and 1.39 μm are due to OH (after [10])

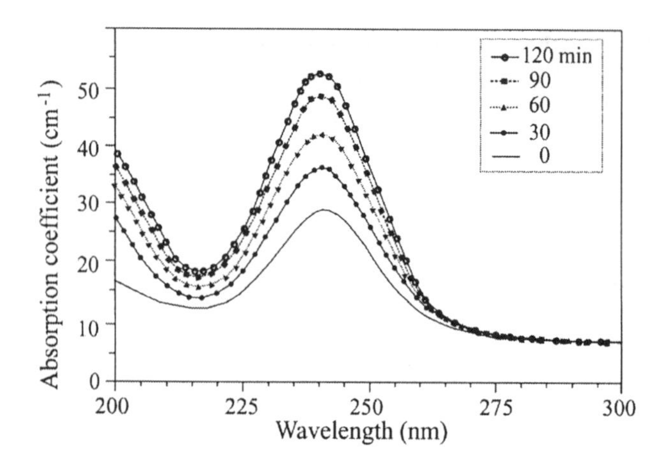

Figure 2.9. Optical absorption spectra of a germanosilicate preform rod heated in a H_2 atmosphere at 500°C for different times (after [66])

2.2.4.2 Flame brushing

Flame brushing is a simple and effective technique for enhancing the photosensitivity in germanosilicate fiber [68]. The photosensitization introduced through flame brushing is achieved with a negligible loss at the important high transmission communications windows. The region of the optical waveguide to be photosensitized is brushed repeatedly by a flame fueled with hydrogen and a small amount of oxygen, reaching a temperature of ~1700°C, with the photosensitization process taking approximately 20 minutes. At these temperatures, the hydrogen diffuses into the fiber core very quickly and reacts with the germanosilicate glass to produce GODCs, creating a strong absorption band at 240 nm and rendering the core highly photosensitive. The flame-brush technique has been used to increase the photosensitivity of standard telecommunications fiber by a factor greater than 10, achieving changes in the index of refraction $>10^{-3}$ [68]. A comparison may be made between 'standard' and flame brushed fiber, for which index changes of 1.6×10^{-4} and 1.75×10^{-3} have been realized, respectively, under similar laser writing conditions.

The enhanced photosensitivity techniques of flame brushing and hydrogen loading follow the same concept. In both cases, hydrogen is used in a chemical reaction with germanosilicate glass to form GODCs that are responsible for the photosensitivity. The formation of Bragg gratings in flame brushed germanosilicate fibers undoubtedly involves both thermal and photolytic mechanisms, except in this case, the thermally driven chemical reactions occur simultaneously as the hydrogen diffuses into the core at elevated temperatures. Subsequent UV irradiation bleaches the GODC band giving rise to index changes. There are several advantages in enhancing fiber photosensitivity by flame brushing. The increased photosensitivity in the fiber is permanent as opposed to hydrogen loading where the fiber loses its photosensitivity as the hydrogen diffuses out of the fiber. It allows strong Bragg gratings to be fabricated in standard telecommunications fibers, that typically exhibit no intrinsic photosensitivity. Localization of photosensitivity due to the relatively small flame can be used to brush the fiber. However, one major drawback in this technique is that the high temperature flame weakens the fiber, having serious implications for the long-term stability of any device fabricated using this approach.

2.2.4.3 Codoping

The addition of various codopants in germanosilicate fiber has also resulted in photosensitivity enhancement. In particular, co-doping with boron can lead to a saturated index change ~4 times larger than that obtained in pure germanosilicate fibers [23]. A comparison of the relative photosensitivity of four different types of fibers including boron-co-doping is given in Table 2.2.

Table 2.2. Relative photosensitivity for four different fibers (After [23])

Fiber type	Fiber Δn	Saturated index modulation	Maximum reflectivity for 2-mm gratings	Time for reflectivity to saturate
Standard fiber ~4 mol% Ge	0.005	3.4×10^{-5}	1.2%	2 hours
High index fiber ~20 mol% Ge	0.03	2.5×10^{-5}	45%	~ 2 hours
Reduced fiber ~ 10 mol% Ge	0.01	5×10^{-5}	78%	~ 1 hour
Boron co-doped fiber ~ 15 mol% Ge	0.003	7×10^{-5}	95%	~ 10 minutes

The fibers were irradiated with a power intensity of 1 W/cm^2 from a frequency doubled CW argon ion laser until the grating reflectivity saturated. The results showed that the fiber containing boron had an enhanced photosensitivity. This fiber was much more photosensitive than the fiber with higher germanium concentration and without boron co-doping. In addition, saturated index changes were higher and achieved faster than for any of the other fibers. This suggests that there is an additional mechanism operating in the boron co-doped fiber that enhances the photoinduced refractive index changes. The germanium-boron co-doped fiber was fabricated with a germanium composition of 15 mol%. In the absence of boron this fiber would have a refractive index difference of 0.025 (Δn) between the core and cladding. However, when the preform was drawn into fiber, the measure value for Δn dropped to 0.003. It appears that the addition of boron reduces the core index of refraction. This result is not surprising, as it is known that the addition of boron oxide to silica can result in a compound glass that has a lower index of refraction than that of silica [69]. Studies have shown that boron-doped silica glass system results in lower refractive index values when the glass is quenched, while subsequent thermal annealing causes the refractive index to increase. This is consistent with the fact that Δn dropped from 0.025 to 0.003 when the preform was drawn into fiber, since fibers are naturally quenched during the drawing process. This effect is assumed to be due to a build-up in thermo-elastic stresses in the fiber core resulting from the large difference in thermo-mechanical properties between the boron-containing core and the silica cladding. It is well known that tension reduces the refractive index through the stress-optic effect. Ultraviolet absorption measurements of the fiber between 200 and 300 nm showed only the characteristic GODC peak at 240 nm. The boron co-doping did not affect the peak absorption at 240 nm, or the shape of the 240 nm peak and no other absorption peaks were observed in this wavelength range [23]. The absorption measurements suggest that boron co-doping does not enhance the fiber photosensitivity through production of GODCs, as in the case of hydrogen loading and flame brushing techniques. Instead, it is believed that boron co-doping increases the photosensitivity of the fiber by allowing photoinduced stress relaxation to occur. In view of the stress-induced refractive index changes known to occur in boron-doped silica fibers, it seems likely that the refractive index increases through photoinduced stress relaxation initiated by the breaking of the wrong-bonds by UV light.

2.2.5 Photosensitivity at other writing wavelength

It has recently been demonstrated that Bragg grating devices can be inscribed in telecommunication fibers using ArF excimer UV radiation at 193 nm [63,70,71] and with far greater writing efficiency. An immediate advantage of using 193 nm is a reduction of laser-induced damage when using a phase mask and the higher spatial resolution in diffraction-limited applications, such as point-by-point writing. More recently, photosensitivity has been demonstrated in fibers exposed to high-energy 351 nm, 334 nm and 157 nm laser wavelengths [72,73-75].

2.2.5.1 Irradiation at 193 nm

A comparison of fiber Bragg gratings written using KrF (248 nm) and ArF (193 nm) excimer laser pulses indicates that use of the latter leads to stronger reflectivity gratings [63]. The decision to investigate photosensitivity at shorter wavelengths was provoked by mounting evidence that the dominant spectroscopic absorption changes associated with photosensitivity were occurring at wavelengths shorter than 200 nm [76,65].

Absorption measurements from 190-400 nm indicate that bleaching is induced with pulse energy densities of 120 mJ/cm^2 at 248 nm and 40 mJ/cm^2 at 193 nm, respectively in non-hydrogenated optical fibers and hydrogenated waveguides. For hydrogenated (<7 days) Ge-doped silica waveguides there is no detectable absorption from the dissolved molecular hydrogen and any UV absorption predominantly results from the tail of the silica UV band gap. Strong absorption changes occur in the doped layer on exposure to UV light. At 248 nm the absorption increase occurs below 190 nm, lacking an absorption band near 242 nm. Conversely, 193 nm excitation results in a distinct absorption band near 242 nm, with an accompanying but weaker feature near 210 nm, in addition to the tail of a strong absorption below 190 nm. The formation of the 242 nm band occurs only in cases of high dosage exposure. Samples hydrogenated for longer time periods give markedly different results, as illustrated in figure 2.10. Absorption bands near 242 nm and at wavelengths shorter than 190 nm are clearly evident, as shown in the insets to the figure. The initial absorption spectra in these cases are similar to those observed in standard optical fiber [76]. 248 nm exposure, figure 2.10(a), also follows the trend observed in optical fibers, where the 242 nm band is quickly and completely bleached after a relatively small UV dose, but the absorption continues to increase significantly at shorter wavelengths with further exposure to the bleaching radiation. Exposure to 193 nm light, figure 2.10(b), results in different absorption changes with two bands appearing early in the exposure, centered near 220 and 260 nm, merging into a strong 225 nm peak at higher UV dose. Little bleaching of the high initial absorption at 193 nm is detected. The strongest absorption increases are achieved in the case shown in figure 2.10(a); i.e. bleaching at 248 nm when there is an initial strong 242 nm band. Comparing writing efficiencies for in-fiber Bragg gratings at 193 nm and 248 nm indicates that a 193 nm grating reaches 80% reflectivity (total UV dose of 1.4 kJ/cm^2 (4-min exposure)), whereas a 248 nm grating of comparable length reaches only 20% after a total dose of 5.4 kJ/cm^2 (6-min exposure). Isochronal annealing experiments reveal no difference in the thermal stability of gratings fabricated at the two wavelengths. The experimental findings of Psaila et al [77] point to the possible existence of two different mechanisms for the creation of photoinduced birefringence, at these wavelengths.

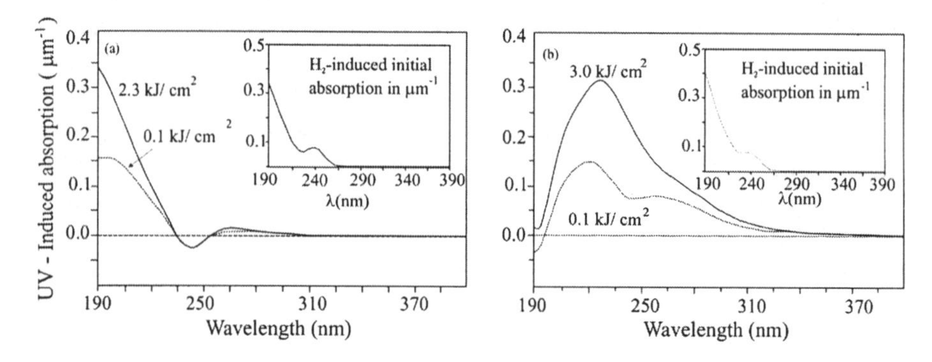

Figure 2.10. Spectra of absorption changes induced by UV light in H2-loaded Ge-doped silica with initial absorption bands (shown in the insets): (a) at 248 nm, (b) at 193 nm. The two curves in each graph correspond to different cumulative UV doses (after [63])

Dyer et al [70] have also reported the formation of high reflectivity gratings in optical fibers using a 193 nm ArF laser irradiated phase mask, producing Type II gratings under single pulse, high energy laser excitation [35]. This mechanism was enhanced by the larger absorption in the doped fiber core at this wavelength. The greater absorption was coupled with an incubation effect in which the damage threshold reduced while the absorption grew with increasing number of pulses, permitting the rapid formation (~10 pulses) of damage gratings at modest fluences, (~400 mJ/cm^2). The formation of Type IIA fiber gratings at 193 nm is also possible, producing highly negative index modulations of ~ $-3x10^{-4}$ [71].

2.2.5.2 Irradiation at 334 and 351 nm

Dianov and co-workers have proposed that if a triplet-state excitation is involved in the photosensitivity of germanosilicate glass, the direct excitation of this state by near-UV 330 nm light should result in the same changes as observed at 240 nm. However, 330 nm cannot directly ionize the defect. The observation of 650 nm luminescence and bleaching of the GODC blue luminescence by CW 351 nm radiation confirmed this [78]. Atkins et al have assigned the luminescence at 650 nm to DIDs [79]. Indeed, the increase in red luminescence has demonstrated a better correlation to the index change than the decrease in blue luminescence, while the relative correlation between the two bands is maintained over a range of excitation intensities, in contrast to 5 eV excitation [80]. The existence of two mechanisms for GODC photodestruction explains the discrepancy between the blue luminescence and refractive index behavior. Direct evidence of a near-UV index change in glass was reported in 1996 with the fabrication of a long period grating in Ge-doped fiber [81]. Starodubov and co-workers have since shown that efficient Bragg grating fabrication at 1550 nm in germanosilicate fibers is possible by use of near-UV light. [72]. Using 334 nm light the side writing of Bragg gratings with an index change of $\sim 10^{-4}$ in Ge-doped fiber was demonstrated. No hydrogenation of the fibers was required. These gratings were shown to have the same temperature stability as gratings fabricated with 240 nm light.

2.2.5.3 Irradiation at 157 nm

Herman et al [75] have demonstrated a new photosensitivity response of optical fibers and slab waveguides to light at 157 nm, from a F_2 excimer laser. Strong photosensitivity responses were anticipated because of the close proximity of 7.9-eV photons to the bandgap of the germanosilicate glass at 7.1 eV. The photosensitivity response of two single-mode fiber types, firstly, a high Ge-doped fiber (8%-GeO_2) and secondly, standard telecommunication fiber (3%-GeO_2 Corning SMF-28) was examined. Both fiber types were also soaked in 3atm hydrogen in excess of two weeks. It was found that rates of index change are several times larger at fluences of 100-450 J/cm^2 and orders of magnitude faster at lower fluence, in comparison with the results of Albert et al for 193 nm irradiation [15]. The 157 nm fluence dependence suggests a single-photon dependence of index change in departure with the two-photon response noted at 193 nm in Albert's work for the same fiber type. The 157 nm induced index changes were ~10 fold faster for hydrogen loaded fiber, while higher germanium content had little effect on the rates.

2.2.6 Mechanism of photoinduced refractive index change

The precise origins of photosensitivity and the accompanying refractive index change have yet to be fully understood. It does become clear that no single model may explain all the experimental results, as there are several microscopic mechanisms at work. There is substantial experimental evidence supporting the mechanism put forward by Hand and Russell [17]. The resultant color-centers are responsible for changes in the UV absorption spectrum of the glass and the refractive index change follows the Kramers-Kronig relationship [17]. Many experiments [29,76,82,83,84] support the GeE' center model for the photosensitivity and it is certainly the mechanism responsible for the original self-organized gratings [4]. However, the color-center model does not completely explain all the experimental observations

[26,34] and an alternative model, based on glass densification induced by photoionization of the Ge defects [85], also has experimental support [86]. We have seen earlier that the influence of the laser-writing wavelength and power, fiber types and processing, lead to many possible reaction pathways.

Measurements of the spectral changes accompanying UV irradiation and grating inscription have shown bleaching of the 240 nm band and the growth of absorption features at shorter wavelengths [29,87,76,82], in particular at 195 nm. Kramers-Kronig analysis of these data yields values for the refractive index change in close agreement with those inferred from measurements on photoinduced gratings, providing support for the color-center model. It is also consistent with the same model that the bleaching of the 240 nm band can be reversed subsequently by heating to 900°C [29]; a grating written, thermally erased and re-written in the same section of fiber exhibited essentially the same properties each time. Malo et al have shown [88], however, that annealing standard germanosilicate telecommunications fiber in air at 1200°C can remove its photosensitivity irrecoverably. Conversely, Cordier et al have presented the results of a TEM investigation of gratings UV-written in a fiber preform. These show microstructural changes aligned with the grating fringes interpreted as densification resulting from strain relaxation induced by the creation of the GeE' centers [89]. These authors argue that a greater spectral absorption range than the ~165 to 300 nm considered by Atkins et al [29] must be included in Kramers-Kronig analysis if the photoinduced refractive index is to be accurately determined. In the sections that follow, we will describe the various mechanisms associated with the changes in the index of refraction. Accumulating all the experimental findings, it becomes obvious that there are two main mechanisms that are involved in photosensitivity (at least for the most common germanosilicate fiber), which are described by the color-center and the compaction model.

2.2.6.1 Color-center model

The color-center model has received a great deal of attention, and whereas its contribution to explaining photosensitivity in germanosilicate fibers is considered complementary to other phenomena, there is increasing evidence that it is most applicable to hydrogenated germanosilicate fibers, where the formation of microscopic defects occurs. Any change in the refractive index, i.e. through the formation of a grating, is associated with the photoinduced change in absorption through the Kramers-Kronig relation [14]:

$$\Delta n_{eff}(\lambda) = \frac{1}{2\pi^2} P \int_0^{\infty} \frac{\Delta\alpha_{eff}(\lambda)}{1-(\lambda/\lambda')^2} d\lambda \tag{2.1}$$

where P is the principle part of the integral, λ is the wavelength and $\Delta\alpha_{eff}(\lambda)$ is the effective change in the absorption coefficient of the defect. Equation (2.1) may be used to calculate the index change induced by bleaching of the absorption bands. The boundaries are set to λ_1 and λ_2, the limits of the spectral range within which absorption changes take place and λ' is the wavelength for which the refractive index is calculated. The validity of equation (2.1) requires that λ' is much greater than the upper and lower bound limits. This relationship demonstrates that the index change produced in the infrared/visible region of the spectrum, by the photoinduced processing, results from a change in the absorption spectrum of the glass in the UV/far UV spectral region. Measuring the Bragg grating reflectivity enables the evaluation of the effective index change that may then be compared to the value calculated from equation (2.1).

In this model, first proposed by Hand and Russell [17], photoinduced changes in the material properties of the glass introduce new localized electronic excitations and transitions of defects. It is precisely these color-center defects, because of their strong optical absorption, that are proposed to give rise to the change in the refractive index associated with photosensitivity. The bleachable wrong bond defects, which initially absorb the light, are transformed into defects that are more polarizable by virtue of the fact that that their electronic transitions take place at longer wavelengths e.g. Ge(1) centers, or to have stronger transitions e.g. Ge(2), GeE' [90]. The observation of weak birefringence induced in low birefringence fibers, by two-photon absorption, indicates that oriented defects are produced [39], in accord with this model. Further to this, the color-center model presumes that the refractive index at a point is related only to the number density and orientation of defects in that region, determined purely by their electronic absorption spectra. Any nuclear displacement arising from the photoinduced process is limited to a few atoms and only weakly coupled to long range displacements of the atoms in the network. Thus only the electronic properties of the defects produced are important. The permanence and thermal annealing properties of the photoinduced index change are attributed to the slow kinetics of the reverse process.

Recently, Hosono et al [87] have found a strong correlation between the optical absorption band peaking at 6.4 eV and the GeE' centers. These results point to compatibility with the Kramers-Kronig mechanism. This particular defect has been strongly linked with photosensitivity in hydrogenated fibers [91]. Tsai et al [84] reported similar thermal stability between photoinduced

GeE' centers and Δn, suggesting an association through the Kramers-Kronig relation and the color-center model (figure 2.16) [29,92]. This is supported by the power-law growth of both Δn [28,31] and the concentration of GeE' centers [93]. By taking into account the contributions from deep UV absorption bands, there is strong evidence [29,92] that this model can explain a large part of the measured magnitude of the index change for Type I gratings.

The color-center model cannot satisfactorily explain the behavior of all fiber types and their dopants. Silica fibers doped with P, Sn or Ta display evidence that the corresponding changes in the UV absorption spectra affect grating formation. This contrasts with observations made on hydrogen-loaded, rare-earth doped, aluminosilicate fibers, where the Kramers-Kronig analysis performed over the wavelength range 190-800 nm fails to account for the refractive index changes. It is useful to explain where discrepancies arise in computing the index change. A knowledge, in principle, of the entire spectrum from zero to infinite wavelength is required. This information is not of course available and theoretical extrapolation has to be used. Given that the color-center model is strictly local, its assessment through the Kramers-Kronig relation requires that during grating fabrication one measures the UV-induced loss spectrum for each exposure interval and at each place along the grating, for the spectral range covering the color-centers' absorption. Assuming that these measurements can be realized with a spatial resolution high enough to match the grating pitch, the Kramers-Kronig analysis would then give, for each exposure, the true form of the periodic refractive index change along the fiber axis.

2.2.6.2 Compaction/densification model

The compaction model is based on laser irradiation-induced density changes that result in refractive index changes. Irradiation by laser light at 248 nm at intensities well below the breakdown threshold has been shown to induce thermally reversible, linear compaction in amorphous silica, leading to refractive index changes [94,95]. Figure 2.11 shows the variation of thin-film (100 nm) amorphous silica samples grown on Si wafers irradiated a KrF excimer laser. At an accumulated dose of 2 kJ/cm^2, there is an obvious reduction in the film thickness (approximately 16%) and a corresponding evolution of refractive index during laser irradiation. Annealing reverses the compaction and the original thickness and pre-irradiated refractive index value is recovered. Continued accumulation of UV irradiation beyond this reversible compaction regime leads to irreversible compaction, until the film is entirely etched after a total accumulated dose of 17 kJ/cm^2. An approximately linear relationship has been found between the index of refraction and the density change. The reversibility in compaction, coincident with the creation of defects, conforms to results taken of implanted fused silica. The evolution of the refractive index, figure 2.11 shows a rapid increase approaching 20% of the equilibrium value. The Lorentz-Lorenz relation for the refractivity (derived from the Clausius Mausotti relation) is

$$R = \frac{(n^2 - 1)}{\rho(n^2 + 2)} \tag{2.2}$$

with ρ being the specific gravity and n the refractive index, one obtains by differentiation,

$$\Delta n = -\frac{(n^2 + 2)(n^2 - 1)}{6n}\left[1 - \frac{\Delta R}{(R\,\Delta V/V)}\right]\frac{\Delta V}{V} \tag{2.3}$$

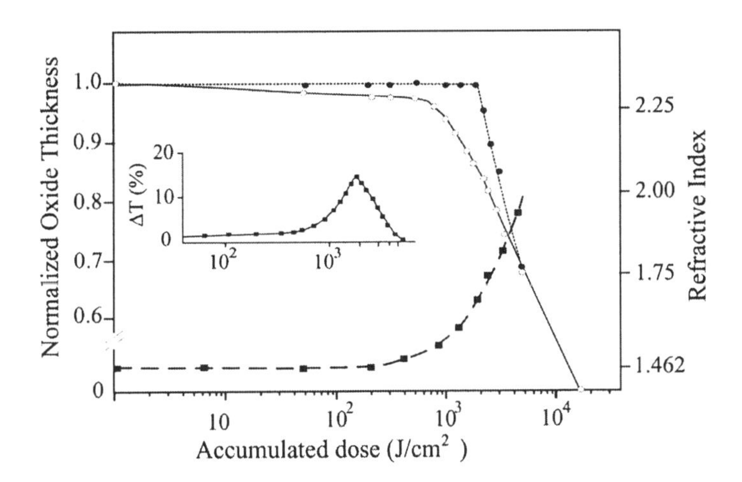

Figure 2.11. Observed compaction of a 100 nm thick oxide as a function of accumulated UV dose (open circle); the same sample after annealing for 1 hour at 950°C in vacuum (solid circle). Inset: difference between these two results. Evolution of the refractive index during irradiation (solid square) (after [94])

If $(V/R)\Delta R/\Delta V<1$, then compaction results in the observed increase in refractive index. The first term in parenthesis relates to structural volume changes, whereas the second term, at least as far as SiO_2 is concerned, to the volume of the oxygen ion. The material density can also be described in terms of the number of polarizable oscillators per unit volume, whereas changes in the glass molar refractivity can be expressed through the material macroscopic polarizability, i.e. the sum of the polarizabilities of these oscillators. Therefore equation (2.3) may be expressed as $\Delta n \propto (\Delta V/V - \Delta\alpha/\alpha)$ [96]. Clearly there is competition between these two terms, affecting the refractive index. For example, the refractive index increase caused by a volume expansion (density decrease) would be larger in the absence of a simultaneous increase in the material polarizability. The measured and predicted values of the refractive index change are found to agree to within 10%, linking index variations to laser-induced structural variations. Fiori and Devine hinted at the possibility that a breakage of high- to low-order membered ring structures in the glass and hence volume, with a reduction in the mean ring sizes to a limit of two-membered rings. This hypothesis has been supported by Raman spectra of compressed amorphous silica [97]. Beyond the two-member limit reversible compaction gives way to irreversible compaction and etching that results from sub-and direct-band gap network defects [98,99]. Measurement of the refractive index variation in hydrostatically compressed silica, produces results in very good agreement with laser-compacted, amorphous silica, confirming that compaction of amorphous silica proceeds through internal structural rearrangements and not primarily through a process of defect creation [95].

The densification process in bulk Si may be extended to describe the UV-induced densification processes in germanosilicate optical fibers, which is critical to understanding the formation dynamics of fiber Bragg gratings. Poumellec [100] has investigated Bragg grating formation in germanosilicate glass using an optical microscope, via interferometric microscopy, leading to the conclusion that densification occurs, accounting for 7% of the UV-induced refractive index increase.

Limberger et al [18] studied compaction and photoelastic-induced index changes in fiber Bragg gratings. They found that the tension on the core of single mode fibers is strongly increased by the formation of a Bragg grating, in contradiction with the stress relief model [90]. This tension increase lowers the refractive index because of the photoelastic effect. On the other hand, the compaction of the core network results in an increased refractive index. The two contributions were evaluated from axial stress measurements, from the determined index modulation amplitude, and from the mean index change of the Bragg gratings. The total Bragg grating index modulation had a positive mean value, explained by a structural modification of the germanosilicate core network into a more compact configuration. The mean index change was observed to be at least 20% smaller than the index modulation amplitude. The total Bragg grating index modulation was found to be smaller than the compaction-induced index modulation by 30%-35% because of the photoelastic effect. It was argued that given that the color-center model cannot account for index changes larger than ~$4x10^{-4}$ [92], a structural modification leading to compaction of the glass matrix must be the main contribution (inelastic) to the observed index change

Douay et al have produced a comprehensive study into densification and its involvement in photosensitivity in silica based optical fibers and glass [38]. A comparison of densification in hydrogen loaded and non-hydrogen loaded preform slices was carried out. It was found that the modulated refractive index was thermally reversible, unlike the mean core index, in most non-hydrogen loaded germanosilicate and aluminosilicate fibers, except for highly doped germanosilicate fibers. Photoelastic densification could account for 40% of the photosensitivity of non-hydrogen-loaded germanosilicate or aluminosilicate plates. Unlike the enhancement of UV photosensitivity via hydrogen loading, no increase in densification following hydrogenation was observed in germanosilicate plates. Given that hydrogenation considerably increases the UV-induced excess loss below 220 nm without strong saturation effects, it was concluded that the color-center model accounts for a large part of the photosensitivity in the hydrogenated germanosilicate plates. The refractive index modulation in non-hydrogen loaded germanosilicate was thermally reversible whereas the mean index was not. If one assumes that the heating-induced increase in the mean refractive index arises from thermal compaction of the germanosilicate fiber core and that densification accounts for a non-negligible part of the UV-induced, refractive index modulation, then thermal reversibility of the modified change implies that the heating-induced-compaction of the core does not prevent further UV-induced densification.

It is not quite clear when and for what parameters compaction is important. Under certain experimental conditions compaction plays a major role in the UV-induced index of refraction, a role that was assumed played by the defect in the color-center model. It is the belief of the authors that compaction is certainly one of the major mechanisms in explaining UV photosensitivity. However, its exact contribution under various experimental conditions has to be further investigated.

2.2.6.3 Stress relief model

The stress-relief model [90] is based on the hypothesis that the refractive index change arises from the alleviation of built-in thermoelastic stresses in the fiber core. The fiber optic core in a germanosilicate fiber is under tension because of the difference in the thermal expansion of the core and the cladding as the glass is cooled below the fictive temperature (glass transition temperature) during fiber drawing. This means that since the temperature of the fiber decreases rapidly there is a point where the structure (including defects) is frozen in. Through the stress-optic effect, it is known that tension reduces the refractive index and is therefore expected that stress-relief will increase the refractive index. It is proposed that during UV irradiation the wrong-bonds break and promote relaxation in the tensioned glass hence reducing frozen-in thermal stresses in the core [90]. Although there is an abundance of breakable wrong bonds in germanosilicate core fibers; this is not the case for pure silica core fibers, which are not photosensitive in the UV (in agreement with the model).

Recently, Fonjallaz et al [101] have reported the measurement of axial stress modifications in fiber Bragg gratings and have shown that tension in the core of single-mode germanosilicate fibers is greatly increased during Bragg grating formation. A strong increase of the tension has also been observed in reference [18], in contradiction to the stress relief model [90]. Finally, the thermal reversibility of grating inscription cannot be explained by this model [76].

2.2.6.4 Other models

The electron charge migration model is applied to explain photorefractivity in BSO or BGO materials [102]. An electric field created by charge migration acts on the index either by the Pockels effect, where the index is proportional to the electric field, or by the Kerr effect, where the index is proportional to the square of the electric field. This is a very weak effect and cannot explain results for grating fabrication via the transverse holographic technique and the use of pulsed UV laser sources. The dipole model is based on the formation of built-in periodic space-charge electric fields by the photo-excitation of defects, however the number of defects must be several orders of magnitude higher than estimated from experiments to account for experimental observations [103]. Other models include the ionic migration model [104] and Soret effect [105], for which there is no supporting evidence.

2.3 PROPERTIES OF FIBER BRAGG GRATINGS

In this section the various properties that are characteristic of fiber Bragg gratings will be described in detail and this will involve the discussion of a diverse range of topics. This will begin by examining the measurable wavelength dependent properties, such as the reflection and transmission spectral profiles, for a number of simple and complex grating structures. The dependence of the grating wavelength response to externally applied perturbations, such as temperature and strain, is also investigated.

2.3.1 Simple Bragg grating

In its simplest form a fiber Bragg grating consists of a periodic modulation of the refractive index in the core of a single mode optical fiber. These types of uniform fiber gratings, where the phase fronts are perpendicular to the fiber longitudinal axis with grating planes having constant period (figure 2.12), are considered the fundamental building blocks for most Bragg grating structures. Light guided along the core of an optical fiber will be scattered by each grating plane. If the Bragg condition is not satisfied, the reflected light from each of the subsequent planes becomes progressively out of phase and will eventually cancel out. Additionally, light that is not coincident with the Bragg wavelength resonance will experience very weak reflection at each of the grating planes because of the index mismatch, this reflection accumulates over the length of the grating. As an example, a 1mm grating at 1.5 μm with a strong Δn of 10^{-3} will reflect ~0.05% of the off-resonance incident light. Where the Bragg condition is satisfied the contributions of reflected light from each grating plane add constructively in the backward direction to form a back-reflected peak with a center wavelength defined by the grating parameters.

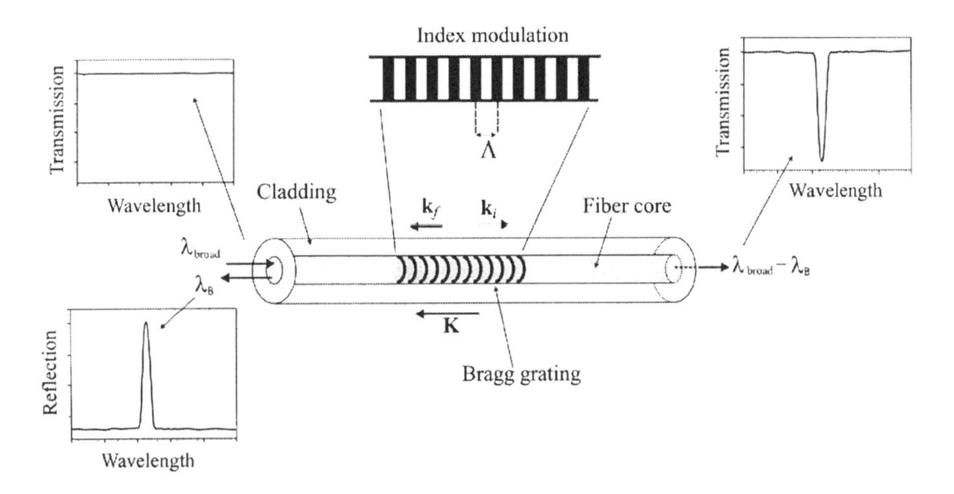

Figure 2.12. Illustration of a uniform Bragg grating with constant index of modulation amplitude and period. Also shown are the incident, diffracted, and grating wave vectors that have to be matched for momentum to be conserved

The Bragg grating condition is simply the requirement that satisfies both energy and momentum conservation. Energy conservation requires that the frequency of the incident radiation and the reflected radiation is the same ($\hbar\omega_f = \hbar\omega_i$). Momentum conservation requires that the incident wavevector, $\mathbf{k}_i$, plus the grating wavevector, $\mathbf{K}$, equal the wavevector of the scattered radiation $\mathbf{k}_f$, this is simply stated as

$$\mathbf{k}_i + \mathbf{K} = \mathbf{k}_f \tag{2.4}$$

where the grating wavevector, $\mathbf{K}$, has a direction normal to the grating planes with a magnitude $2\pi/\Lambda$ (Λ is the grating spacing shown in figure 2.12). The diffracted wavevector is equal in magnitude, but opposite in direction, to the incident wavevector. Hence the momentum conservation condition becomes

$$2\left(\frac{2\pi n_{eff}}{\lambda_B}\right) = \frac{2\pi}{\Lambda} \tag{2.5}$$

which simplifies to the first order Bragg condition

$$\lambda_B = 2 n_{eff} \Lambda \tag{2.6}$$

where the Bragg grating wavelength, λ_B, is the free space center wavelength of the input light that will be back reflected from the Bragg grating, and n_{eff} is the effective refractive index of the fiber core at the free space center wavelength.

2.3.2 Uniform Bragg grating

Consider a uniform Bragg grating formed within the core of an optical fiber with an average refractive index n_0. The index of refractive profile can be expressed as

$$n(z) = n_0 + \Delta n \cos\left(\frac{2\pi z}{\Lambda}\right) \tag{2.7}$$

where Δn is the amplitude of the induced refractive index perturbation (typically 10^{-5} to 10^{-3}) and z is the distance along the fiber longitudinal axis. Using the coupled mode theory [6] the reflectivity of a grating with constant modulation amplitude and period is given by the following expression

$$R(l,\lambda) = \frac{\Omega^2 \sinh^2(sl)}{\Delta k^2 \sinh^2(sl) + s^2 \cosh^2(sl)} \tag{2.8}$$

where $R(l, \lambda)$ is the reflectivity which is a function of the grating length l, and wavelength λ. Ω is the coupling coefficient, $\Delta k = k - \pi/\lambda$ is the detuning wave-vector, $k = 2\pi n_0 / \lambda$ is the propagation constant and finally $s^2 = \Omega^2 - \Delta k^2$. For sinusoidal variation in the index perturbation the coupling coefficient, Ω, is given by

$$\Omega = \frac{\pi \Delta n}{\lambda} M_{power} \tag{2.9}$$

where M_{power} is the fraction of the fiber mode power contained by the fiber core. In the case where the grating is uniformly written through the core, M_{power} can be approximated by $1-V^{-2}$, where V is the normalized frequency of the fiber. The normalized frequency of the fiber is given by $(2\pi/\lambda) a\sqrt{n_{co}^2 - n_{cl}^2}$ where a is the core radius, and n_{co} and n_{cl} the core and cladding indices, respectively. At the center wavelength of the Bragg grating the wavevector detuning is $\Delta k = 0$, therefore the expression for the reflectivity becomes

$$R(l, \lambda) = \tanh^2(\Omega l) \tag{2.10}$$

The reflectivity increases as the induced index of refraction change increases. Similarly as the length of the grating increases so does the resultant reflectivity. Figure 2.13 shows a calculated reflection spectrum as a function of wavelength of uniform Bragg grating. The side lobes of the resonance are due to multiple reflections to and from opposite ends of the grating region. The sine spectrum arises mathematically through the Fourier transform of a harmonic signal having finite extent; an infinitely long grating would transform to an ideal delta function response in the wavelength domain.

A general expression for the approximate full-width-half maximum bandwidth of a grating is given by [106]

$$\Delta\lambda = \lambda_B s \sqrt{\left(\frac{\Delta n}{2n_0}\right)^2 + \left(\frac{1}{N}\right)^2} \tag{2.11}$$

where N is the number of the grating planes. The parameter $s\sim 1$ for strong gratings (with near 100% reflection) whereas $s\sim 0.5$ for weak gratings.

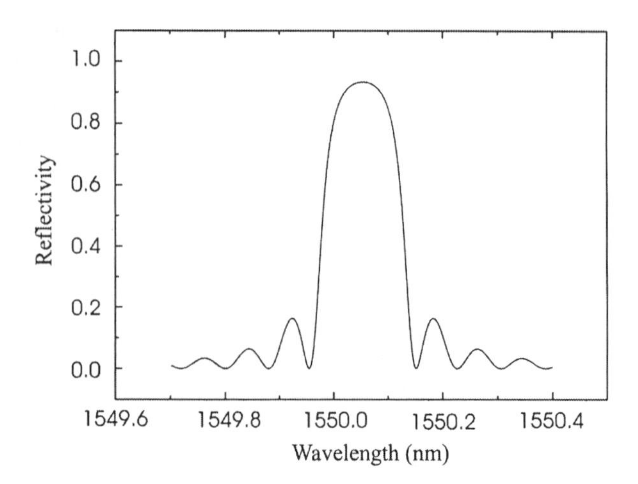

Figure 2.13. A typical reflection spectrum of a Bragg grating center at 1550 nm as a function of wavelength

2.3.3 Phase and group delay of uniform gratings

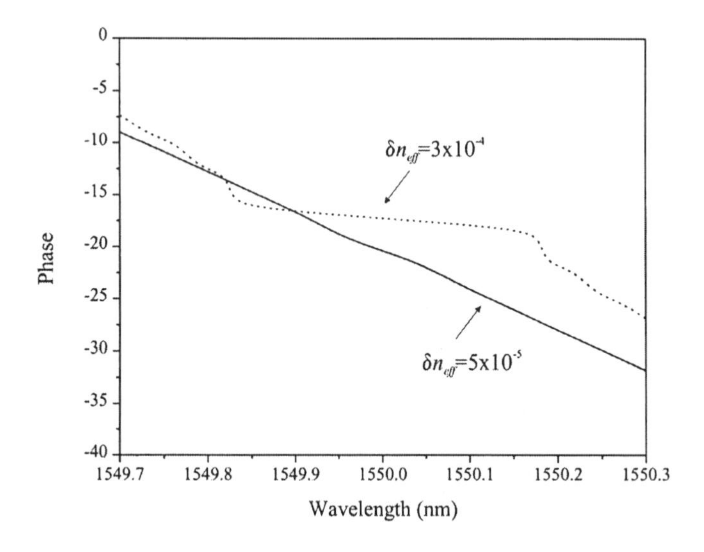

Figure 2.14. Typical phase response in reflection from a uniform-period Bragg grating as a function of wavelength. The strong grating ($\delta n_{eff}=3\times10^{-4}$) has a phase change that is almost constant for the same bandwidth when compared with the weaker grating ($\delta n_{eff}=5\times10^{-5}$)

Figure 2.14 shows the phase response of two uniform-period Bragg gratings as a function of wavelength. The two gratings have the same length

(1cm): however they have different index perturbation change, namely a "strong" grating with $\delta n_{eff} = 3\times10^{-4}$ and a "weak" grating with $\delta n_{eff} = 5\times10^{-5}$. It appears that the phase change decreases with increase index of refraction perturbation.

The group delay of the same two gratings is shown in figure 2.15. Clearly strong dispersion (change of group delay with wavelength) is noticable at the edge of the band stop and increases with increasing index perturbation change, although limited to a small bandwidth. The group delay is minimum at the center of the band.

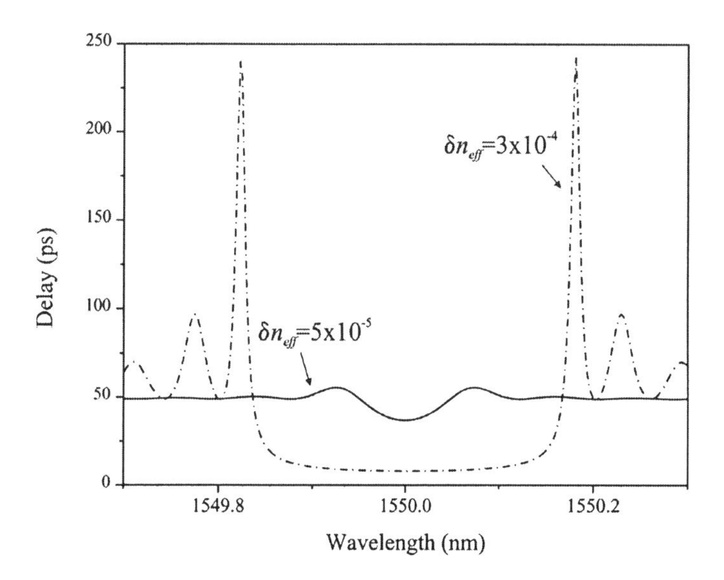

Figure 2.15. Typical group delay response in reflection from a uniform-period Bragg grating as a function of wavelength. For the strong grating the group delay in the center of the band is constant, while at the band edges it increases rapidly with increasing bandwidth confinement

2.3.4 Strain and temperature sensitivity of Bragg gratings

The Bragg grating resonance, which is the center wavelength of back reflected light from a Bragg grating, depends on the effective index of refraction of the core and the periodicity of the grating. The effective index of refraction, as well as the periodic spacing between the grating planes, will be affected by changes in strain and temperature. Using equation 2.6 the shift in the Bragg grating center wavelength due to strain and temperature changes is given by

$$\Delta\lambda_B = 2\left[\Lambda\frac{\partial n_{eff}}{\partial l} + n_{eff}\frac{\partial \Lambda}{\partial l}\right]\Delta l + 2\left[\Lambda\frac{\partial n_{eff}}{\partial T} + n_{eff}\frac{\partial \Lambda}{\partial T}\right]\Delta T \tag{2.12}$$

The first term in the above equation represents the strain effect on an optical fiber. This corresponds to a change in the grating spacing and the strain-optic induced change in the refractive index. The above strain effect term may be expressed as [40]

$$\Delta\lambda_B = \lambda_B\left[1 - \frac{n^2}{2}\left[p_{12} - \nu(p_{11} + p_{12})\right]\right]\varepsilon_z \tag{2.13}$$

where p_{11} and p_{12} are components of the strain optic tensor, ν is the Poisson's ratio and $\varepsilon_z = \delta l/l$. For a typical germanosilicate fiber there is a 1.2 pm shift in the center wavelength of the grating as a result of applying 1 $\mu\varepsilon$ to the Bragg grating. Experimental results of a Bragg center wavelength shift with applied stress on a 1555.1 nm grating are shown in figure 2.16.

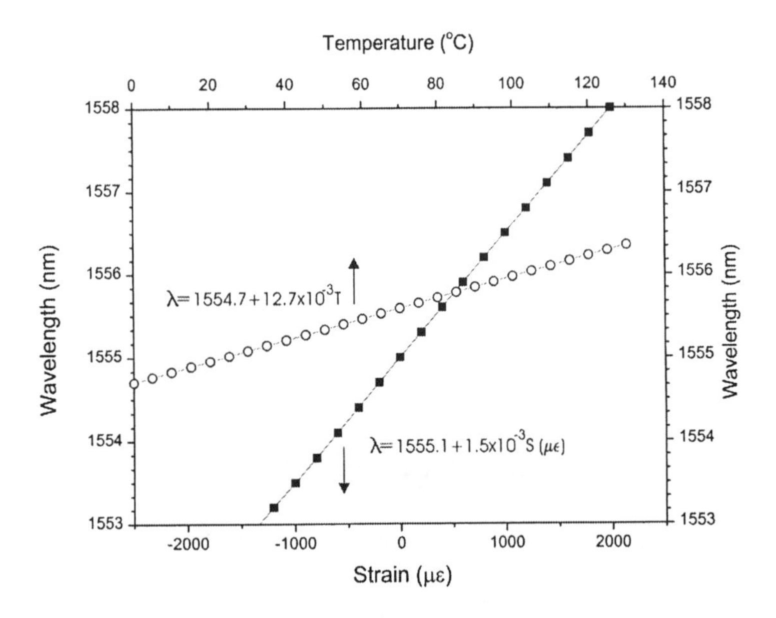

Figure 2.16. Peak reflection from the Bragg grating; under applied stress (square symbols), at different temperatures (open circles). The Bragg grating formed the output coupler of an erbium doped fiber laser

The second term in the equation 2.12 represents the effect of temperature on an optical fiber. A shift in the Bragg wavelength due to thermal expansion changes the grating spacing and the index of refraction. This fractional wavelength shift for a temperature change ΔT may be written as [40]

$$\Delta\lambda_B = \lambda_B(\alpha_\Lambda + \alpha_n)\Delta T \tag{2.14}$$

where $\alpha_\Lambda = (1/\Lambda)(\partial\Lambda/\partial T)$ is the thermal expansion coefficient for the fiber (approximately 0.55×10^{-6} for silica). The quantity $\alpha_n = (1/n)(\partial n/\partial T)$ represents the thermo-optic coefficient and its approximately equal to 8.6×10^{-6} for the germania-doped, silica-core fiber. Clearly the index change is by far the dominant effect. From equation 2.14 the expected sensitivity for a ~1550 nm Bragg grating is approximately 13.7 pm/°C. Figure 2.16 shows experimental results of a Bragg grating center wavelength shift as a function of temperature. It now becomes apparent that any change in wavelength associated with the action of an external perturbation to the grating, is the sum of strain and temperature terms. Therefore, in sensing applications where only one perturbation is of interest, the deconvolution of temperature and strain becomes necessary.

2.3.5 Other properties of fiber gratings

When a grating is formed under conditions for which the modulated index change is saturated under UV exposure, then the effective length will be reduced as the transmitted signal is depleted by reflection. As a result, the spectrum will broaden appreciably and depart from a symmetric sinc or Gaussian shape spectrum, whose width is inversely proportional to the grating length. This is illustrated in figure 2.17 (a) and (b). In addition, the cosinusoidal shape of the grating will distort into a waveform with steeper sides. A second-order Bragg line (figure 2.17(c)) will appear from the new harmonics in the Fourier spatial spectrum of the grating [107].

Another interesting feature, which is observed in strongly reflecting gratings with large index perturbations, is the small-shape spectral resonance on the short wavelength side of the grating centerline. This is due to self-chirping from $\Delta n_{eff}(z)$. Such features do not occur if the average index change is held constant or adjusted to be constant by a second exposure of the grating. A Bragg grating will also couple dissimilar modes in reflection and transmission, provided the following two conditions are satisfied, namely phase matching and sufficient mode overlap in the region of the fiber that contains the grating. The phase matching condition, which ensures a coherent exchange of energy between the modes, is given by [107]:

$$n_{eff} - \frac{\lambda}{\Lambda_z} = n_e \tag{2.15}$$

where n_{eff} is the modal index of the incident wave and n_e is the modal index of the grating-coupled reflected or transmitted wave. It should be pointed out that the above equation allows for a tilted or blazed grating by adjusting the grating pitch along the fiber axis Λ_z.

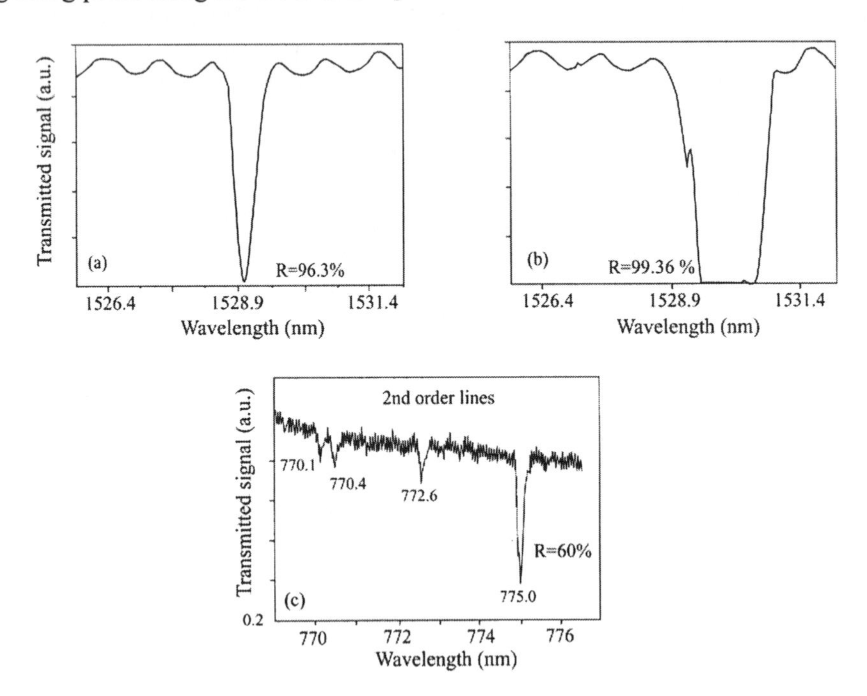

Figure 2.17. A strongly reflecting grating with a large index change (a) becomes saturated and (b) the spectrum broadens under continuous exposure because the incident wave is completely reflected before reaching the end of the grating. The strongly saturated grating is no longer sinusoidal, and the peak index regions are flattened whereas the valleys in the perturbation index distribution are sharpened. As a result second order Bragg reflection lines (c) are observed at about one-half the fundamental Bragg wavelength and at other shorter wavelengths for higher order modes (after [107])

2.3.5.1 Cladding and radiation mode coupling

Bragg gratings written in a highly photosensitive fiber, or in fiber that has been hydrogenated, have a very pronounced transmission structure on the short wavelength side of the Bragg peak (figure 2.18). This feature is only observable in the transmission spectrum (when viewed in reflection only the main peak appears) and therefore this structure must result from light leaving the side of the fiber – to analyze it one must take into account radiation mode coupling. Usually radiation mode coupling, which is routinely observed from surface relief gratings made by physically etching the core of a polished optical fiber, is a smooth function of wavelength. However, the transmission spectrum of the Bragg grating consists of multiple sharp peaks that modulate this coupling and it is a direct consequence of the cylindrical cladding-air interface. Dipping the cladding into glycerin, which in effect eliminates the cladding-air interface, may eliminate this issue. Nevertheless, the cladding mode radiation-related problems become very serious with large excess losses at wavelengths shorter than the peak reflection wavelength. As a result, highly reflective chirped gratings have lower reflectivity at shorter wavelengths when the signal is coupled from the longer wavelength side of the fiber grating. There are several approaches to avoiding the radiation mode effect. One proposed method to counter this problem is the suppression of the normalized refractive index modulation, associated with this coupling by having a uniform photosensitive region across the cross-section plane of the optical fiber [108]. From the orthogonality principle of the modes, the overlap of the modal fields and the grating index modulation would be zero in this case. The LP_{01} (fundamental) mode will therefore not couple into any of the cladding modes. Since the LP_{01} mode only has significant field distribution over the core and the part of the cladding immediately next to the core, it is usually sufficient to have only this part of the optical fiber photosensitive. Although it is possible to introduce a photosensitive cladding around a photosensitive core, it is, however very difficult to obtain the same photosensitivity over both cladding and core.

2.3.5.2 Apodization of fiber gratings

The main peak in the reflection spectrum of a finite length Bragg grating with uniform modulation of the index of refraction is accompanied by a series of sidelobes at adjacent wavelengths. It is important in some applications to lower and if possible to eliminate the reflectivity of these sidelobes, or to apodize the reflection spectrum of the grating. For example, in dense wavelength division multiplexing (DWDM) it is important to have very high rejection of the non-resonant light in order to eliminate cross talk between information channels and therefore apodization becomes absolutely necessary. Apodized fiber gratings can have very sharp spectral responses, with channel spacings down to 100GHz. For applications such as add-drop filters, or in demultiplexers the grating response to less than –30 dB from the maximum reflection is important. Another benefit of apodization is the improvement of the dispersion compensation characteristic of chirped Bragg grating, for which the group delay becomes linearized and the modulation associated with the presence of side-lobes is eliminated. For a 10Gb/s signal the level of the modulation should be less than 100ps, with ±10ps being considered an acceptable value. In practice, apodization is accomplished by varying the amplitude of the coupling coefficient along the length of the grating. The apodization of fiber Bragg gratings using a phase mask with variable diffraction efficiency has been reported by Albert et al [109]. Bragg gratings, with sidelobe levels 26dB lower than the peak reflectivity, were fabricated in standard telecommunication fibers. This represents a reduction of 14dB in the sidelobe levels compared to uniform gratings with the same bandwidth and reflectivity. Figure 2.19 shows the spectral reflection response of an apodized and an unapodized fiber Bragg grating reflector reported by the same group [110], where a 20 dB reduction in the sidelobe levels was achieved.

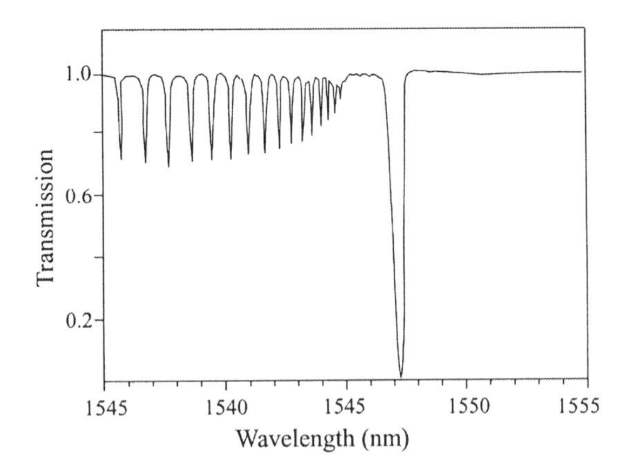

Figure 2.18. Schematic transmission profile for a strong fiber Bragg grating, showing loss to radiation modes on the short wavelength side, sharply modified by the cladding mode structure (after [107])

2.3.6 Types of fiber Bragg gratings

There are a several distinct types of fiber Bragg grating structures such as the common *Bragg reflector*, the *blazed Bragg grating*, and the *chirped Bragg grating*. These fiber Bragg gratings are distinguished either by their grating pitch (spacing between grating planes) or tilt. The most common fiber Bragg grating is the *Bragg reflector*, which has a constant pitch. The *blazed grating* has phase fronts tilted with respect to the fiber axis, that is the angle between the grating planes and the fiber axis is less than 90^0. The *chirped grating* has an aperiodic pitch, displaying a monotonic increase in the spacing between grating planes. A brief overview of these Bragg gratings along with some of their applications will be presented solely for the purpose of establishing the grating properties.

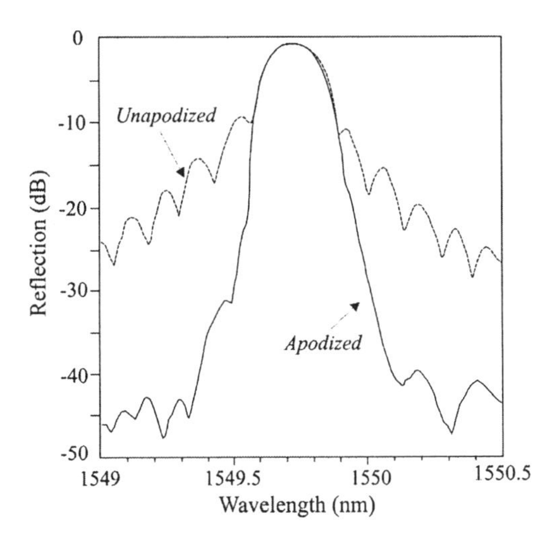

Figure 2.19. Reflection spectrum of fiber Bragg gratings photoimprinted with a uniform diffrating phase mask and with a phase mask having a Gaussian profile of diffraction efficiency (after [110])

2.3.6.1 Common Bragg reflector

The *common Bragg reflector*, the simplest and most used fiber Bragg grating, is illustrated in figure 2.12. Depending on the parameters such as grating length and magnitude of induced index change, the Bragg reflector can function as a narrow-band transmission or reflection filter or a broadband mirror. In combination with other Bragg reflectors, these devices can be arranged to function as band-pass filters. Two such configurations are shown in figure 2.20. The first configuration modifies a broadband spectrum, utilizing the Bragg gratings to remove discrete wavelength components, whereas the second arrangement incorporates the Bragg gratings as highly reflecting mirrors to construct a fiber Fabry-Perot cavity.

Bragg reflectors are considered to be excellent strain and temperature sensing devices because the measurements are wavelength-encoded. This eliminates the problems of amplitude or intensity fluctuations that exist in many other types of fiber-based sensor systems. Each Bragg reflector can be designated its own wavelength-encoded signature, therefore, a series of gratings can be written in the same fiber, each having a distinct Bragg resonance signal. This configuration can be used for wavelength division multiplexing or quasi-distributed sensing [111]. Gratings have also proven to be very useful components in tunable fiber or semiconductor lasers [112,113], serving as one or both ends of the laser cavity (depending on the laser configuration). Varying the Bragg resonance feedback signal to the grating tunes the laser wavelength. Using this approach Ball and Morey [114] demonstrated a continuously tunable, single-mode, erbium fiber laser with two Bragg reflectors configured in a Fabry-Perot cavity. Continuous tunability, without mode hopping, was achieved when both the gratings and enclosed fiber were stretched uniformly. Bragg grating fiber lasers can also be used as sensors where the Bragg reflector serves the dual purpose of tuning element and sensor [115]. A series of Bragg reflectors having distinct wavelength encoded signatures can be multiplexed in a fiber laser sensor configuration for multipoint sensing [116,117].

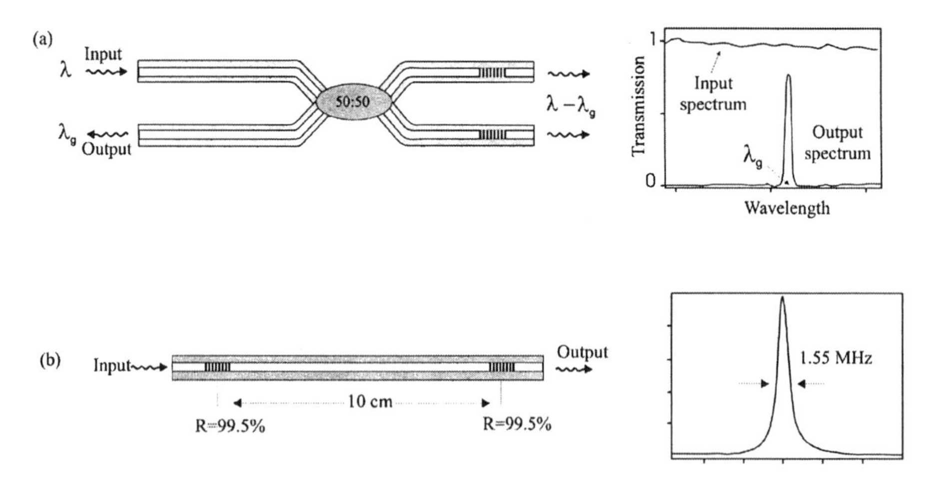

Figure 2.20. Fiber-optic bandpass filters using Bragg reflectors (a) filter arranged in a Michelson-type configuration and (b) filter arranged in a Fabry Perot-type configuration

2.3.6.2 Blazed Bragg gratings

Tilting (or blazing) the Bragg grating planes at angles to the fiber axis, shown in figure 2.21(a), will result in light that is otherwise guided in the fiber core, being coupled into loosely bound, guided-cladding or radiation modes. The tilt of the grating planes and strength of the index modulation determines the coupling efficiency and bandwidth of the light that is tapped out. The criterion to satisfy the Bragg condition of a blazed grating is similar to that of the Bragg reflector that was analyzed earlier. Figure 2.21(b) also illustrates the vector diagram of the Bragg condition (energy and momentum conservation) for the blazed grating. Here the wavevector of the grating **K** is incident at an angle, θ_b, with respect to the fiber axis. The magnitudes of the incident $\mathbf{v}_i$ and the scattered, $\mathbf{v}_s$, wavevectors must be equal ($v = |\mathbf{v}_i| = |\mathbf{v}_s|$). Simple trigonometry shows that the scattered wavevector must be at an angle $2\theta_b$ with respect to the fiber axis. Applying the law of cosines to the momentum diagram gives

$$|\mathbf{v}_i|^2 + |\mathbf{v}_s|^2 - 2|\mathbf{v}_i||\mathbf{v}_s|\cos(\pi - 2\theta_b) = |\mathbf{K}|^2 \tag{2.16}$$

this reduces to $\cos(\theta_b)=|\mathbf{K}|/2v$ and shows that the scattering angle is restricted by the Bragg wavelength and the effective refractive index. It is clear from the above equation that for blazed gratings not only different wavelengths emerge at different angles, but different modes of the same wavelength also emerge at slightly different angles due to their different propagation constants.

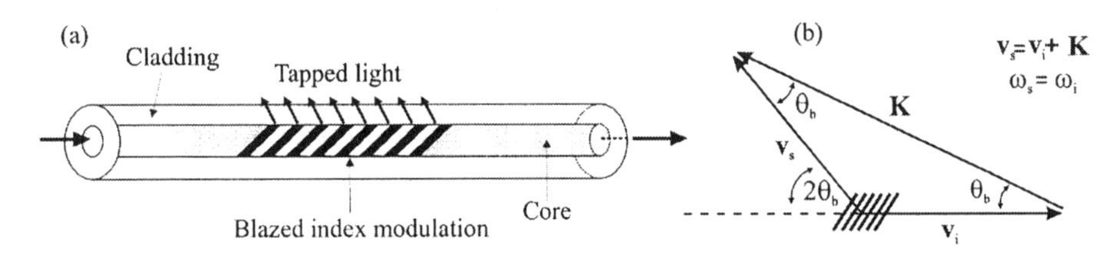

Figure 2.21. (a) A schematic diagram of a blazed grating. (b) A vector diagram for the Bragg condition of a blazed grating

Figure 2.22 shows the output coupling of 488 and 514.5 nm light from an argon ion laser. The green argon ion wavelength has two modes and the blue wavelength has three modes that propagate in the fiber. These wavelengths as well as their modes are well separated and resolvable, thus the grating tap acts as a spectrometer and mode discriminator. Meltz and Morey [118] have achieved out-coupling efficiencies as high as 21% at 488 and 514.5 nm.

Erbium doped fiber amplifiers are now an integral part of long haul, high-bit-rate communication systems and are finding applications in areas of wide bandwidth amplification. Kashyap et al [119] demonstrated the use of multiple blazed gratings to flatten the gain spectrum of erbium doped fiber amplifiers. A gain variation of ±1.6 dB over a bandwidth of 33 nm in a saturated erbium-doped, fiber amplifier was reduced to ±0.3 dB. This is important in fiber communications that use several signals at different wavelengths and gives a uniform signal-to-noise ratio at the receiver output. Another interesting application of blazed gratings is in mode conversion.

Mode converters are fabricated by inducing a periodic refractive index perturbation along the fiber length with a periodicity that bridges the momentum mismatch between the modes to allow phase matched coupling between the selected modes. Different grating periods are used for mode conversion at different wavelengths. Hill et al [9] demonstrated efficient mode conversion between forward propagation LP_{01} and LP_{11} modes.

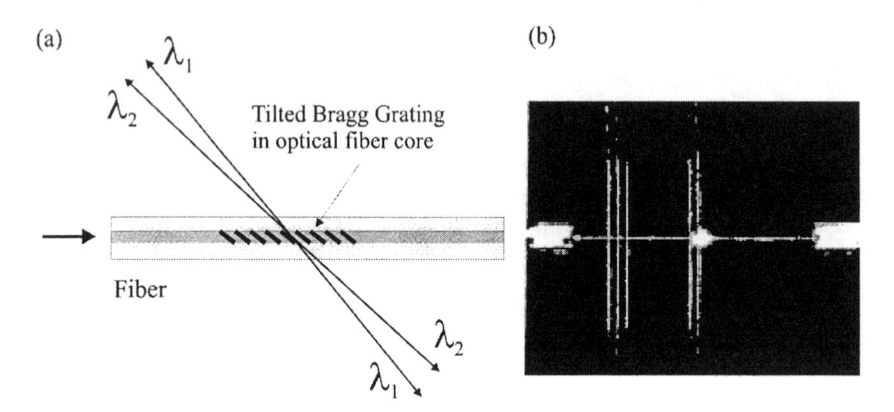

Figure 2.22. Illustration of separated wavelength tapped out at different angles. Image of the radiation out-coupled at the 488 and 514.5 nm from a fiber Bragg grating tap (after [118])

2.3.6.3 Chirped Bragg grating

One of the most interesting Bragg grating structures with immediate applications in telecommunications is the chirped Bragg grating. This grating has a monotonically varying period, as illustrated schematically in figure 2.23. There are certain characteristic properties offered by monotonically varying the period of gratings that are considered advantages for specific applications in telecommunication and sensor technology, such as dispersion compensation and the stable synthesis of multiple wavelength sources [120]. These types of gratings can be realized by axially varying either the period of the grating Λ or the index of refraction of the core or both. From equation 2.6 we have

$$\lambda_B(z) = 2n_{eff}(z)[\Lambda_0 + \Lambda_1 z] \tag{2.17}$$

where the variation in the grating period is assumed linear. Λ_0 is the starting period and Λ_1 is the linear change (slope) along the length of the grating. Thus such a grating structure may be considered to be made up of a series of smaller length uniform Bragg gratings, increasing in period. If such a structure is designed properly one may realize a broadband reflector. Typically the linear chirped grating has associated with it a chirped value/unit length (Λ_1) and the starting period. For example a chirped grating 2 cm in length may have a starting wavelength at 1550 nm and a chirped value of 1 nm/cm. This implies that the end of the chirped grating will have a wavelength period corresponding to 1552 nm. Chirped gratings have been written in optical fibers using various methods [121-123]. These fabrication techniques will be discussed in the next section.

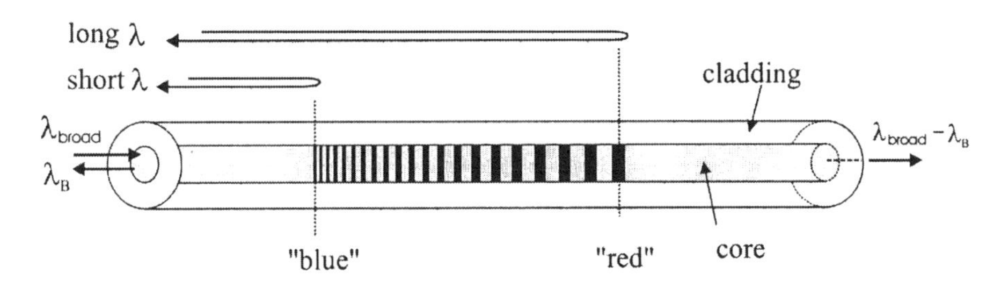

Figure 2.23. A schematic diagram of a chirped grating with an aperiodic pitch. For forward propagating light as shown, long wavelengths travel further into the grating before being reflected

In optically amplified, long haul, high-bit-rate communication systems the main limitation to data transmission is pulse broadening caused by chromatic dispersion. The pulse broadening can be eliminated by incorporating an element having a dispersion of opposite sign and equal magnitude to that of the optical fiber link. Traditionally, optical fibers displaying the correct negative dispersion characteristics have been incorporated into telecommunication lines. However, the modal field diameter of the compensating fiber rarely matches that of the standard guide, therefore, splicing between different fiber sections requires pre-fusion preparation (heating of the fiber to produce diffusion of the dopants in the guiding core until the modal field overlap is optimized). This can be achieved with a single fusion-splicing device; however, it is advantageous if this can be avoided, particularly if a suitable in-line, in-fiber component is available. In a chirped grating the resonant frequency is a linear function of the axial position along the grating, so that different frequencies present in the pulse are reflected at different points and thus acquire different delay times (figure 2.23). It is now possible to compress temporally broadened pulses.

2.3.7 Photosensitivity types of fiber Bragg gratings

In section 2.2 the different types of Bragg gratings were classified into three distinct categories of photosensitivity-types, Type I, Type IIA and Type II. The final grating type depended upon the initial writing conditions (laser power and wavelength, CW or pulsed energy delivery) and fiber properties. Here the spectral properties of these types of gratings will be discussed.

2.3.7.1 Type I Bragg gratings

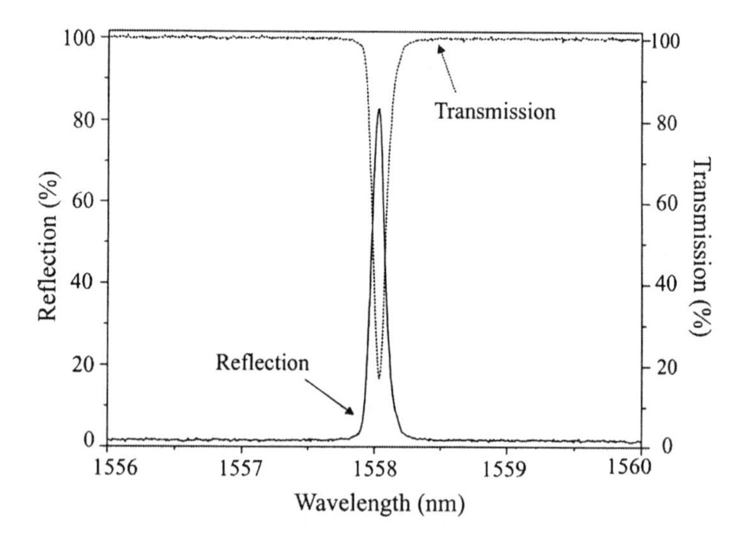

Figure 2.24. A spectral response of a Type I Bragg grating with a uniform period. Reflection and transmission spectra of a broadband light source in the region of where the Bragg condition is satisfied

Type I Bragg gratings refer to gratings that are formed in normal photosensitive fibers under moderate intensities. A typical spectral response of a uniform period, Type I Bragg grating is shown in figure 2.24. It is interesting to point out that the reflection spectra of the guiding mode is complementary to the transmission signal implying that there is negligible loss due to absorption or reflection into the cladding. This is a fundamental characteristic of a Type I Bragg grating. Furthermore, due to the photosensitivity type of the Bragg grating, the grating itself has a characteristic behavior with respect to temperature erasure. Type I gratings can be erased at relatively low temperatures, approximately 200°C. Nevertheless, Type I gratings are the most utilized Bragg gratings and operate effectively from -40 to +80°C, a temperature range that satisfactorily covers most telecommunication and sensor applications.

2.3.7.2 Type IIA Bragg gratings

Type IIA fiber Bragg gratings appear to have the same spectral characteristics as Type I gratings. The transmission and reflection spectra are complimentary, rendering this type of grating indistinguishable from Type I in a static situation. However, due to the different mechanism involved in fabricating these gratings there are some distinguishable features noticeable under dynamic conditions observed either in the initial fabrication or the temperature erasure of the gratings. Type IIA gratings are inscribed through a long process, following Type I grating inscription. At approximately 30 minutes of exposure (depending of the fiber type and exposure fluence) the Type IIA grating is fully developed. Clearly Type IIA gratings are not very practical to fabricate. Although the mechanism of the index change is different from Type I, occurring through compaction of the glass matrix, the behavior subject to external perturbations is the same for both grating types. Irrespective of the subtleties of the index change on a microscopic scale, the perturbations act macroscopically and therefore the wavelength response remains the same. However, when the grating is exposed to high ambient temperature a noticeable erasure is observed only at temperatures as high as 500°C. A clear advantage of the Type IIA gratings over the Type I is the dramatically improved temperature stability of the grating, this may prove very useful if the system has to be exposed in high ambient temperatures (as may be the case for sensor applications).

2.3.7.3 Type II Bragg gratings

Type II Bragg gratings are the most distinct of all grating types and are formed under very high, single-pulse fluence (>0.5 J/cm^2) [35]. A typical transmission and reflection spectrum of a Type II grating is shown in figure 2.25. The reflection appears to be broad and several features over the entire spectral profile are believed to be due to non-uniformities in the excimer beam profile that are strongly magnified by the highly nonlinear response mechanism of the glass core. Type II gratings pass wavelengths longer than the Bragg wavelength, whereas shorter wavelengths are strongly coupled into the cladding, as is observed for etched or relief fiber gratings [124], permitting their use as effective wavelength selective taps.

Type II gratings were first demonstrated in an experiment carried out to study the relationship between pulse energy and grating strength, where a series of single-pulse gratings were produced with a UV excimer laser beam

[35]. The index of refraction change was estimated from the reflection spectrum using coupled mode theory with the result summarized in figure 2.4. It is apparent that there is threshold at pulse energy of 650 mJ/cm^2, above which the induced index modulation increases dramatically. Below the threshold point, the index modulation grows linearly with energy density, whereas above this point the index modulation appears to saturate. The gratings formed with low index of refraction modulation were Type I. The behavior observed in figure 2.4 suggests that there is a critical level of absorbed energy, which triggers off a highly non-linear mechanism, initiating dramatic changes in the optical fiber. Examination of a Type II grating with an optical microscope revealed a damaged track at the core-cladding interface that is unique to this grating type, strongly suggesting that it may be responsible for the large index change. The fact that this damage is localized on one side of the core suggests that most of the UV light has been strongly absorbed locally.

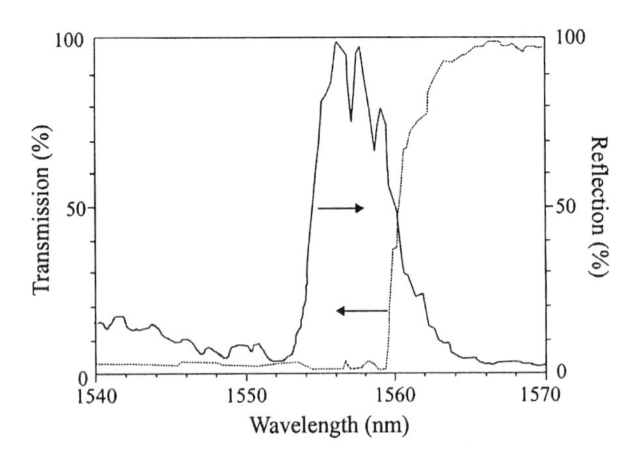

Figure 2.25. Reflection and transmission spectra of a Type II grating. At wavelengths below the Bragg wavelength 1556 nm the light is strongly coupled into the cladding (after [35])

Results of stability tests have shown Type II gratings to be extremely stable at elevated temperatures [35]. At 800°C over a period of 24 hours, no degradation in grating reflectivity was evident. At 1000°C, most of the grating disappears after 4 hours, implying that the localized fusion has been thermally "washed out". This superior temperature stability can be usefully utilized for sensing applications in hostile environments. One of the most attractive features of Type II gratings is that (as with Type I gratings) highly reflective gratings can be formed in just a few nanoseconds, the duration of a single excimer pulse. This is of great practical importance for large-scale mass production of strong gratings during the fiber drawing process before application of the protective polymer coating. Although the concept of fabrication of single pulse Type I and Type II Bragg gratings during the fiber drawing process has been successfully demonstrated [125,126], the quality of in-line gratings must be improved. One distinct advantage of producing fiber Bragg gratings during the draw process is that in-line fabrication avoids potential contact with the pristine outer surface of the glass, whereas off-line fabrication requires a section of the fiber to be stripped off its UV absorbing polymer coating, in order for the grating to be exposed. This drastically weakens the fiber at the site of the grating due to surface contamination, even if the fiber is subsequently recoated.

2.3.8 Novel Bragg grating structures

In this section several atypical grating structures that have immediate applications to telecommunications and sensing applications are included.

2.3.8.1 Superimposed multiple Bragg gratings

Othonos and co-workers [127] demonstrated the inscription of several Bragg gratings at the same location on an optical fiber. This is of interest as a device in fiber communications, lasers and sensor systems, because multiple Bragg gratings at the same location basically perform a comb function and this device is ideally suited for multiplexing and demultiplexing signals. All the gratings are written at the same location of the fiber, lending this approach to optical integrated technology, where the issue of size is always a concern. This can also be used for material detection where the multiple Bragg lines can be designed to match the signature frequencies of a given material. A narrow linewidth KrF excimer laser was used in an interferometric setup to inscribe the different Bragg gratings on the same fiber location. The fiber, AT&T Accutether, was first hydrogen loaded to enhance its photosensitivity. Figure 2.26 shows the reflectivity for seven superimposed Bragg gratings. The first grating was written at 1550.05 nm and reached a reflectivity of ~100% within 15sec (at 30Hz) of UV exposure and had a linewidth of 0.25 nm. After adjusting the interferometer to write at a different Bragg wavelength, the second grating was written at 1542.6 nm with approximately the same characteristics. Each time a new grating was inscribed, the reflectivity of the existing gratings was reduced. Nevertheless, (figure 2.27), even after superimposing five gratings, the individual grating reflectivities were higher than 60%. Additionally, the center wavelength of the existing Bragg gratings shifted to longer wavelengths each time a new grating was inscribed due to change of the effective index of refraction, for example, the first grating shifted to 1550.975 nm by the time the last grating was inscribed. The shift in wavelength of the first grating after writing all seven gratings corresponds to an effective index of refraction increase of 0.86×10^{-3}.

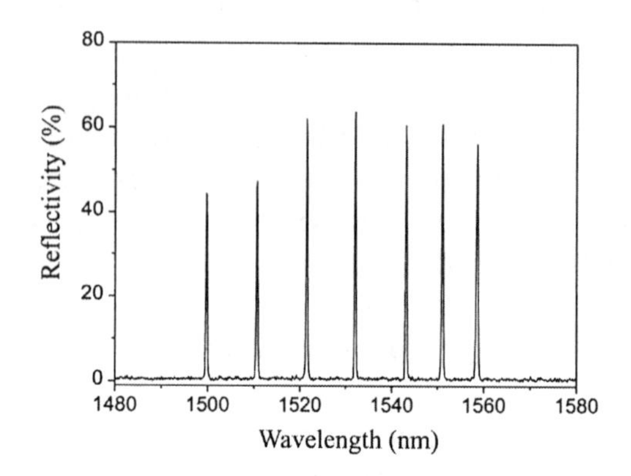

Figure 2.26. Reflection spectrum for seven Bragg gratings superimposed at the same location on an optical fiber. The seven gratings cover a span of 60 nm ranging form 1500 to 1560 nm (after [127])

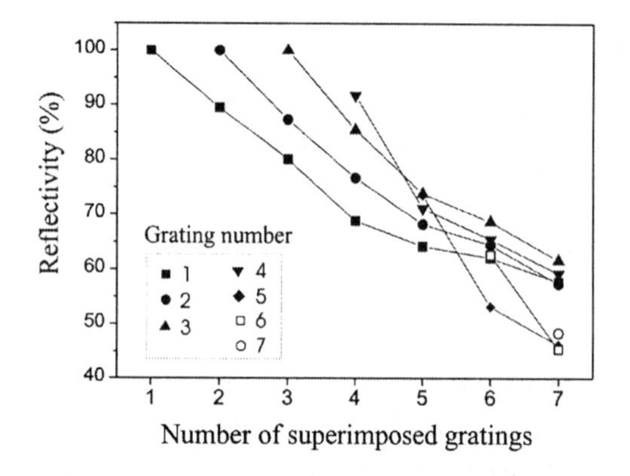

Figure 2.27. Reflectivities for each Bragg grating as a function of the number of gratings superimposed on the same location. The lines through the symbols are a guide to the eye (after [127])

2.3.8.2 Superstructure Bragg gratings

The superstructure Bragg grating refers to a grating fiber structure fabricated with a modulated exposure over the length of the grating [128]. One such approach used by Eggleton et al [128] was to translate the UV writing beam along a fiber and phase mask assembly while the intensity of the beam was modulated. An excimer-pumped, dye laser with a frequency doubler was used to produce 2.0 mJ at 240 nm. Hydrogenated, single-mode, boron-codoped fiber was placed in near contact with a phase mask, and the ultraviolet light was focused through the phase mask into the fiber core by a cylindrical lens, exposing a length of approximately 1mm. To fabricate a 40 mm long superstructure, the excimer laser was periodically triggered at intervals of 15 sec to produce bursts of 150 short at a repetition rate of 10 Hz, while the ultraviolet beam was translated at a constant velocity of 0.19 mm/s along mask. The resulting period of the grating envelope was approximately 5.65 mm, forming seven periods of the superstructure. The reflection spectrum of this grating structure is shown in figure 2.28. These superstructure gratings can be used as comb filters for signal processing, and for increasing the tunability of the fiber laser-grating reflector.

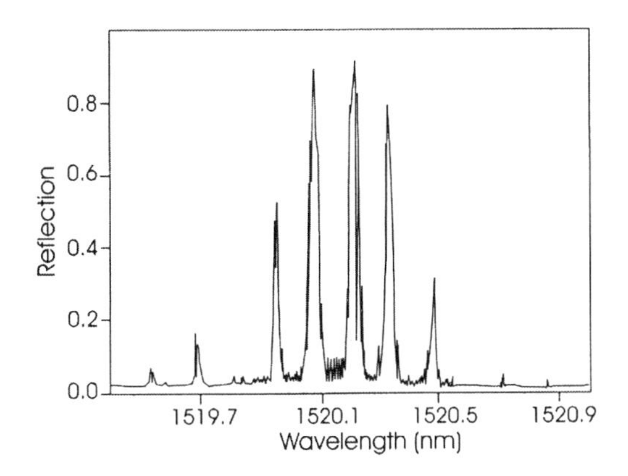

Figure 2.28. Reflection spectrum from grating supersturcture fabricated by translating the UV writing beam along a fiber and phse-mask assemble while the intensity of the beam was modulated (after [128])

2.3.8.3 Phase shifted Bragg gratings

Bragg gratings generally act as narrow-band reflection filters centered at the Bragg wavelength because of the stop band associated with a one-dimensional periodic medium. Many applications, such as channel selection in a multi-channel communication system, would benefit if the fiber grating could be designed as a narrow-band transmission filter. Although techniques based on Michelson and Fabry-Perot interferometers have been developed for this purpose [129], their use requires multiple gratings and may introduce additional losses. A technique commonly used in distributed feedback (DFB) semiconductor lasers [130] can be used to tailor the transmission spectrum to suit specific requirements. This approach relies on introducing a phase shift across the fiber grating, whose location and magnitude can be adjusted to design a specific transmission spectrum. It is a generalization of an idea first proposed by Haus and Shank [131] in 1976. The principle of the phase-shift was demonstrated by Alferness et al [132] in periodic structures made from semiconductor materials where a phase shift was introduced by etching a larger spacing at the center of the device. This forms the basis of single-mode phase-shifted semiconductor DFB lasers. A similar device may be constructed in optical fibers using various techniques:

- Phase masks, in which phase-shift regions have been written into the mask design [133].
- Post-processing of a grating by exposure of the grating region to pulses of UV laser radiation which has been described in reference [134],
- And post-fabrication processing using localized heat treatment that has also been reported [135].

Such processing produces two gratings out of phase with each other, which act as a wavelength selective Fabry-Perot resonator, allowing light at the resonance to penetrate the stop-band of the original grating. The resonance wavelength depends on the size of the phase change. One of the most obvious applications includes production of very narrow-band transmission and reflection filters. Moreover, multiple phase shifts can be introduced to produce other devices such comb filters, or to obtain single-mode operation of DFB fiber lasers.

2.3.9 Thermal delay of fiber Bragg gratings

Although all grating types written in non-hydrogenated fiber are stable at room temperature over many years, since the formation of gratings involves laser-induced excitation of glass into a metastable state the gratings will decay over time at elevated temperatures. The extent to which this decay occurs depends on the fiber and grating type. One approach to stabilize the grating is to preanneal at a temperature that exceeds the anticipated serviceable temperature of the grating-based device – called accelerated aging. There is however, no single, universal stabilization anneal for all grating types as the refractive index stability depends on the fiber properties and the laser writing conditions. The thermally induced decay under accelerated test temperatures implies that the UV-induced defects are not thermodynamically stable, having sites that are reversible in nature followed by decay of the associated refractive index. Each individual site has an associated activation energy barrier that must be overcome in order to cause the index decay. The amorphous nature of glass results in a distribution of activation energies, the components of which can decay with discrete time constants. One observes an initial rapid decay followed by a slowing but non-zero rate [1].

2.4 INSCRIBING BRAGG GRATINGS IN OPTICAL FIBRES

Here we will describe various techniques used in fabricating standard and complex Bragg grating structures in optical fibers. Depending on the fabrication technique Bragg gratings may be labeled as internally or externally written. Although internal Bragg gratings may not be considered very practical or useful, nevertheless it is important to consider them, thus obtaining a complete historical perspective. Externally written Bragg gratings, that is gratings inscribed using techniques such as interferometric, point-by-point and phase-mask overcome the limitations of internally written gratings and are considered far more useful. Although most of these inscription technique were initially considered difficult, due to the requirements of sub-micron resolution and thus stability, today they are well controlled and the inscription of Bragg grating using these techniques is considered routine.

2.4.1 Internally inscribed Bragg gratings

Internally inscribed Bragg gratings were first demonstrated in 1978 by Hill and his co-workers [4,5] in a simple experimental setup shown in figure 2.29. An argon ion laser was used as thc sourcc, oscillating on a single longitudinal mode at 514.5 nm (or 488 nm) exposing the photosensitive fiber by coupling light into its core. Isolation of the argon ion laser from the back-reflected beam was necessary to avoid instability furthermore the pump laser and the fiber were placed in a tube for thermal isolation. The incident laser light interfered with the 4% reflection (from the cleaved end of the fiber) to initially form a weak standing wave intensity pattern within the core of the fiber. At the high intensity points the index of refraction in the photosensitive fiber changed permanently. Thus a refractive index perturbation having the same spatial periodicity as the interference patter was formed. These types of gratings normally have a long length (tens of centimeters) in order to achieve useful reflectivity values due to the small index of refraction changes.

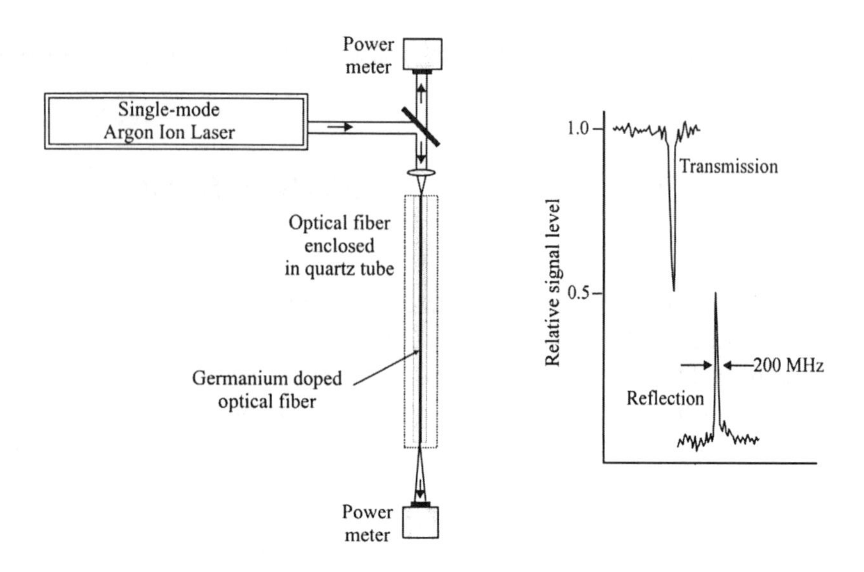

Figure 2.29. A typical apparatus used in generating self-induced Bragg gratings using an argon ion laser. A typical reflection and transmission characteristics of these types of gratings are shown in the graph

2.4.2 Inscribing Bragg gratings interferometrically

The interferometric fabrication technique, which is an external writing approach for inscribing Bragg gratings in photosensitive fibers was first demonstrated by Meltz and coworkers [8]. In that experiment, an incident UV light beam was split into two beams that were subsequently recombined to form an interference pattern that side-exposed a photosensitive fiber, inducing a permanent refractive index modulation in the core. Bragg gratings in optical fibers have been fabricated using both amplitude-splitting and wavefront-splitting interferometers. These two different types of interferometers will be examined next.

2.4.2.1 Amplitude-splitting interferometer

An amplitude-splitting interferometer was used by Meltz et al [8] to fabricate fiber Bragg gratings in an experimental arrangement similar to the one shown in figure 2.30. An excimer-pumped dye laser operating at a wavelength in the range of 486-500 nm was frequency doubled using a non-linear crystal. This provided a UV source in the 244 nm band with adequate coherence length (a critical parameter in this inscription technique). The UV radiation was split into two beams of equal intensity that were recombined to produce an interference pattern, normal to the fiber axis. A pair of cylindrical lenses focused the light onto the fiber and the resulting focal line was approximately 4mm long by 124μm wide. A broad band source was also used in conjunction with a high-resolution monochromator to monitor the reflection and transmission spectra of the grating. The graph in figure 2.30 shows the reflection and complementary transmission spectra of the grating formed in a 2.6 μm diameter core, 6.6 mol % GeO_2-doped fiber after 5 minute exposure to a 244 nm interference pattern with an average power of 18.5 mW. The length of the exposed region was estimated to be between 4.2 and 4.6 mm.

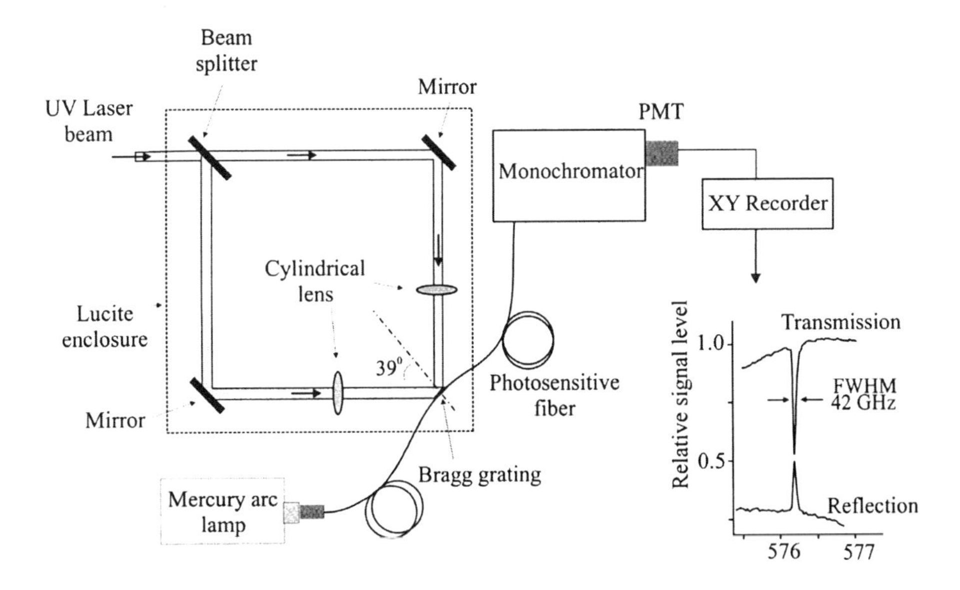

Figure 2.30. An amplitude splitting interferometer used by Meltz et al [8], which demonstrated the first externally fabricated Bragg grating in optical fiber. The reflection and transmission spectra for a 4.4mm long Bragg grating fabricated with this apparatus is also shown

In a conventional interferometer such as the one shown in figure 2.30 the UV writing laser light is split into equal intensity beams that subsequently recombine after having undergone a different number of reflections in each optical path. Therefore, the interfering beams (wavefronts) acquire different (lateral) orientations. This results in a low quality fringe pattern for laser beams having low spatial coherence. This problem may be eliminated by including a second mirror in one of the optical paths as shown in figure 2.31, which in effect compensates for the beam splitter reflection. Since now the total number of reflections is now the same in both optical arms, the two interfering beams at the fiber are identical.

The interfering beams are normally focused to a fine line matching the fiber core using a cylindrical lens placed outside the interferometer. This results in higher intensities at the core of the fiber, thereby improving the grating inscription. In an interferometer system as the one shown in figures 2.30 and 2.31 the interference fringe pattern period Λ, depends on both the irradiation wavelength λ_w and the half angle between the intersecting UV beams φ (figure 2.31). Since the Bragg grating period is identical to the period of the interference fringe pattern, then the fiber grating period is given by

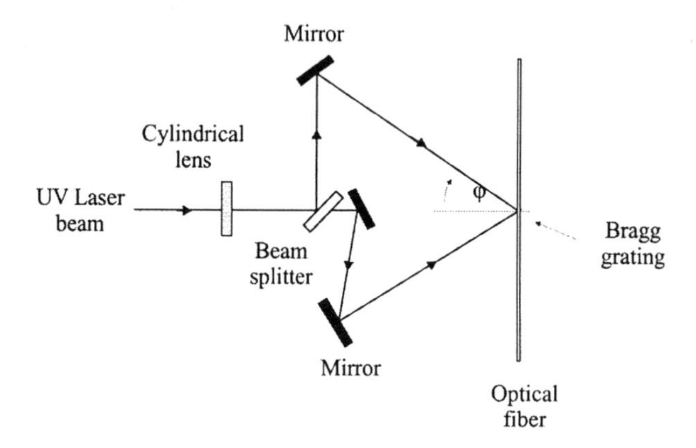

Figure 2.31. An improved version of the amplitude splitting interferometer, where an additional extra mirror is used to achieve an equal number of reflections, thus eliminating the different lateral orientations of the interfering beams. This type of interferometer is applicable to fabrication systems where the source spatial coherence is low, such as with excimer lasers

$$\Lambda = \frac{\lambda_w}{2\sin\varphi} \tag{2.18}$$

where, λ_w is the UV wavelength and φ is the half angle between the intersection UV beams (figure 2.31). Given the Bragg condition, $\lambda_B=2n_{eff}\Lambda$, the Bragg resonance wavelength, λ_B, can be represented in terms of the UV writing wavelength and the half angle between intersecting UV beams as

$$\lambda_B = \frac{n_{eff}\lambda_w}{\sin\varphi} \tag{2.19}$$

where n_{eff} is the effective core index. From Equation 2.19 one can easily see that the Bragg grating wavelength can be varied either by changing λ_w [136]

and/or φ. The choice of λ_w is limited to the UV photosensitivity region of the fiber, however there is no restriction for the choice of the angle φ.

One of the advantages of the interferometric method is the ability to introduce optical components within the arms of the interferometer, allowing the wavefronts of the interfering beams to be modified. In practice, incorporating one or more cylindrical lenses into one or both arms of the interferometer produces chirped gratings with a wide parameter range [2]. The most important advantage offered by the amplitude-splitting interferometric technique is the ability to inscribe Bragg gratings at any wavelength desired. This is accomplished by changing the intersecting angle between the UV beams. This method also offers complete flexibility for producing gratings of various lengths, which allows the fabrication of wavelength narrowed or broadened gratings. The main disadvantage of this approach is a susceptibility to mechanical vibrations. Sub-micron displacements in the position of mirrors, beam splitter or other optical mounts in the interferometer during UV irradiation will cause the fringe pattern to drift, washing out the grating from the fiber. Furthermore, because the laser light travels long optical path lengths, air currents, which affect the refractive index locally, can become problematic, degrading the formation of a stable fringe pattern. In addition to the above shortcomings, quality gratings can only be produced with a laser source that has good spatial and temporal coherence and excellent wavelength and output power stability.

2.4.2.2 Wavefront-splitting interferometers

Wavefront-splitting interferometers are not as popular as amplitude-splitting interferometers for grating fabrication but, however, they offer some useful advantages. Two examples of wavefront-splitting interferometers used to fabricate Bragg gratings in optical fibers are the prism interferometer [137,138] and Lloyd's interferometer [139]. The experimental set-up for fabricating gratings with Lloyd interferometer is shown in figure 2.32. This interferometer consists of a dielectric mirror, which directs half of the UV beam to a fiber that is perpendicular to the mirror. The writing beam is centered at the intersection of the mirror surface and fiber. The overlap of the direct and deviated portions of UV beam creates interference fringes normal to the fiber axis. A cylindrical lens is usually placed in front of the system to focus the fringe pattern along the core of the fiber. Since half of the incident beam is reflected, interference fringes appear in a region of length equal to half the width of the beam. Secondly, since half the beam is folded onto the other half, interference occurs, but the fringes may not be of high quality. In the Lloyd arrangement, the folding action of the mirror limits what is possible. It requires a source with a coherence length equal to at least the path difference introduced by the fold in the beam. Ideally the coherence and intensity profile should be constant across the writing beam, otherwise, the fringe pattern and thus the inscribed grating will not be uniform. Furthermore diffraction effects at the edge of the dielectric mirror may also cause problems with the fringe pattern.

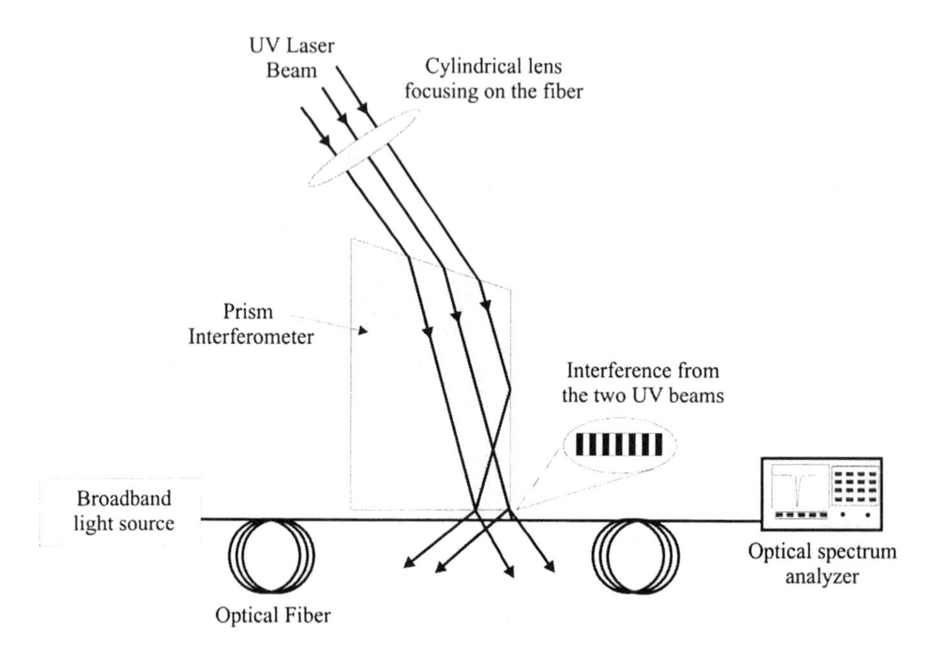

Figure 2.32. A schematics of the Lloyd wavefront-splitting interferometer

A schematic of the prism interferometer is shown in figure 2.33. The prism is made from high homogeneity, ultraviolet-grade, fused silica allowing for good transmission characteristics. In this setup the UV beam is spatially bisected by the prism edge and half the beam is spatially reversed by total internal reflection from the prism face. The two beam halves are then recombined at the output face of the prism, giving a fringe pattern parallel to the photosensitive fiber core. A cylindrical lens placed just before the setup helps in forming the interference pattern on a line along the fiber core. The interferometer is intrinsically stable as the path difference is generated within the prism and remains unaffected by vibrations. Writing times of over 8 hours have been reported with this type of interferometer. One disadvantage of this system is the geometry of the interference. The interferogram is formed by folding the beam onto itself, hence different parts of the beam must interfere, thus requiring a UV source with good spatial coherence.

A key advantage of the wavefront-splitting interferometer is the requirement for only one optical component, greatly reducing sensitivity to mechanical vibrations. In addition, the short distance where the UV beams are separated reduces the wavefront distortion induced by air currents and temperature differences between the two interfering beams. Furthermore, this assembly can be easily rotated to vary the angle of intersection of the two beams for wavelength tuning. One disadvantage of this system is the limitation on the grating length, which is restricted to half of the beam width. Another disadvantage is the range of Bragg wavelength tunability, which is restricted by the physical arrangement of the interferometers. As the intersection angle increases, the difference between beam path lengths increases, therefore, the beam coherence length limits the Bragg wavelength tunability.

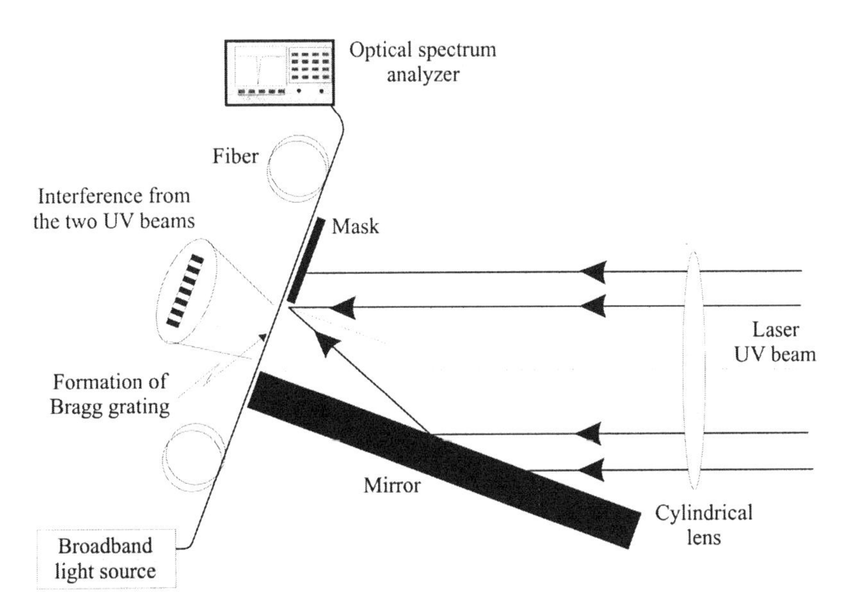

Figure 2.33. Schematic of the prism wavefront-splitting interferometer

2.4.2.3 Laser source requirements

Laser sources used for inscribing Bragg gratings via the above interferometric techniques must have good temporal and spatial coherence. The spatial coherence requirements can be relaxed in the case of the amplitude split interferometer by simply making sure that the total number of reflections are the same in both arms. This is especially critical in the case where a laser with low spatial coherence, like an excimer laser, is used as the source of UV light. The temporal coherence has to be at least the length of the grating in order for the interfering beams to have a good contrast ratio thus resulting in good quality Bragg gratings. The above coherence requirement together with the UV wavelength range needed (240-250 nm) forced researchers initially to use very complicated laser systems.

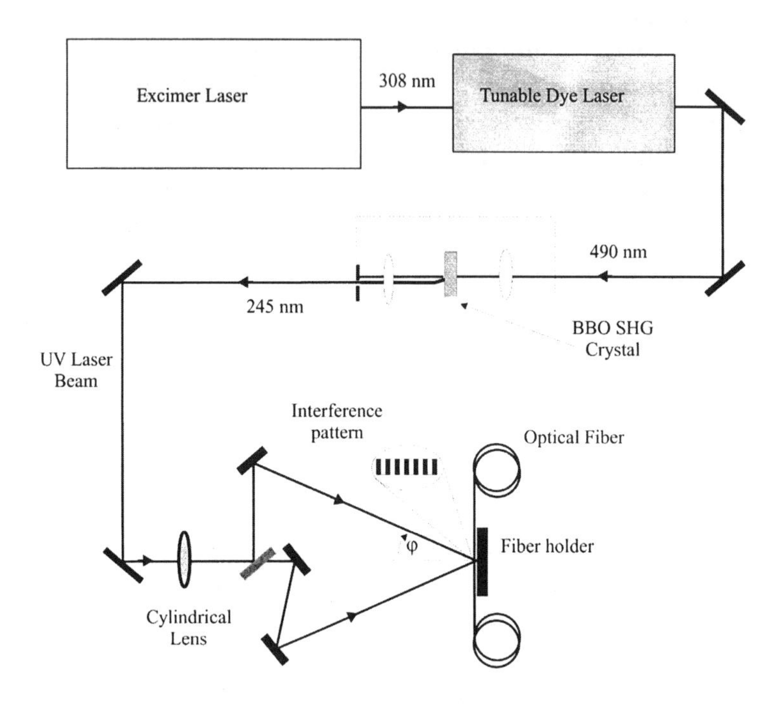

Figure 2.34. Experimental set-up of an excimer pump dye laser with a frequency doubled BBO crystal for generating UV light at 245 nm for inscribing Bragg gratings in an interferometer

One such system consists of an excimer pumped tunable dye laser, operating in the range of 480 to 500 nm. The output from the dye laser is focussed on a non-linear crystal to double the frequency of the fundamental light (figure 2.34). Typically this arrangement provides 10-20 nsec pulses (depending on the excimer pump laser) of approximately 3-5 mJ, with excellent temporal and spatial coherence. An alternative to this elaborate and often troublesome set-up is a specially designed excimer laser that has a long temporal coherence length. These spectrally narrow line-width excimer lasers may operate for extended periods of time on the same gas mixture with little changes in their characteristics. Commercially available narrow line-width excimer systems are complicated oscillator amplifier configurations, which make them extremely costly. Othonos and Lee [140] developed a low cost simple technique where an existing KrF excimer laser may be retrofitted with a spectral narrowing system for inscribing Bragg gratings in a side written interferometric configuration. In that work a commercially available KrF excimer laser (Lumonics Ex-600) was modified to produce a spectrally narrow laser beam (figure 2.35) with a linewidth of approximately 4×10^{-12} m. This system was used to inscribe successfully Bragg gratings in photosensitive optical fibers [140]. An alternative to the above system which is becoming very popular is the intracavity frequency-doubled argon ion laser [49] that uses Beta-Barium Borate (BBO). This system efficiently converts high-power visible laser wavelengths into deep ultraviolet (244 and 248 nm). The characteristics of these lasers include unmatched spatial coherence, narrow line-width and excellent beam pointing stability make such systems very successful in inscribing Bragg gratings in optical fibers [141].

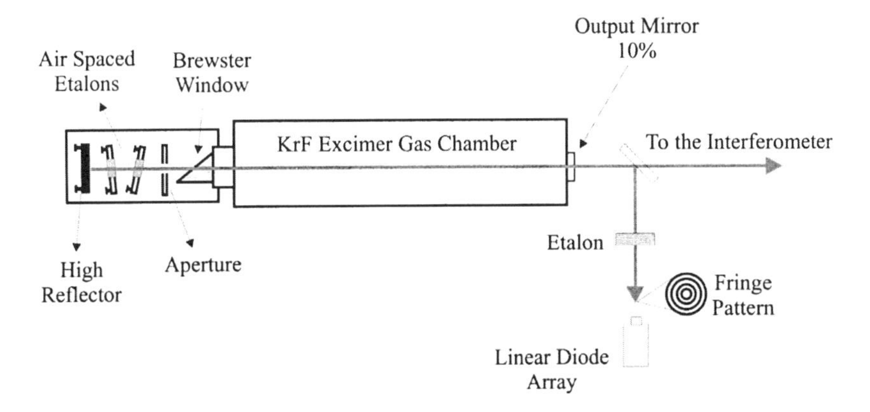

Figure 2.35. A schematic of a narrow linewidth, excimer laser system (SLN-KrF), consisting of two air-spaced etalons and an intracavity aperture placed between the KrF excimer gas chamber and the high reflector (after [140])

2.4.3 Phase mask technique

One of the most effective methods for inscribing Bragg gratings in photosensitive fiber is the phase mask technique [11]. This method employs a diffractive optical element (phase mask) to spatially modulate the UV writing beam (figure 2.36). Phase masks may be formed holographically or by electron-beam lithography. Holographically induced phase masks have no stitch error, which is normally present in the electron-beam phase masks. However, complicated patterns can be written into the electron-beam-fabricated masks (quadratic chirps, Moire patterns etc.). The phase mask grating has a one-dimension surface-relief structure fabricated in a high quality fused silica flat transparent to the UV writing beam. The profile of the periodic surface-relief gratings is chosen such that when a UV beam is incident on the phase mask, the zero-order diffracted beam is suppressed to less than a few percent (typically less than 5%) of the transmitted power. In addition the diffracted plus and minus first orders are maximized; each containing typically more than 35% of the transmitted power. A near field fringe pattern is produced by the interference of the plus and minus first order diffracted beams. The period of the fringes is one half that of the mask. The interference pattern photo-imprints a refractive index modulation in the core of a photosensitive optical fiber placed in contact with or in close proximity to and immediately behind the phase mask (figure 2.36). A cylindrical lens may be used to focus the fringe pattern along the fiber core.

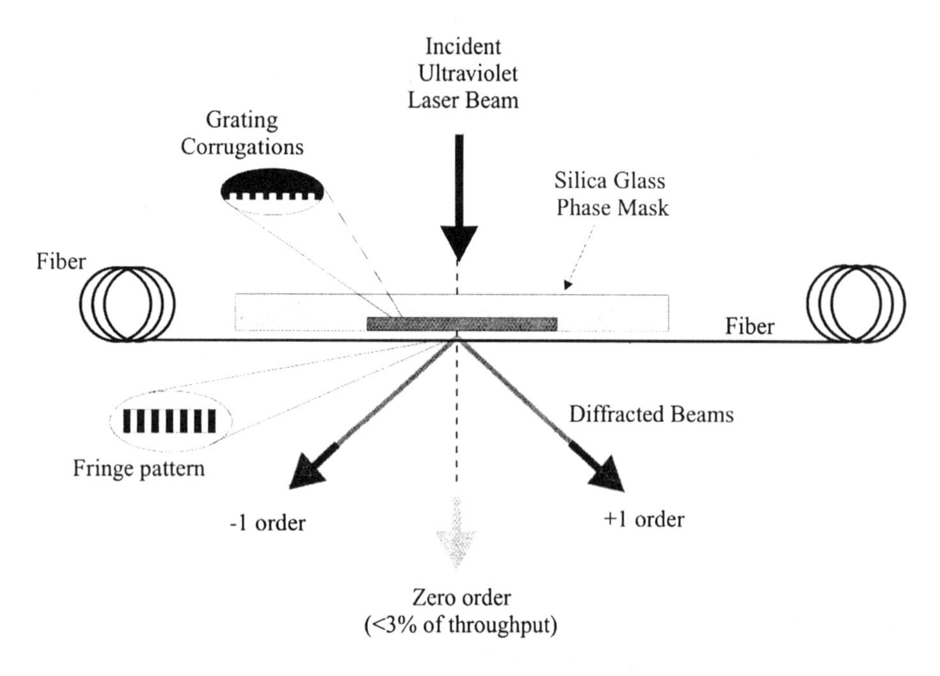

Figure 2.36. Schematic of the phase-mask geometry for inscribing Bragg gratings in optical fibers. The plus and minus first-order diffracted beam interfere at the fiber core, placed at a known distance from the mask

The phase-mask greatly reduces the complexity of the fiber grating fabrication system. The simplicity of using only one optical element provides a robust and an inherently stable method for reproducing fiber Bragg gratings. Since the fiber is usually placed directly behind the phase mask in the near field of the diffracting UV beams, sensitivity to mechanical vibrations and therefore stability problems are minimized. Low temporal coherence does not effect the writing capability (as opposed to the interferometric technique) due to the geometry of the problem.

KrF excimer lasers are the most common UV sources used to fabricate Bragg gratings with a phase-mask. The UV laser sources typically have low spatial and temporal coherence. The low spatial coherence requires the fiber to be placed in near contact to the grating corrugations on the phase mask in order to induce maximum modulation in the index of refraction. The further the fiber is placed from the phase mask, the lower the induced index modulation, resulting in lower reflectivity Bragg gratings. Clearly, the separation of the fiber from the phase mask is a critical parameter in producing quality gratings. However, placing the fiber in contact with the fine grating corrugations is not desirable due to possible damage to the phase mask. Othonos and Lee [142] demonstrated the importance of spatial coherence of UV sources used in writing Bragg gratings, employing the phase-mask technique. Improving the spatial coherence of the UV writing beam not only improves the strength and quality of the gratings inscribed by the phase mask technique, it also relaxes the requirement that the fiber has to be in contact with the phase mask.

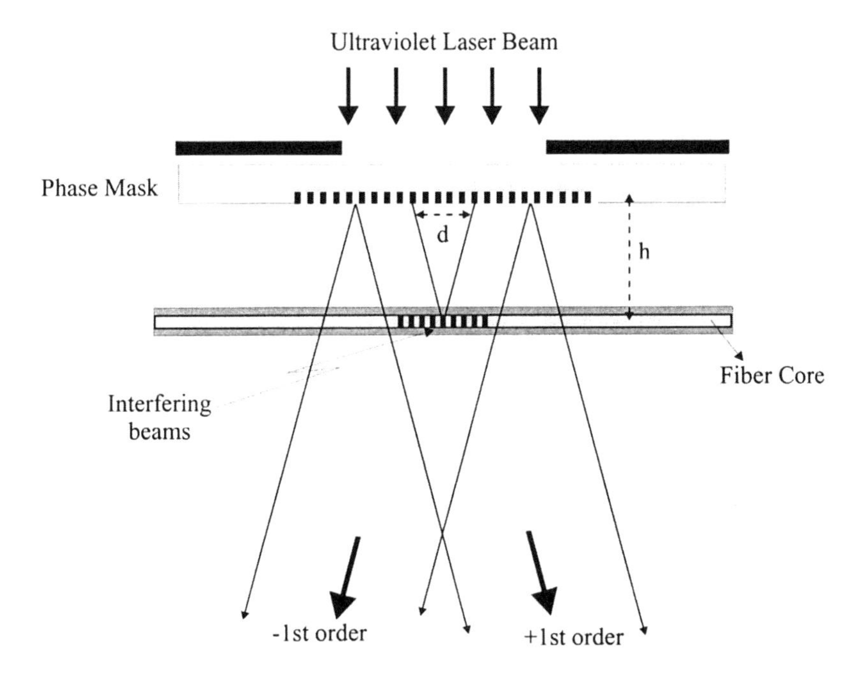

Figure 2.37. A simple schematic of the phase-mask geometry for inscribing Bragg gratings in optical fibers. The plus and minus first-order diffracted beams interfere at the fiber core, which is placed at a distance, h, from the mask

To understand the significance of spatial coherence in the fabrication of Bragg gratings using the phase mask technique, it is helpful to consider a simple schematic diagram (figure 2.37). Consider the fiber core to be at a distance h from the phase mask. The transmitted *plus* and *minus* first orders that interfere to form the fringe pattern on the fiber emanate from different part of the mask (referred to as distance d in figure 2.37). Since the distance of the fiber from the phase mask is identical for the two interfering beams, the requirement for temporal coherence is not critical for the formation of a high contrast fringe pattern. On the other hand, as the distance h increases, the separation d between the two interfering beams emerging from the mask increases. In this case, the requirement for good spatial coherence is critical for the formation of a high contrast fringe pattern. As the distance h extends beyond the spatial coherence of the incident UV beam, the interference fringe contrast will deteriorate, eventually resulting in no interference at all. The importance of spatial coherence was also demonstrated by Dyer et al [143] who used a KrF laser irradiated phase mask to form gratings in polyimide film.

One of the advantages of not having to position the fiber against the phase mask is the freedom to be able to angle the fiber relative to the mask forming blazed gratings. Placing one end of the exposed fiber section against the mask and the other end at some distance form the mask, it is possible to change the induced Bragg grating center wavelength. From simple geometry (see inset in figure 2.38) a general expression for the tunability of the Bragg grating center wavelength can be derived, given by:

$$\lambda_B = 2n\Lambda\sqrt{1+\left(\frac{r}{l}\right)^2} \tag{2.20}$$

where Λ is the period of the fiber grating, r is the distance from one end of the exposed fiber section to the phase mask, and l is the length of the phase grating. For a fixed phase mask period, changing r will result in blazed gratings with a changing center Bragg wavelength. In the experiments described in Othonos and Lee [142] a phase mask with Λ=0.531μm (l=10000 μm) was utilized resulting in λ_B=1558 nm at r=0 (the fiber placed parallel to the phase mask). Figure 2.38 shows the theoretical curve for the tunability of the inscribed Bragg grating as a function of distance r. The experimental values for the peak reflectivities of the Bragg gratings are also shown for different r values.

A variation to the phase mask scheme with the fiber in near contact to the mask (as described above) has been demonstrated [144]. This technique is based on a UV transmitting silica prism. The -1 and +1 orders are internally reflected within a rectangular prism, as shown in figure 2.39 and interfere at the fiber. This non-contact technique is flexible and allows quick changes of the Bragg wavelength.

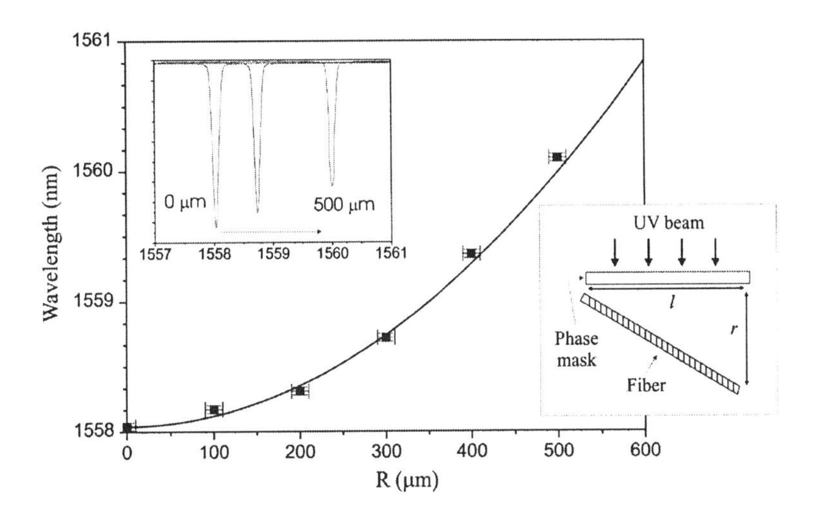

Figure 2.38. Experimental and calculated results of tuning a Bragg grating resonance by tilting the writing fiber with respect to the phase-mask. This tuning technique results in blazed gratings

2.4.4 Point-by-point fabrication of Bragg gratings

The point-by-point techniquc [145] for fabricating Bragg gratings is accomplished by inducing a change in the index of refraction one step at a time along the core of the fiber. Each grating plane is produced separately by a focused single pulse from an excimer laser. A single pulse of UV light from an excimer laser passes through a mask containing a slit. A focusing lens images the slit onto the core of the optical fiber from the side, as shown in figure 2.40, and the refractive index of the core in the irradiated fiber section increases locally. The fiber is then translated through a distance Λ corresponding to the grating pitch in a direction parallel to the fiber axis and the process is repeated to form the grating structure in the fiber core. Essential to the point-by-point fabrication technique is a very stable and precise submicron translational system.

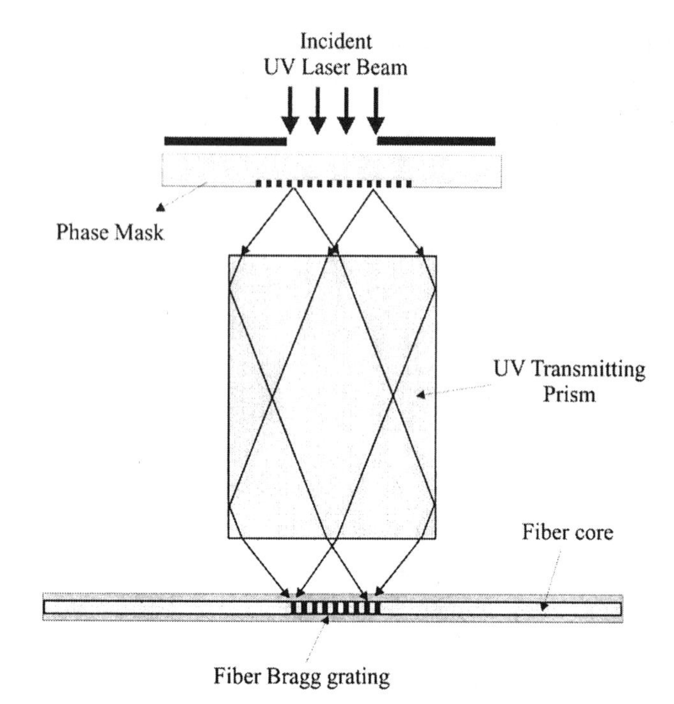

Figure 2.39. Non-contact technique interferometric phase mask technique for generating fiber Bragg gratings

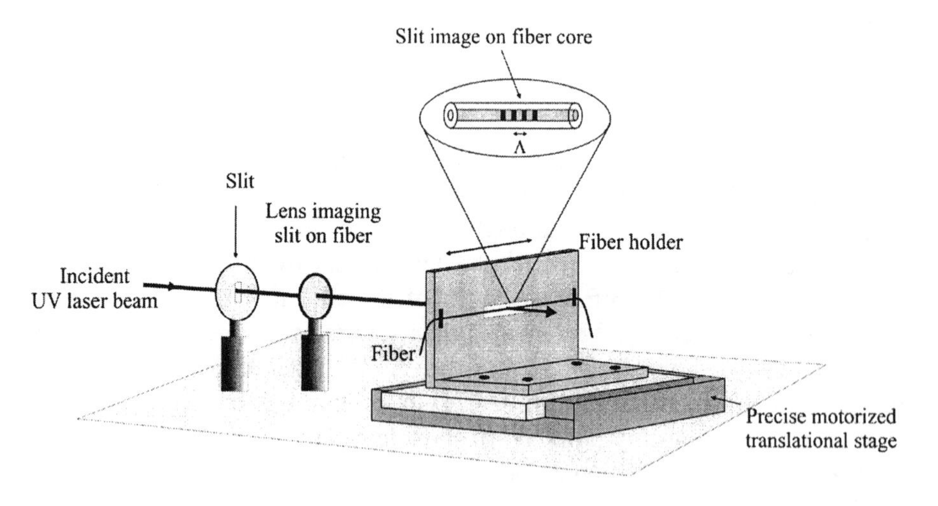

Figure 2.40. A schematic of a setup for fabricating Bragg gratings using the point-by-point technique

The main advantage of the point-by-point writing technique lies in its flexibility to alter the Bragg grating parameters. Because the grating structure is built up a point at a time, variations in grating length, grating pitch, and spectral response can easily be incorporated. Chirped gratings can be produced accurately simply by increasing the amount of fiber translation each time the fiber is irradiated. The point-by-point method allows for the fabrication of spatial mode converters [146] and polarization mode converters or rocking filters [9] that have grating periods, Λ, ranging from tens of micrometers to tens of millimeters. Because the UV pulse energy can be varied between points of induced index change, the refractive index profile of the grating can be tailored to provide any desired spectral response.

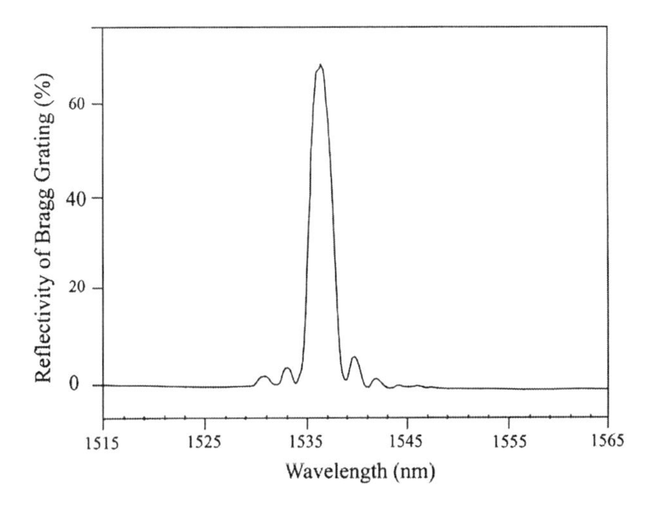

Figure 2.41. The plot at the upper-right corner illustrates a reflection spectrum of a third-order Bragg grating fabricated using the point-by-point method (after [145])

One disadvantage of the point-by-point technique is that it is a tedious process. Because it is a step-by-step procedure, this method requires a relatively long process time. Errors in the grating spacing due to thermal effects and/or small variations in the fiber strain can occur. This limits the gratings to very short lengths. Typically, the grating period required for first order reflection at 1550 nm is approximately 530 nm. Because of the submicron translation and tight focusing required, first order 1550 nm Bragg gratings have yet to be demonstrated using the point-by-point technique. Malo et al [145] have only been able to fabricate Bragg gratings which reflect light in the 2nd and 3rd order that have a grating pitch of approximately 1μm and 1.5μm respectively. Figure 2.41 shows the reflection spectrum of the 3rd order Bragg grating fabricated using the point-by-point method. This 3rd order grating is made up of 225 index perturbations with grating period Λ, of 1.59μm resulting in a grating length of 360μm. The grating has a peak reflectivity of 70% at 1536 nm and a full width half maximum of 2.7 nm.

2.4.5 Mask image projection

In addition to the above well know techniques for fabricating fiber Bragg gratings, high resolution mask projection has been demonstrated [147] as a means of inscribing Bragg gratings in optical fiber using excimer laser pulses. The mass projection system consists of a excimer laser source generating a UV beam which is incident on a transmission mask. In Mihailov and Gower's [147] experiments the transmission mask consisted of a series of UV opaque line-spaces. The transmitted beam was imaged onto the fiber core by a multicomponent fused silica high-resolution system having a demagnification of 10:1. In their work, gratings with periods of 1,2,3,4, and 6 μm have been written in single mode Ge-doped fiber using mask-imaging techniques. Because of the simplicity of the source and setup, the recording of coarse period gratings by mask imaging exposures in some cases may be more flexible than other techniques. Complicated grating structures (blazed, chirped, etc.) can be readily fabricated with this method by implementing a simple change of mask.

2.4.6 Special fabrication processing

Here a brief description of several techniques used in inscribing special Bragg grating structures will be given. These include, for example, the fabrication of fiber gratings using a single laser pulse and the inscription of gratings during fiber drawing.

2.4.6.1 Single pulse inscription

Fiber Bragg gratings typically require exposure periods of seconds to minutes (depending on photosensitivity of the fiber) and the precise overlap of the intensity patterns of hundreds or thousands of individual exposures. Successful multi-shot interferometric exposures require the spatial stability of the interference pattern relative to points along the fiber axis to be less than the Bragg period. This strenuous stability requirement demands the suppression of air currents, vibration, thermal excursions and laser pointing instability over a period corresponding to the exposure time. This holds whether a pulsed or a CW laser is used and as a result fabrication of a useable grating is quite difficult. Clearly the fabrication of fiber Bragg gratings by a short laser pulse (such as a few nanosecond pulse from an excimer laser) will satisfy many of the stability requirements. Askins et al [125] first demonstrated the inscription of Bragg reflection gratings in a Ge-doped silica core optical fibers by interfering two beams of a single 20ns pulse from a KrF excimer laser. Following the work by Askins et al and Archambault et al [35] on the single-pulse fiber Bragg grating fabrication using the interferometric technique, Malo and co-workers [148] demonstrated the inscription of single-pulse fiber Bragg gratings using the phase-mask technique.

2.4.6.2 Inscription during fiber drawing

One of the major drawbacks in inscribing a Bragg grating is that a section of the fiber must be stripped of its UV absorbing polymer coating in order for the fiber to be exposed. This process weakens the fiber at the site of the grating, due to surface contamination, even if the fiber is subsequently recoated. This problem can be overcome by writing the gratings during the fiber drawing process, just before the fiber is coated [126]. Dong et al [126] demonstrated Bragg grating inscription during fiber drawing using a line-narrowed KrF excimer laser with pulse width of 20 ns.

2.4.6.3 Long fiber Bragg gratings

Many of the applications of fiber Bragg gratings require complex structures that often depend on having a relatively long length typically a few centimeters. A method of writing very long gratings has been described by Martin et al [149] where a UV beam is scanned over a long phase-mask in a fixed position relative to the fiber. The idea behind this approach is to keep the mask and fiber held together, placing the fiber directly behind the mask and to move the writing beam along the mask and fiber assembly. As long as the mask and fiber do not move relative to each other, the phase of the fringes created in the fiber core remains determined by the mask, regardless of the writing beam position, resulting in gratings that can be as long as the mask itself. Another method is to use specially designed long phase-masks where the complex structure already has been implemented in the mask itself [150]. Complex grating structures have also been demonstrated by Loh et al [151] where a UV beam was scanned over a phase-mask while moving the fiber slightly with a piezoelectric transducer.

2.4.6.4 Chirped gratings

There has been a great deal of effort in producing chirped gratings. The motivation behind this drive has been twofold. Firstly, given the fact that although strong, uniform-period gratings can provide a wide bandwidth, this is accompanied by substantial, unavoidable losses on the short-wavelength side of the Bragg resonance. Ideally, gratings would be fabricated with the desired bandwidth and minimal losses. Secondly, it is to exploit applications in telecommunications, for dispersion compensation in high-bit-rate transmission systems and in laser cavities. Chirped gratings may be fabricated using all the techniques referred to thus far for inscribing conventional Bragg gratings. The first report of chirped grating fabrication was by Byron et al [121] who used a conventional two-beam, UV interferometer to produce a uniform-period fringe pattern in a tapered photosensitive. A far more flexible and controllable approach to chirped grating fabrication relies on two-beam interference and is based on the use of dissimilar curvatures in the interfering wavefronts. This is accomplished by placing two lenses in the two arms of the interferometer [1]. The phase-mask is probably one of the most precise and controlled techniques for inscribing chirped gratings. In this technique the mask may be divided into subsections, each having its own period, that is progressively changing in a linear or other functional form depending on the type of chirped grating required. The number of sections depends on the chirp value and the length of the grating. Several other methods have been described in which a uniform phase-mask may be used, for example the "stretch-and-write" method mentioned earlier, producing a piecewise chirped grating.

2.4.6.5 Phase-shifted gratings

The phase-shifted gratings may be fabricated either directly or via post-processing after the grating has been written. In the direct approach, the phase-mask may be constructed with a phase-shift incorporated into the corrugated structure of the mask [133]. The phase-mask approach is probably the most controlled and precise way of fabricated phase-shifted gratings in optical fibers. Another approach is to inscribe the first section of the grating on the fiber, move the fiber by an amount corresponding to the required phase-shift and then inscribe the second part of the grating structure. Other post-processing techniques [134] include exposing a grating region to pulses of UV laser radiation to alter the index of refraction thus in effect inserting a phase-shift at that particular area, or localized heat treatment [135] to achieve a similar result.

2.4.6.6 Apodization of gratings

The word apodization has its roots from the Greek word "*αποδος – apodos*" meaning without-foot (footless). When this word is applied to fiber Bragg grating design, it means the suppression of the spectral side lobes observed due to the fine length of the gratings. Apodization [109, 110, 152], in practice, corresponds to the modulation of the index of refraction of the fiber grating with much larger period than the period of the grating. This may be accomplished using various techniques, for example, exposing the optical fiber to the interference pattern from two non-uniform beams. An alternative approach uses a phase-mask with apodization achieved by varying the exposure time along the length of the grating, either from a double exposure, or by scanning a small writing beam or using a variable diffraction efficiency phase-mask.

2.5 SIMULATIONS OF SPECTRAL RESPONSE FROM BRAGG GRATINGS

There have been many models developed to describe the behavior of Bragg gratings in optical fibers [1]. The most widely used of the techniques has been the coupled-mode theory where the counter-propagating fields inside the grating structure, obtained by convenient perturbation of the fields in the unperturbed waveguide are related by coupled differential equations. Here a simple T-Matrix formalism will be presented for solving the coupled-mode equations for a Bragg grating structure [1], thus obtaining its spectral response.

2.5.1 Coupled-mode theory and the T-matrix formalism

The spectral profile from a Bragg grating structure may be simulated using the T-Matrix Formalism. For this analysis, two counter-propagating plane waves are considered confined to the core of an optical fiber in which an intracore uniform Bragg grating of length l and uniform period Λ exist. This is illustrated in figure 2.42. The electric fields of the backward and forward waves can be expressed as $E_a(x,t)=A(x)\exp[i(\omega t-\beta x)]$ and $E_b(x,t)=B(x)\exp[i(\omega t+\beta x)]$ respectively where β is the wave propagation constant. The complex amplitudes A(x) and B(x) of these electric fields obey the coupled-mode equations [152]

$$\frac{dA(x)}{dx} = i\kappa B(x)\exp[-i2(\Delta\beta)x]$$

$$(0 \le x \le l) \qquad (2.21)$$

$$\frac{dB(x)}{dx} = -i\kappa^* A(x)\exp[i2(\Delta\beta)x]$$

where $\Delta\beta=\beta-\beta_0$ is the differential propagation constant (β_0 is π/Λ, and Λ is the grating period) and κ is the coupling coefficient. For uniform gratings, κ is constant and its related to the index modulation depth. For a sinusoidal modulated refractive index, the coupling coefficient is real and it is given by $\pi\Delta n/\lambda$.

Assuming that there are both forward and backward inputs to the Bragg grating, and boundary conditions $B(0)=B_0$ and $A(l)=A_1$, closed form solutions for A(x) and B(x) are obtained from the above. Following these assumptions, the closed form solutions for x-dependencies of the two waves are $a(x)=A(x)\exp(-i\beta x)$ and $b(x)=B(x)\exp(i\beta x)$. Therefore the backward output (reflected wave), a_0, and the forward output (transmitted wave), b_1, from the grating can be expressed by means of the scattering matrix

$$\begin{bmatrix} a_0 \\ b_1 \end{bmatrix} = \begin{bmatrix} S_{11} & S_{12} \\ S_{21} & S_{22} \end{bmatrix} \cdot \begin{bmatrix} a_1 \\ b_0 \end{bmatrix} \qquad (2.22)$$

with $a_1=A_1\exp(i\beta l)$ and $b_0=B_0$, and

$$S_{11} = S_{22} = \frac{is\exp(-i\beta_0 l)}{-\Delta\beta\sinh(sl) + is\cosh(sl)},$$

$$(2.23)$$

$$S_{12} = S_{21}\exp(2i\beta_0 l) = \frac{\kappa\sinh(sl)}{-\Delta\beta\sinh(sl) + is\cosh(sl)}$$

where $s = \sqrt{|\kappa|^2 - \Delta\beta^2}$. Based on the scattering-matrix expression in (5.3), the T-matrix for the Bragg grating is [153]:

$$\begin{bmatrix} a_0 \\ b_0 \end{bmatrix} = \begin{bmatrix} T_{11} & T_{12} \\ T_{21} & T_{22} \end{bmatrix} \cdot \begin{bmatrix} a_1 \\ b_1 \end{bmatrix} \qquad (2.24)$$

where

$$T_{11} = T_{22}^* = \exp(-i\beta_0 l)\frac{\Delta\beta \sinh(sl) + is\cosh(sl)}{is},$$

$$T_{12} = T_{21}^* = \exp(-i\beta_0 l)\frac{\kappa \sinh(sl)}{is} \tag{2.25}$$

(a)

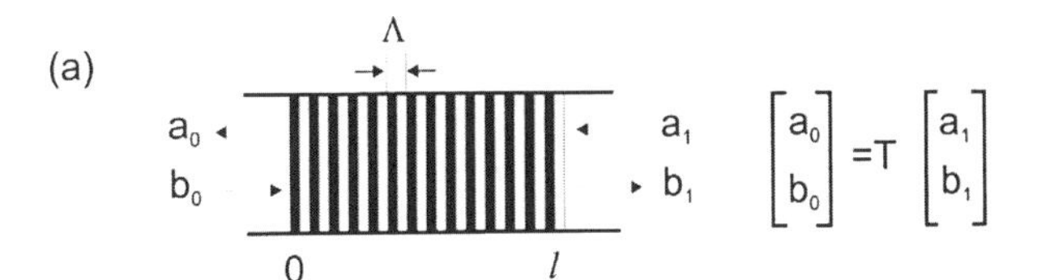

(b)

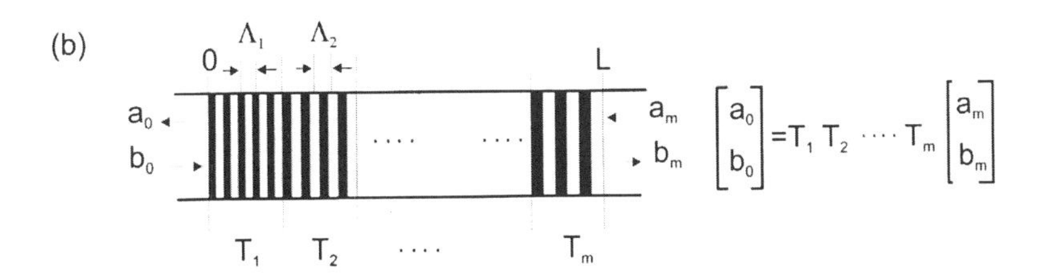

Figure 2.42. Illustration of the T-matrix model: (a) a single uniform Bragg grating and (b) a series of gratings with different periods back to back

The T-matrix relates the input and output of the Bragg grating and is ideal for analyzing a cascade of gratings (figure 2.42). Figure 2.42(b) shows a series of gratings back-to-back with a total length, L. This grating structure is made up of "m" Bragg grating segments. Each segment has a different period Λ_m and has its own T-matrix T_m . The total grating structure may be expressed as

$$\begin{bmatrix} a_0 \\ b_0 \end{bmatrix} = T_1 \cdot T_2 \cdots T_{m-1} \cdot T_m \cdot \begin{bmatrix} a_m \\ b_m \end{bmatrix}, \tag{2.26}$$

and the spectral reflectivity of the grating structure is given by $|a_0(\lambda)/b_0(\lambda)|^2$. From the phase information, the delay for the light reflected back from the grating may also be obtained [1]. It should be noted that this model does not take into account cladding mode-coupling losses. Next the reflectivity spectral response of a few Bragg grating structures will be calculated using this T matrix formalism.

2.5.2 Uniform index of refraction Bragg gratings

2.5.2.1 Grating length dependence

The reflection spectral response for uniform Bragg gratings is calculated using the T-matrix formalism described above. The objective of this set of simulations is to demonstrate how the spectral response of a grating is affected as the length of the grating is altered. The index of refraction change is assumed uniform over the grating length: however the value of the change is reduced accordingly with increasing grating length such that the maximum grating reflectivity remains constant. Figure 2.43 shows the spectral profile of three uniform Bragg gratings. Clearly, as seen from the various plots, the bandwidth of the gratings decreased with increasing length. The 1cm long uniform grating had a bandwidth approximately 0.15 nm, the 2 cm long grating was 0.074 nm and finally 4cm long grating was 0.057 nm. Theoretically Bragg gratings may be constructed with extremely small bandwidths by simply increasing the grating length. However, in practice such devices are not easy to manufacture. The error associated with the spacing between the periods of a grating (during manufacture) is cumulative, therefore with increasing grating length the total error will increase, resulting in out-of-phase periods (leading to broadening of the Bragg grating). Furthermore, if a long period Bragg grating is constructed the effects of the environment have to be considered very carefully. For example, any strain or temperature fluctuations on any part of the grating will cause the periods to move out of phase resulting in broadening of the Bragg grating.

2.5.2.2 Index of refraction dependence

Figure 2.44 shows a set of simulations assuming a uniform Bragg grating of 2 cm in length. In these simulations the index of refraction changes were varied. For the first grating with $\Delta n=0.5\times10^{-4}$ the reflectivity is 90% and the bandwidth is approximately 0.074 nm. Reducing the change of the index of refraction to half the value of the first grating ($\Delta n = 0.25\times10^{-4}$), the reflectivity decreases to 59% and the bandwidth to 0.049 nm. A further decrease in the index of refraction change ($\Delta n = 0.1\times10^{-4}$) results in a reflectivity of 15% and a bandwidth of 0.039 nm. It appears that the bandwidth approaches a minimum value and remains constant for further reduction in the index of refraction change.

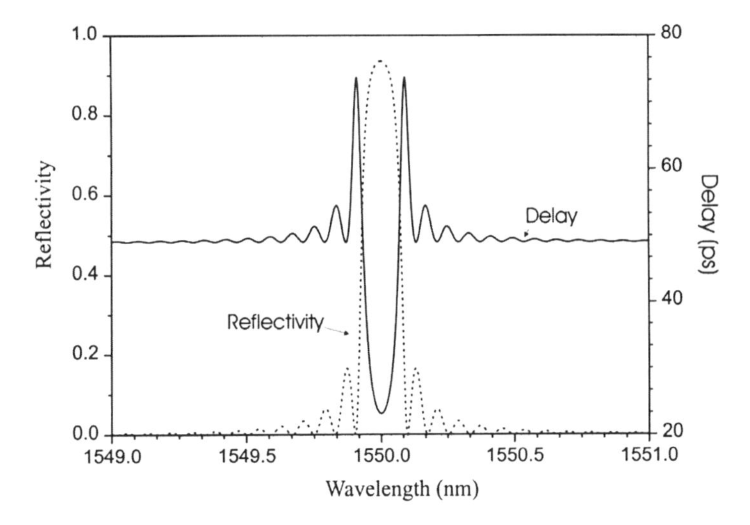

Figure 2.43. Spectral profiles for uniform Bragg gratings. The various spectral profiles correspond to different grating lengths, (1cm, solid; 2cm, dashed and 4cm, dotted)

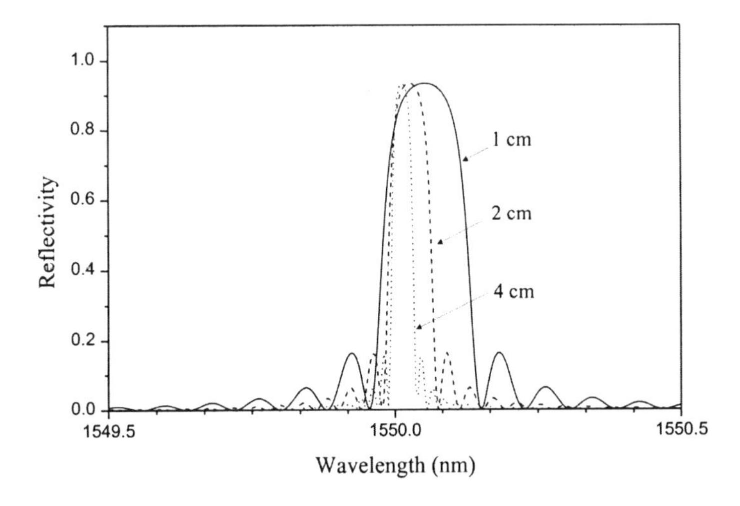

Figure 2.44. Spectral reflectivity response from uniform Bragg grating 2cm in length for different indexes of refraction. The solid line corresponds to 0.5×10^{-4} index of refraction change

2.5.2.3 Time delay dependence

Figure 2.45 shows the delay, τ, calculated from the derivative of the phase with respect to the wavelength for a uniform grating of length ~10mm.

The design wavelength for this grating was 1550 nm and the index of refraction of the fiber was set at $n_{eff} = 1.45$. Figure 2.45 also shows the reflectivity spectral response of the same Bragg grating. Clearly both reflectivity and delay are symmetric about the wavelength λ_{max}. The dispersion is zero near λ_{max} for uniform gratings and becomes appreciable near the band edges and side lobes of the reflections spectrum where it tends to vary rapidly with wavelength.

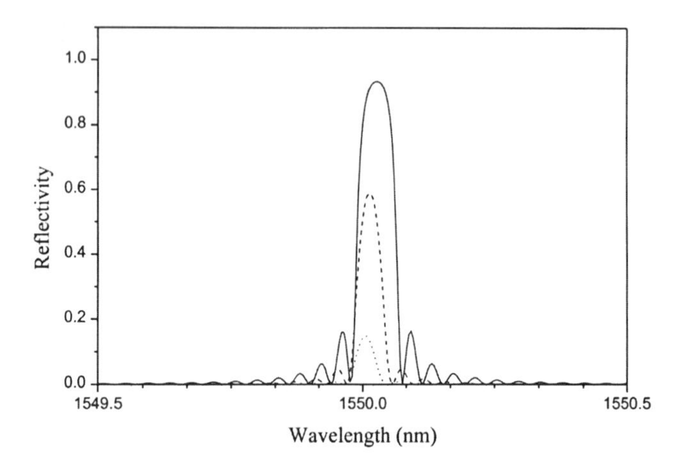

Figure 2.45. Calculated group delay (solid line) and reflectivity (dash line) for uniform weak Bragg grating with an index change of 10^{-4} and an approximate length of 10mm long

2.5.3 Chirped Bragg gratings

The simulations shown in fig. 2.46(a) and 2.46(b) show chirped Bragg grating structures. Three different reflection spectra are shown in fig. 2.46(a), corresponding to three different chirped values, 0, 0.2 and 0.4 nm over the entire length of the grating. In these simulations, all the gratings are assumed 10 mm in length with a constant index of refraction change 1×10^{-4}. With increasing chirp value the reflectivity response becomes broader and the reflection maximum decreases. In these simulations the chirp gratings are approximated with a number of progressively increasing period gratings whose total length amounts to the length the chirp grating. The number of "steps" (the number of smaller gratings) assumed in the calculations is 100 (simulations indicated that calculations with more than 20 steps will give approximately the same result). Fig. 2.46(b) shows the spectral response from Bragg gratings with very large chirp values (1, 4 and 8 nm over the 10mm length of the grating). Clearly with increasing chirp value, it is possible to span a very large spectral area, with a reduction in the maximum reflectivity of the grating. This problem may be overcome by increasing the index of refraction modulation.

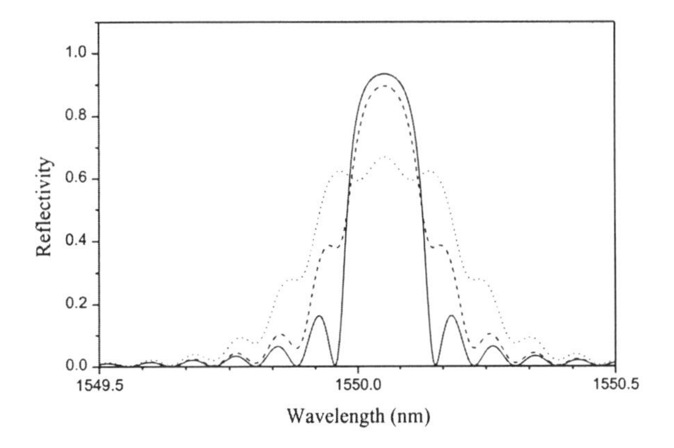

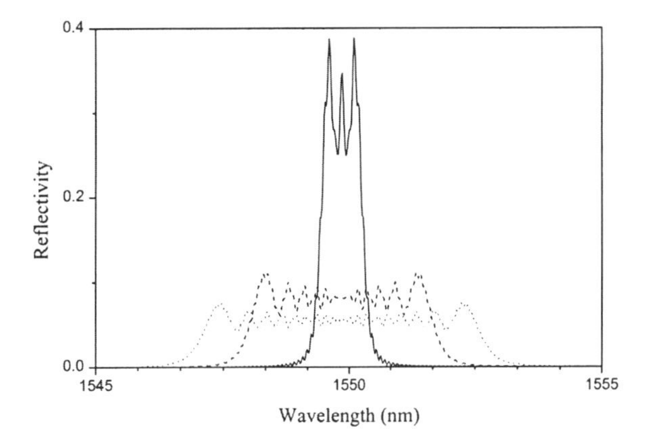

Figure 2.46. (a) Spectral reflectivity response from different Bragg gratings showing the effect of chirping. All gratings are 10mm and the index of refraction change is assumed 1×10^{-4} for all. The solid curve corresponds to a 0 chirp, the dashed and dotted curves correspond to 0.2 and 0.4 nm chirp respectively (where the chirp value is over the length of the grating). (b) Spectral reflectivity response from highly chirped Bragg gratings for the following chirp values of 1 nm, 4 nm and 8 nm over the 10 mm length of the gratings

2.5.4 Apodization of the spectral response of Bragg gratings

The reflection spectrum of a finite length Bragg grating with uniform modulation of the index of refraction is accompanied by a series of sidelobes at adjacent wavelengths. It is very important to minimize and if possible eliminate the reflectivity of these sidelobes, (or apodise the reflection spectrum of the grating) in devices where high rejection of the nonresonant light is required. An additional benefit of apodization is the improvement of the dispersion compensation characteristics of chirped Bragg gratings [154]. In practice apodization is accomplished by varying the amplitude of the coupling coefficient along the length of the grating. A method used to apodize the response consists in exposing the optical fiber to the interference pattern formed by two non-uniform ultraviolet light beams [155]. In the phase mask technique, apodization can also be achieved by varying the exposure time along the length of the grating, either from a double exposure or by scanning a small writing beam or using a variable diffraction efficiency phase mask. In all these apodization techniques, the variation in coupling coefficient along the length of the grating comes from local changes in the intensity of the UV light reaching the fiber.

Figure 2.47 shows two plots of anodized gratings with a Gaussian shape index of refraction change. The index of refraction changes are shown in the upper inset, which are plotted against the length of the grating. The full width half maximum of the index of refraction profile is doubled for one of the curves (the dotted curve) resulting in a larger spectral reflection response. It is interesting to point out that in both the anodized Bragg gratings, the sidelobes have been eliminated.

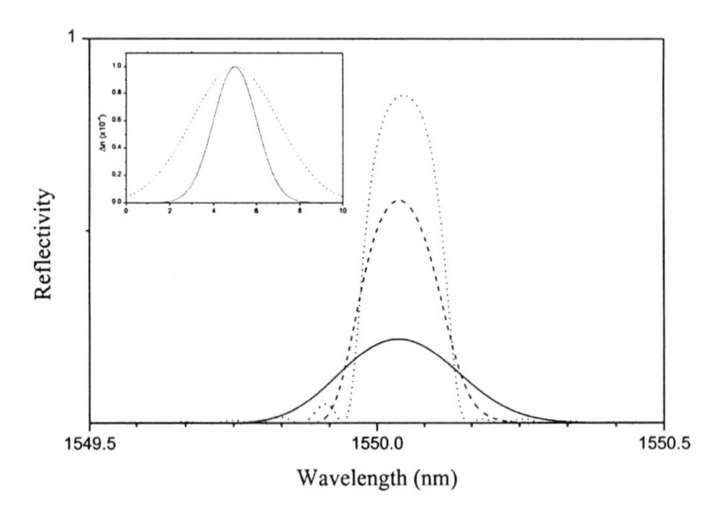

Figure 2.47. Spectral reflectivity response from Gaussian apodized Bragg gratings. The apodization profiles plotted against the length of the gratings (10 mm) are shown in the inset at the upper right corner

Apodization of the spectral response of fiber Bragg gratings using a phase mask with variable diffraction efficiency has been reported by Albert et al [109]. Bragg gratings with sidelobe levels 26dB lower than the peak reflectivity have been fabricated in standard telecommunication fibers. This represents a reduction of 14 dB in the sidelobe levels compared to uniform gratings with the same bandwidth and reflectivity. Figure 2.19 shows the spectral reflection response of an anodized and an unapodized fiber Bragg grating reflector reported by the same group [110] where they have achieved a reduction of 20 dB in the sidelobe levels.

A technique for cosine apodization that was obtained by repetitive, symmetric longitudinal stretching of the fiber around the center of the grating while the grating was written has been demonstrated [156]. This apodization scheme is applicable to all types of fiber gratings, written by direct replication by a scanning or a static beam, or by use of any other interferometer and is independent of length. The simplicity of this technique allows the rapid production of fiber gratings required for wavelength-division-multiplexed systems (WDM) and dispersion compensation.

2.6 APPLICATIONS OF BRAGG GRATINGS

Fiber Bragg gratings have emerged as important components in a variety of lightwave applications. Their unique filtering properties and versatility as in-fiber devices are illustrated by their use in wavelength stabilized lasers, fiber lasers, remotely pumped amplifiers, Raman amplifiers, phase conjugators, wavelength converters, passive optical networks, wavelength division multiplexers (WDMs) demultiplexers, add/drop multiplexers, dispersion compensators and gain equalizers [1]. Fiber Bragg gratings are also considered excellent sensor elements, suitable for measuring static and dynamic fields, such as temperature, strain and pressure [111]. The principle advantage is that the measurand information is wavelength-encoded (an absolute quantity) thereby making the sensor self-referencing, rendering it independent of fluctuating light levels and the system immune to source power and connector losses that plague many other types of optical fiber sensors. It follows that any system incorporating Bragg gratings as sensor elements is potentially interrupt-immune. Their very low insertion loss and narrowband wavelength reflection offers convenient serial multiplexing along a single monomode optical fiber. There are further advantages of the Bragg grating over conventional electrical strain gauges, such as linearity in response over many orders of magnitude, immunity to electromagnetic interference (EMI), light weight, flexibility, stability, high temperature tolerance and even durability against high radiation environments (darkening of fibers). Moreover, Bragg gratings can easily be embedded into materials to provide local damage detection as well as internal strain field mapping with high localization, strain resolution and measurement range. The Bragg grating is an important component for the development of smart structure technology, with applications also emerging in process control and aerospace industries. This section will describe, in brief, some applications of fiber gratings.

2.6.1 Fiber Bragg grating diode laser

A fiber Bragg grating may be coupled to a semiconductor laser chip to obtain a fiber Bragg diode laser [157]. A semiconductor laser chip may be antireflection coated on the output facet and coupled to a fiber with a Bragg grating, as illustrated in figure 2.48. If this Bragg grating reflects at the gain bandwidth of the semiconductor material, it is possible to obtain lasing at the Bragg grating wavelength. The grating bandwidth can be narrow enough to force single frequency operation with a linewidth of much less than a 1GHz. High output powers, up to tens of mW have been obtained with these types of lasers. An added advantage of these systems is their temperature sensitivity which is approximately 10% of that of the semiconductor laser, thus reducing temperature-induced wavelength drift. These fiber grating semiconductor laser sources have been used to generate ultrashort mode-locked soliton pulses up to 2.9GHz [158].

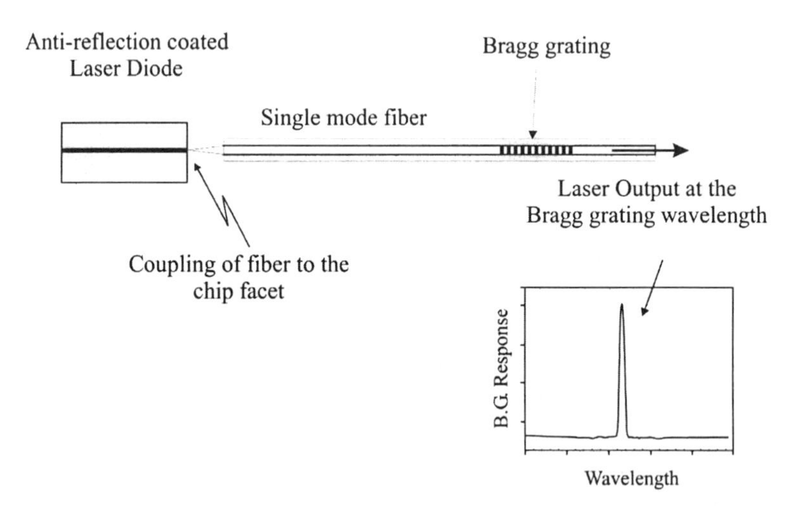

Figure 2.48. External fiber grating semiconductor laser. Semiconductor laser chip is antireflection coated on the output face and coupled to a fiber with a Bragg grating, forcing oscillation at the Bragg grating wavelength

A manufacture problem in DBR lasers is the precise control of the laser wavelength. Routine production of DBR lasers with a wavelength specified to better than 1 nm is difficult. On the other hand Bragg gratings can be manufactured precisely (to better than 0.1 nm) to the wavelength required. With anti-reflection coating on the semiconductor chip, the lasing wavelength may be selected from anywhere in the gain bandwidth by choosing the appropriate fiber Bragg grating. Clearly, such an approach will increase the yield from the semiconductor wafers. In addition, since each laser has to be coupled to a fiber, the Bragg grating may be written after the packaging process has proved to be successful, thus reducing the time spent on unsuccessful products.

2.6.2 Fiber Bragg grating lasers

The majority of Bragg grating fiber laser research has been on erbium-doped lasers, due to its potential in communication and sensor applications. The characteristic broadband gain profile of the erbium-doped fiber around the 1550 nm region makes it an extremely useful tunable light source. Employing this doped fiber in an optical cavity as the lasing medium, along with some tuning element, results in a continuously tunable laser source over its broad gain profile. In fact, a tunable erbium doped fiber with an external grating was reported by Reekie et al [159] in 1986. Since then, several laser configurations have been demonstrated with two or more intracavity gratings [45,159-163].

A simple Bragg grating tunable Er doped fiber laser was demonstrated by the author and co-worker, where a broad band Bragg mirror and a narrow Bragg grating served as the high reflector and the output coupler respectively [164]. The broadband mirror was constructed from a series of Bragg gratings resulting in the broadband reflector with a bandwidth of approximately 4 nm. It should be noted that with continuous advancements in photosensitivity and writing techniques such a broadband mirror may have any shape and bandwidth desired. The fiber laser consisted of a two-meter long erbium doped fiber with Bragg gratings at each end (broadband and narrowband) providing feedback to the laser cavity. The output coupler to the fiber laser cavity was a single grating with approximately 80% reflectivity and 0.12 nm linewidth. Figure 2.49 show the broadband fluorescence obtained from the Er doped fiber laser system before lasing threshold is reached. The spectrum is the typical characteristic broadband gain profile from an erbium-doped fiber spanning a range of several tens of manometers, namely between 1.45 and 1.65 μm. Superimposed on the gain profile is a broadband peak at 1550 nm, corresponding to the reflection of the fluorescence from the broadband Bragg mirror and within this peak, there is a notch at 1550 nm corresponding to the narrow Bragg grating. With increasing incident pump power, the losses in the fiber laser cavity are overcome and lasing begins. At pump powers just above the threshold value, the notch due to the Bragg grating begins to grow in the positive direction and as the pump power increases further, the laser grows even stronger by depleting the broadband fluorescence (figure 2.50). In figure 2.50 the output spectrum from the erbium doped fiber laser is shown for various coupled pump powers into the doped fiber, starting below lasing threshold at 0.50 to 1.0 mW and 1.5 mW where the laser line at 1550 nm begins to grow. At 3.0mW of coupled pump power lasing is dominant. This is shown in the inset at the upper left-hand corner where a vertical line at 1550 nm represents the lasing wavelength.

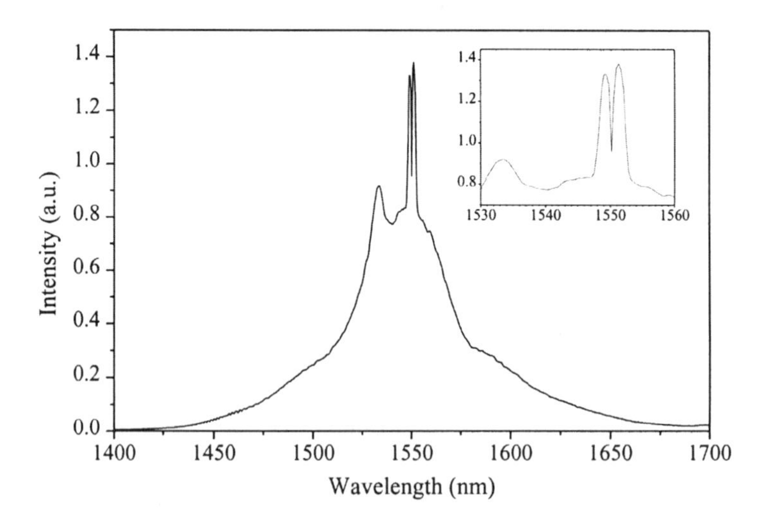

Figure 2.49. Broadband fluorescence obtained from an erbium-doped fiber laser. Superimposed on the gain profile is the broadband mirror at 1550 nm. Within this peak there is a notch at 1550 nm corresponding to the Bragg grating [after 164]

Single frequency Er^{3+} doped Fabry-Perot fiber lasers using fiber Bragg gratings as the end mirrors [112,165] are emerging as an interesting alternative to DFB diode lasers for use in future optical CATV networks and high capacity WDM communication systems [166]. They are fiber compatible, simple, scaleable to high output powers, and have low noise and kilohertz linewidth. In addition, the lasing wavelength can be determined to an accuracy of better than 0.1 nm, which is very difficult to achieve DFB diode lasers.

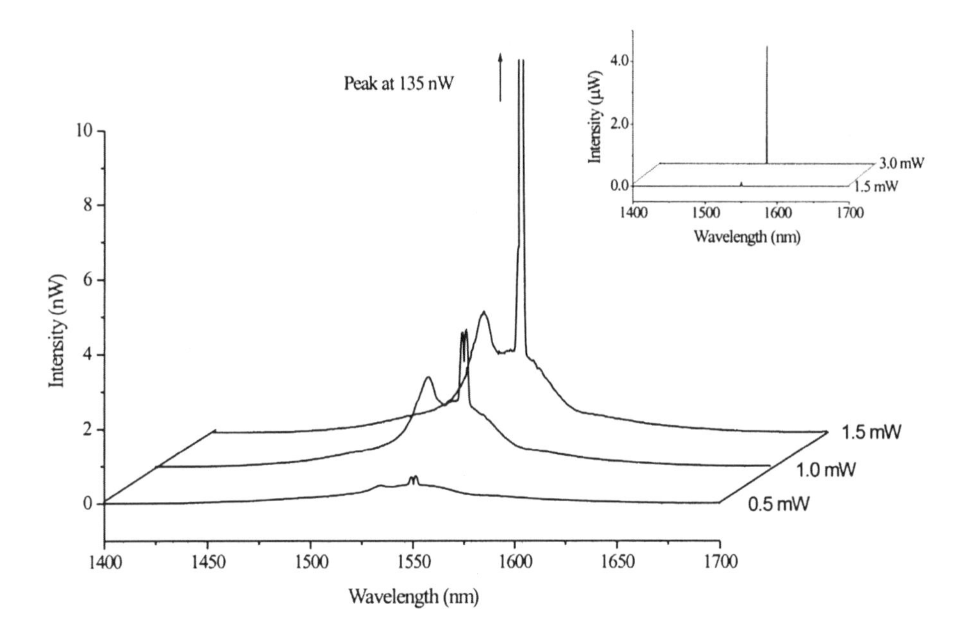

Figure 2.50. Output spectrum from erbium-doped fiber laser at various coupled input powers varying from below the lasing threshold of 0.5mW to nearly lasing at 1.0mW and fully lasing at 1.5mW. The inset at the upper left corner shows the lasing spectrum as a function of coupled pump power of 1.5 and 3.0mW (note that the laser power at 3mW is approximately 40 times stronger than that at 1mW)

Fiber lasers can operate in a single frequency mode provided that the grating bandwidth is kept below the separation between the axial mode spacings. Furthermore, it is ncccssary to kccp the erbium concentration low enough (a few 100 ppm) to reduce ion-pair quenching, which causes a reduction in the quantum efficiency and in addition may lead to strong self-pulsation of the laser [165,166]. The combination of these practical limits implies that the pump absorption of an erbium-doped fiber system can be as low as a few percent resulting in low output lasing power. One solution to this problem is to use the residual pump power to pump an erbium doped fiber amplifier following the fiber laser. However, in such cases the amplified spontaneous emission from the amplifier increases the output noise. Another way to overcome the problem of low pump absorption is by codoping the erbium doped fiber with Yb^{3+}. This increases the absorption at the pump wavelength by more than two orders of magnitude and enables high efficient operation of centimeter long lasers with relatively low Er^{3+} concentration. Kringlebotn et al [167] reported a highly-efficient, short, robust single-frequency and linearly polarized Er^{3+}:Yb^{3+} codoped fiber laser with fiber grating Bragg reflectors, an output power of 19 mW and a linewidth of 300 kHz for 100 mW of 980 nm diode pump power.

One other interesting application of Bragg gratings in fiber lasers makes use of stimulated Raman scattering. The development of fiber Bragg gratings has made possible the fabrication of numerous highly reflecting elements directly in the core of germanosilicate fibers. This technology, coupled with that of cladding-pumped fiber lasers, has enabled the development of Raman fiber lasers. The pump light is introduced through one set of highly reflecting fiber Bragg gratings. The cavity consists of several hundred meters to a kilometer of germanosilicate fiber. The output consists of a set of highly reflecting gratings through Raman order n-1 and the output wavelength of Raman order n is coupled out, by means of a partially reflecting fiber grating (R~20%). The intermediate Raman Stokes orders are contained by sets of highly reflecting fiber Bragg gratings and this power is circulated until it is nearly entirely converted to the next successive Raman Stokes order [168].

2.6.3 Fiber Bragg grating filters and mode converters

Band-pass filters are considered as one of the most fundamental devices in multiwavelength optical networking and in most communications systems where wavelength demultiplexing is required. There may be techniques for fabricating these band-pass filters utilizing fiber Bragg gratings. One approach is based on the interferometric principle where gratings are incorporated into Saganac [169], Michelson [170,171] or Mach-Zehnder [172,173] configurations. Another approach, based on the principle of the Moiré grating resonator [174], has also been applied with uniform period [175] and chirped [176] grating types. Resonant filter structures have been fabricated by introducing a phase shift into the grating by an additional UV exposure, or by using a phase-shifted phase mask. In general the resonant type transmission filters are capable of large wavelength selectivity and are, in principle, simple to manufacture and do not require carefully balanced arms or identical gratings, as in the case of interferometric filters.

A single Bragg grating in a single mode fiber acts as a wavelength selective distributed reflector or a band rejection filter by reflecting wavelengths around the Bragg resonance. However, by placing identical gratings in two lengths of a fiber coupler, as in a Michelson arrangement, a bandpass filter can be made [177]. This filter, shown in fig. 2.20(a), passes only wavelengths in a band around the Bragg resonance and discards other wavelengths without reflections. If the input port is excited by broad-band light and the wavelengths reflected by the gratings arrive at the coupler with identical optical delays, then this wavelength simply returns to the input port. If, however, a path-length difference of $\pi/2$ is introduced between the two arms, then it is possible to steer the reflected wavelength to arrive at the second input port, creating a bandpass filter. In principle, this is a low-loss filter: however, there is a 3dB loss penalty for the wavelengths which are not reflected, unless a Mach-Zehnder interferometer is used to recombine the signal at the output [170]. An efficient band-pass filter was demonstrated by Bilodeau et al [171] using a scheme identical to that presented in reference [170]. The device had back reflection of -30dB. However, all wavelengths out of the pass band suffered from the 3dB loss associated with the Michelson interferometer.

Optical fiber communication systems employing wavelength division multiplexing/demultiplexing (WDM/D) techniques require low loss, compact, stable and reliable components which can be used as wavelength-selective channel dropping or inserting filters. By adding a second coupler to close the legs containing the gratings of fig. 2.20(a), as in a Mach-Zehnder arrangement, a drop-add filter can be made [1] (figure 2.51). The Mach-Zehnder is balanced so that any optical signal in the 1550 nm transmission window launched into the input fiber (port A or C) is coupled into the output fiber (port D or B). The two identical Bragg reflection filters and the input coupler, which has a precise 50:50 splitting ratio at the center wavelength of the reflection gratings, form a Michelson interferometer transmission filter. The path length in each arm of the interferometer is set equal, thus giving maximum transmission. Any signal launched into port A, at the transmission wavelength of the Michelson interferometer, will be coupled into port B. An input optical signal $\Sigma(\lambda_\tau)$, at the transmission wavelength of the Michelson interferometer (λ_τ), can be multiplexed onto an optical signal containing many discrete wavelengths $\Sigma(\lambda_{1..m})$ by launching into Ports A and C respectively, with the output signal $\Sigma(\lambda_{1...\tau..m})$ being transmitted into Port B. Conversely, an optical signal $\Sigma(\lambda_\tau)$ can be demultiplexed into port B from an optical signal containing many discrete wavelengths $\Sigma(\lambda_{1...\tau..m})$ launched into Port A, with the remainder of the signal $\Sigma(\lambda_{1..m})$ being transmitted into Port D.

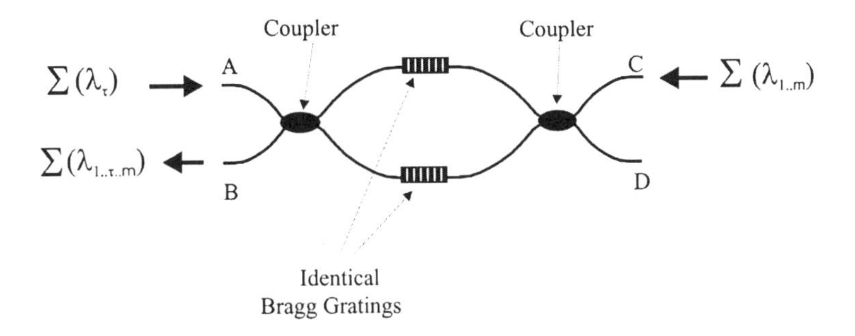

Figure 2.51. Schematic diagram of a drop-add filter formed by adding a second coupler to close the legs containing the grating of a Mach-Zehnder arrangement

Placing two identical Bragg gratings in series on a single mode fiber results in a Fabry-Perot etalon within the fiber core. Fig. 2.20(b) shows one such construction with 99.5% reflectors, a free spectral range of 1.06GHz, and a finesse of 660. With the advancements in the inscription of Bragg gratings in optical fiber, it now is possible to obtain etalons with finesse as high as several thousand.

By tilting or blazing the Bragg grating at angels to the fiber axis, light can be coupled out of the fiber core into loosely cladding modes or to radiation modes outside the fiber. This wavelength selective tap occurs over a rather broad range of wavelengths that can be controlled by the grating and waveguide design. One of the advantages is that the signals are not reflected and thus the tap forms an absorption type filter. An application that was mentioned earlier is that this grating tap can be used as a gain flattening filter for an erbium fiber amplifier.

With a small tilt of the grating planes to the fiber axis (~1°), a reflecting spatial mode coupler can be made such that the grating reflects one guided mode into another. It is interesting to point out that by making long period gratings, one can perturb the fiber to couple to other forward going modes. A wavelength filter based on this effect has been demonstrated by Hill et al [150]. The spatial mode converting grating was written using the point by point technique with a period of 590μm over a length of 60 cm. Using mode strippers before and after the grating makes a wavelength filter. In a similar manner, a polarization mode converter or rocking filter in polarization maintaining fiber can be made. A rocking filter of this type, generated with the point-by-point technique, was demonstrated by Hill and co-workers [9]. In their work they demonstrated an 87 cm long, 85 step rocking filter which had a bandwidth of 7.6 nm and peak transmission of 89%.

2.6.4 Pulse dispersion compensation

One of the main problems that can occur in single mode optical fibers is chromatic dispersion, causing different wavelength components of a data pulse to travel at different group velocities. This causes broadening of the signal pulse and increasing bit-error rates. With increasing network data rates, chromatic dispersion in standard single mode fiber is the main limiting factor in performance. For a low data rate of 2.5Gbit/s, a signal can be transmitted without significant degradation for up to 1000km. However, this distance drops to 60km at 10Gbit/s and to 15km at 20Gbit/s. In addition, a large portion of the already worldwide installed fiber is optimized for transmission at 1.31 μm. This type of fiber exhibits high chromatic dispersion of the order of 17 ps/nm km when used to transmit at the more commonly used telecommunication wavelength of 1.55 μm.

Chirped fiber Bragg gratings provide the means for dispersion compensation. The basic principle of operation of a chirped fiber grating as a dispersion compensating element is that different wavelength components of the broadened pulse are reflected from different locations along the Bragg grating. Therefore, the dispersion imparted by the grating in reflection to a pulse with a given spectral content is equal to twice the propagation delay through the grating length, L, as

$$\tau = \frac{2L}{\nu_g} \tag{2.27}$$

where ν_g is the group velocity of the pulse incident on the grating. Therefore a grating with a linear wavelength chirp of $\Delta\lambda$ nm will have a dispersion of

$$d = \frac{\tau}{\Delta\lambda}(ps/nm) \tag{2.28}$$

Eggleton et al [138] demonstrated dispersion compensation by pulse compression with the use of a chirped grating. Figure 2.52(a) and 2.52(b) show respectively the 21ps input pulse and the compressed 13ps reflected pulse. To show that the observed compression is not due to truncation of the pulse spectrum by the grating, the direction of the grating was reversed which resulted in the pulse stretching to 40ps as shown in figure 2.52(c). Pulse stretching is expected because the sign of the dispersion in the reversed case adds to the dispersion of the optical pulse.

The first practical demonstrations of dispersion compensation used short pulses (figure 2.52) traveling in single mode optical fiber and a chirped Bragg grating to compensate for the pulse broadening arising from negative group delay dispersion and non-linear self phase modulation in the fiber [178]. Specifically short pulses of 1.8 ps were sent through 200m of optical fiber which had a measured group delay dispersion of –100 ps/nm km. These pulses suffered significant dispersive broadening in the fiber as seen in figure 2.52(b). A 50:50 coupler between the fiber and a linearly chirped fiber Bragg grating provided an output for the pulses directly from the fiber and those reflected off the grating. The pulses reflected by the grating were measured temporally by cross-correlation with pulses derived directly from the source. These results indicated that the chirped grating provided satisfactory dispersion compensation, which was in good agreement with numerical simulation.

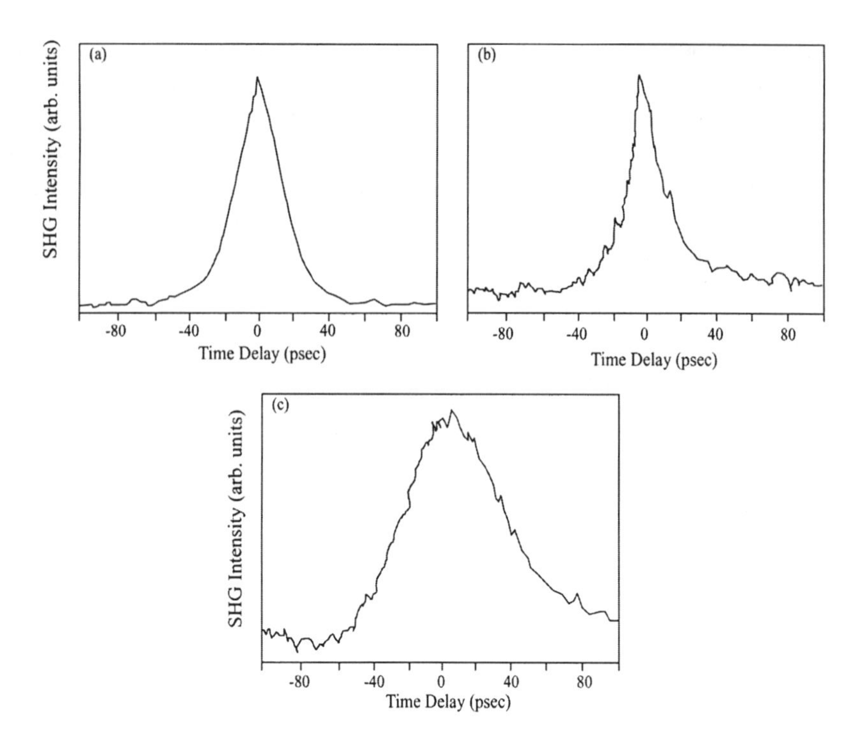

Figure 2.52. Traces of optical pulses before and after reflection from a chirped grating. (a) the initial chirped pulse with duration of 21±1 ps (b) compressed pulse after reflection from a chirped grating with duration of 13±1 ps, (c) stretched pulse after reflection from a reversed chirped grating with duration of 40±1 ps. From Eggleton et al (after [138])

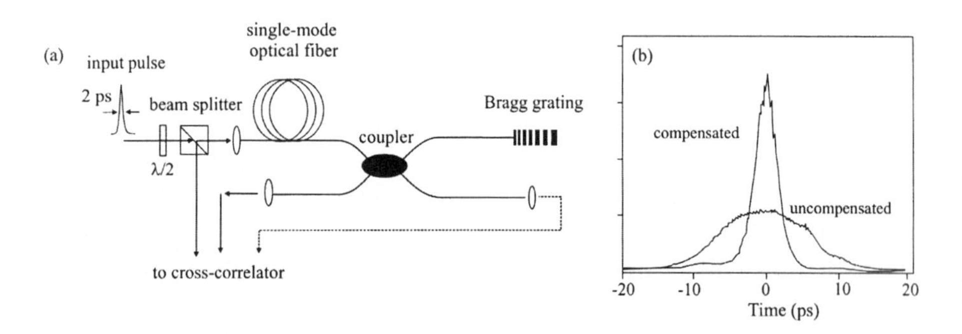

Figure 2.53. (a) Experimental configuration used to investigate dispersion compensation of a linearly chirped fiber Bragg grating. Plot (b) shows a comparison between dispersive pulse and dispersion compensated pulse from the experimental setup shown in (a). (after [178])

2.6.5 Fiber Bragg grating sensors

Fiber Bragg gratings are excellent fiber optic sensing elements. They are integrated into the light guiding core of the fiber and are wavelength encoded, eliminating the problems of amplitude or intensity variations that plague many other types of fiber sensors. Due to their narrow band wavelength reflection they are also conveniently multiplexed in a fiber optic network. Fiber gratings have been embedded in composite materials for smart structures monitoring and tested with civil structures to monitor load levels. They have also been successfully tested as acoustic sensing arrays. Applications for fiber grating sensors should also be emerging in process control and aerospace industries in the near future.

The temperature sensitivity of a Bragg grating occurs principally through the effect on the index of refraction and to a lesser extent through the expansion coefficient. It is noteworthy that temperature sensitivity can be enhanced or mulled by proper bonding to other materials. The maximum operating temperatures may be around 500° C, however this may depend on the fabrication condition of the Bragg grating. For example, type II gratings may operate at high temperatures than type I gratings.

Strain affects the Bragg response directly through the expansion or contraction of the grating elements and through the strain optic effect. Many other physical parameters other than tension can also be measured, such as pressure, flow, vibration acoustics, acceleration, electric, magnetic fields, and certain chemical effects. Therefore, fiber Bragg gratings can be thought of as generic transducer elements. There are various schemes for detecting the Bragg resonance shift, which can be very sensitive. One such scheme involves the injection of a broadband light (generated by super-luminescent diode, edge-emitting LED, erbium-doped fiber super-fluorescent source) into the fiber and determining the peak wavelength of the reflected light. Another scheme involves the interrogation of the Bragg grating with a laser tuned to the sensor wavelength, or by using the sensor as a tuning element in a laser cavity. Detecting small shifts in the Bragg wavelength of a fiber Bragg grating sensor element, which corresponds to changes of the sensing parameter, is important. In a laboratory environment this can be accomplished using a high precision optical spectrum analyzer. In practical applications, this function must be performed using compact, low cost instrumentation. Schemes based on simple broadband optical filtering, interferometric approaches, and fiber-laser approaches allow varying degrees of resolution and dynamic range and should be suitable for most applications.

The most straightforward means for interrogating a Bragg grating sensor is based on a passive broadband illumination of the device. Light with a broadband spectrum which covers that of the Bragg grating sensor is input to the system, and the narrowband component reflected by the grating is directed to a wavelength detection system. Several options exist for measuring the wavelength of the optical signal reflected from a Bragg grating element including a miniaturized spectrometer, passive optical filtering, tracking using a tunable filter, and interferometric detection. The optical characteristics of these filtering options are as shown in figure 2.54. Filtering techniques based on the use of broadband filters allow the shift in the Bragg grating wavelength of the sensor element to be assessed by comparing the transmittance through the filter compared to a direct 'reference' path [179]. A relatively limited sensitivity is obtained using this approach due to problems associated with the use of bulk-optic components and alignment stability. One way to improve this sensitivity is to use a fiber device with a wavelength-dependent transfer function, such as a fiber WDM coupler. Fused WDM couplers for 1550/1570 nm operation are commercially available. This coupler will provide a monotonic change in the coupling ratio between two output filters for an input optical signal over the entire optical spectrum of an Erbium broadband source, and thus has a suitable transfer function for wavelength discrimination over this bandwidth. An alternate means to increasing the sensitivity is to use a filter with a steeper cut-off (figure 2.54) such as an edge filter. However, this can limit the dynamic range of the system. One of the most attractive filter-based techniques for interrogating Bragg grating sensors is based on the use of a tunable passband filter for tracking the Bragg grating signal. Examples of these types of filter include Fabry-Perot filters [180], acousto-optic filters [181], and fiber Bragg grating based filters [182].

2.6.5.1 Tunable filter interrogation

Figure 2.55 shows the configuration used to implement a tunable filter (such as a fiber Fabry-Perot) to interrogate a Bragg grating sensor. The fiber Fabry-Perot can be operated in either a tracking or scanning mode for addressing single or multiple grating elements respectively. In a single sensor configuration, the Fabry-Perot filter with a bandwidth of 0.1 nm is locked to the Bragg grating reflected light (R) using a feedback loop. This is accomplished by dithering the fiber Fabry-Perot resonance wavelength by a small amount (typically 0.01 nm) and using a feedback loop to lock to the Bragg wavelength of the sensor return signal. The fiber Fabry-Perot control voltage is a measure of the mechanical or thermal perturbation of the Bragg grating sensor.

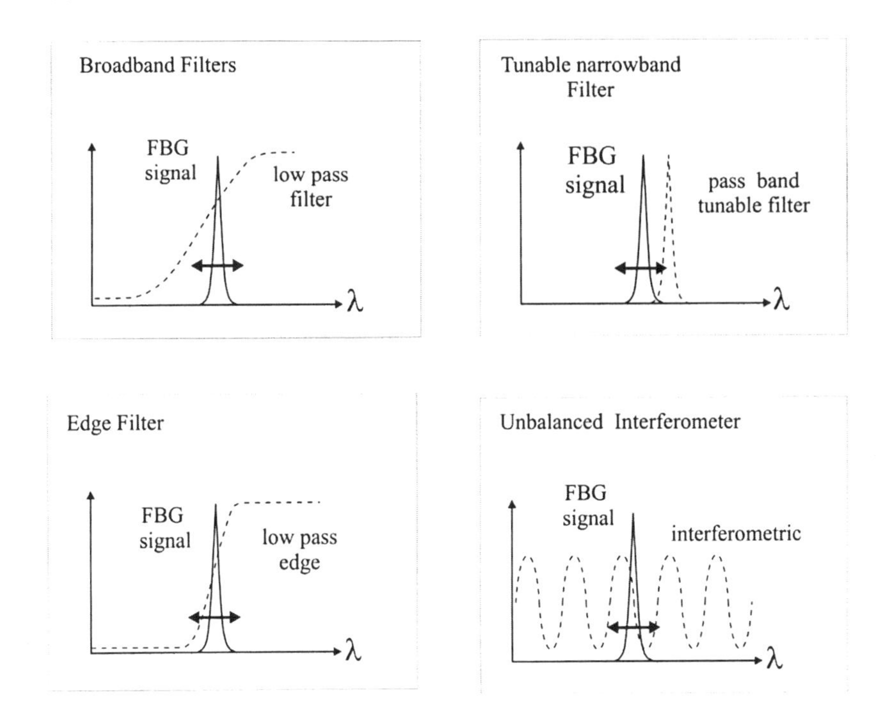

Figure 2.54. Diagram of basic filtering function for processing Fiber Bragg grating return signals

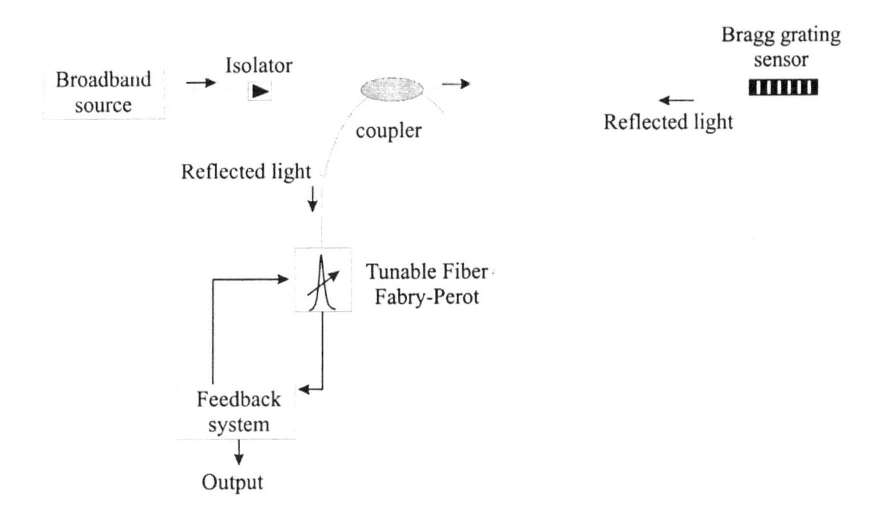

Figure 2.55. Schematic of a tuned filter based interrogation technique for a Fiber Bragg grating sensor

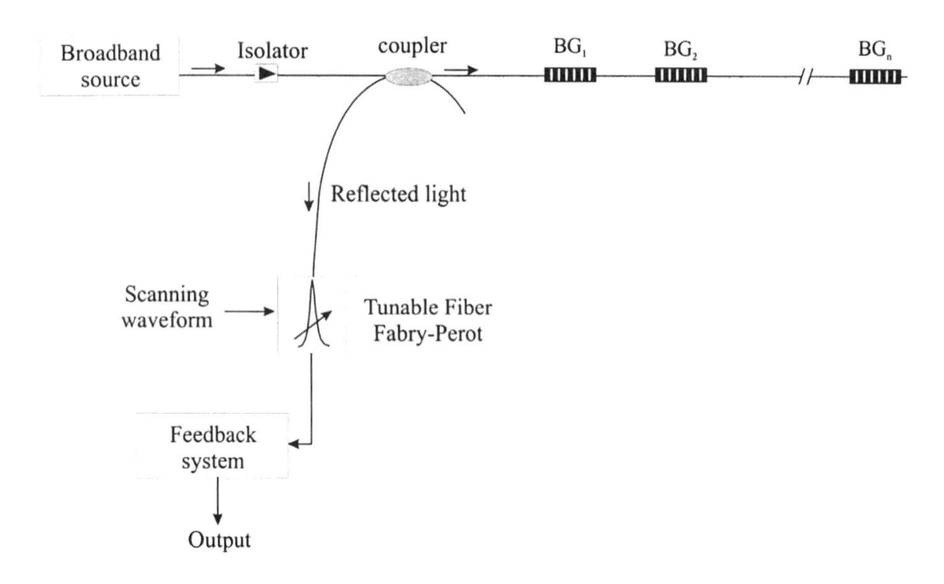

Figure 2.56. Schematic of a multiplexed Fiber Bragg grating sensor array with a scanning Fabry-Perot filter

Operating the fiber Fabry-Perot filter in a wavelength-scanning mode provides a means for addressing a number of fiber Bragg gratings elements placed along a fiber path (figure 2.56). In this mode, the direct Bragg grating sensor spectral returns are obtained from the photodetector output. The minimum resolvable Bragg wavelength shift that can be detected has a limited resolution. By dithering the Fabry-Perot filter transmission, the derivative response to the spectral components in the array output can be obtained, producing a zero crossing at each of the Bragg grating center wavelengths and thus improving the resolution in determining the wavelength shifts and hence the strain (or any other sensing parameter the transducer is designed to measure).

2.6.5.2 Interferometric interrogation

A sensitive technique for detecting the wavelength shifts of a fiber Bragg grating sensor makes use of a fiber interferometer. The principle behind such system is shown in figure 2.57. Light from a broad-band source is coupled along a fiber to the Bragg grating element. The wavelength component reflected back along the fiber toward the source is tapped off and fed to an unbalanced Mach-Zehnder interferometer. This light, in effect, becomes the light source into the interferometer, and wavelength shifts induced by perturbation of the Bragg grating sensor resemble a wavelength modulated source. The unbalanced interferometer behaves as a spectral filter with a raised cosine transfer function. The wavelength dependence on the interferometer output can be expressed as:

$$I(\lambda_b) = A\left[1 + k\cos\left(\frac{2\pi nd}{\lambda_b} + \phi\right)\right] \tag{2.29}$$

where A is proportional to the input intensity and system losses, d is the length imbalance between the fiber arms, n is the effective index of the core, λ_b is the wavelength of the return light from the grating sensor and ϕ is a bias phase offset of the Mach-Zehnder interferometer. Pseudoheterodyne phase modulation is used to generate two quadrature signals with a 90^0 phase shift with respect to each other, thus providing directional information. Wavelength shifts are tracked using a phase demodulation system developed for interferometric fiber optic sensors. In practical applications, a reference wavelength source is used to provide low frequency drift compensation. Strain resolution as low as 0.6 $n\varepsilon/Hz^{-0.5}$ at 500 Hz has been reported [183].

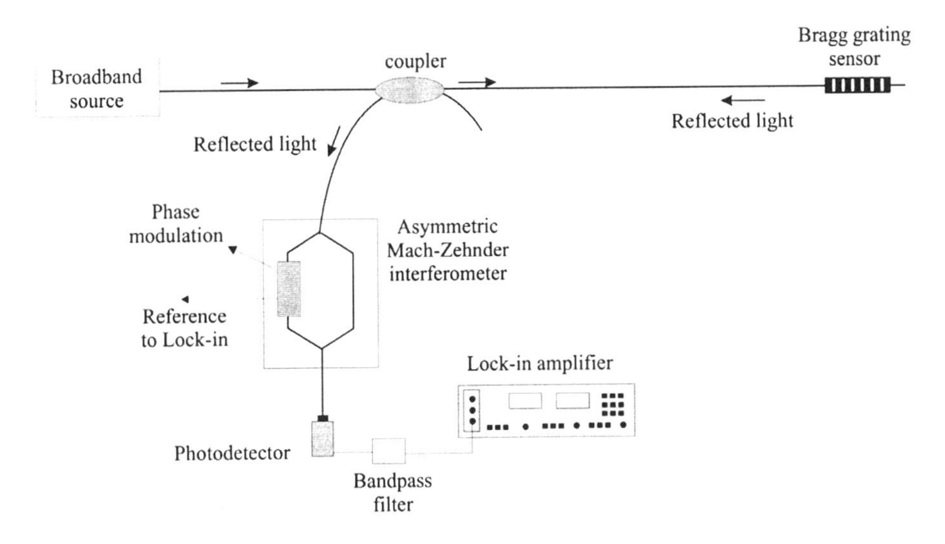

Figure 2.57. Schematic of an interferometer interrogation technique for a Fiber Bragg grating sensor

2.6.5.3 Active laser interrogation

In active interrogation, the fiber Bragg grating sensor is used as an optical feedback element of an optical laser cavity [115]. Compared to the passive broad-band base system, forming a fiber Bragg laser sensor generally provides stronger optical signals, and thus has the potential to provide improved signal to noise performance. The basic concept is shown in figure 2.58. The laser cavity is formed between the mirror and the fiber Bragg grating element, which may located at some sensing point. A gain section within the cavity can be provided via a semiconductor or doped fiber (such as the Erbium doped fiber). Once the laser gain is greater than unity, the fiber laser will lase at the wavelength determined by the fiber Bragg grating wavelength. As the Bragg grating changes its periodicity due to strain or temperature, the lasing wavelength will also shift. Reading of the laser wavelength using filtering, tracking filters or interferometric techniques can then be used to determine induced shifts. This laser sensor configuration is limited, however, to a single fiber Bragg grating element. A means to increase the number of Bragg gratings that can be addressed is to incorporate an additional tuning element within the cavity, which selectively optimizes the gain at certain wavelengths. In this way, a number of fiber Bragg gratings each operating at nominally different wavelengths can be addressed in a sequential manner to form a quasi-distributed fiber laser sensor. By tuning a wavelength selective filter located within the laser cavity over the gain bandwidth the laser selectively lases at each of the Bragg wavelengths of the sensors. Thus strain-induced shifts in the Bragg wavelengths of the sensors are detected by the shift in the lasing wavelengths of the system.

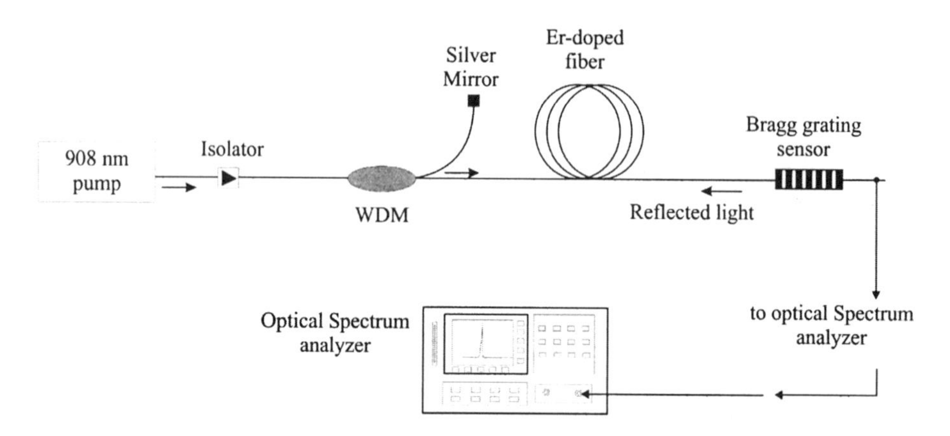

Figure 2.58. A schematic of a fiber laser sensor configuration with fiber Bragg grating elements

An alternative multiplexed fiber laser sensor is based on a single element fiber laser sensor utilizing a WDM (see Tsai et al [61]). Theoretically, since erbium is a homogeneously broadened medium, it will support only one lasing line at any time. To produce several laser lines within a single length of optical fiber, a section of erbium-doped fiber is placed between the successive Bragg gratings. With sufficient pump power and enough separation between the Bragg grating center wavelengths, a multiplexed fiber laser sensor is possible. The maximum number of sensors utilized would depend on the total pump power, the required dynamic range, and finally the gain profile of the active medium. A schematic configuration of the serially multiplexed Bragg grating fiber laser is shown in figure 2.59. One of the drawbacks in such a serial multiplexed configuration is that the cavities are coupled so their respective gains are not independent. In fact, gain coupling is a common effect in such systems.

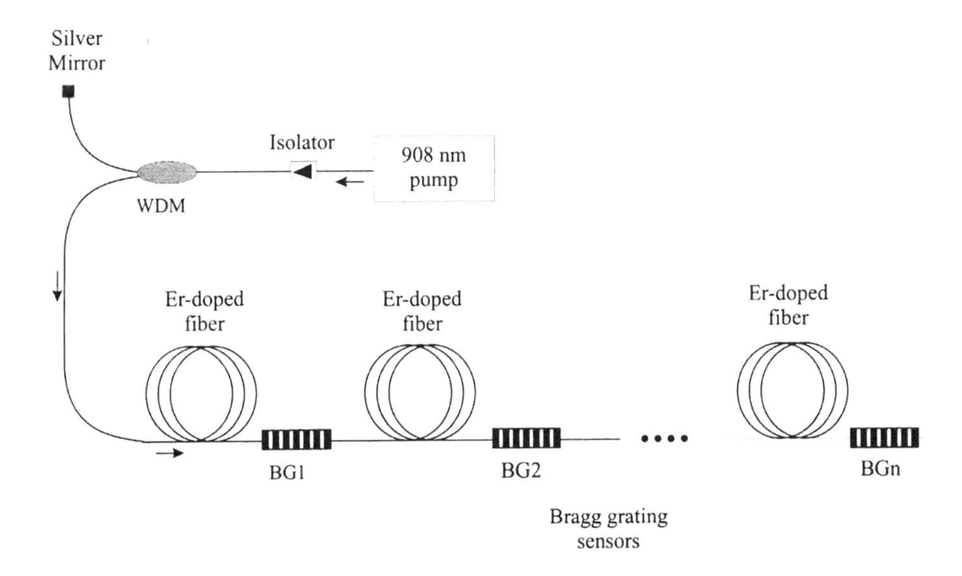

Figure 2.59. Schematic configuration of a serially m multiplexed Bragg grating fiber laser

At the cost of adding more elements in a fiber laser sensor system, an alternative is to multiplex the fiber laser sensor in a parallel configuration. This system, in essence, incorporates several single fiber lasers, one for each fiber Bragg grating.

2.6.5.4 Simultaneous measurements of strain and temperature

Although Bragg gratings are well suited for measuring strain and temperature in a structure, one of the drawbacks is the actual separation of the temperature from the strain component. This complicates the Bragg grating application as a strain or a temperature gauge. On a single measurement of the Bragg wavelength shift, it is impossible to differentiate between the effects of changes in strain and temperature. Various schemes for discriminating between these effects have been developed. These include the use of a second grating element contained within a different material and placed in series with the first grating element [184] and the use of a pair of fiber gratings, surface-mounted on opposite surfaces of a bent mechanical structure [185]. However, these methods have limitations when the wavelength of a large number of fiber gratings needs to be interrogated. Various techniques, such as measuring two different wavelengths or two different optical modes, have been employed. In one other scheme two superimposed fiber gratings having different Bragg wavelengths (850 nm and 1300 nm) have been used simultaneously to measure the strain and temperature (figure 2.60). The change in the Bragg wavelength of the fiber grating due to a combination of strain and temperature can be expressed as follows:

$$\Delta\lambda_B(\varepsilon,\lambda) = \Psi_\varepsilon \Delta\varepsilon + \Psi_T \Delta T \qquad (2.30)$$

In the case where there are two Bragg gratings with different wavelengths (referred to 1 and 2) then the following relation holds:

$$\begin{bmatrix} \Delta\lambda_{B_1} \\ \Delta\lambda_{B_2} \end{bmatrix} = \begin{bmatrix} \Psi_{\varepsilon 1} & \Psi_{T1} \\ \Psi_{\varepsilon 2} & \Psi_{T2} \end{bmatrix} \begin{bmatrix} \Delta\varepsilon \\ \Delta T \end{bmatrix} \qquad (2.31)$$

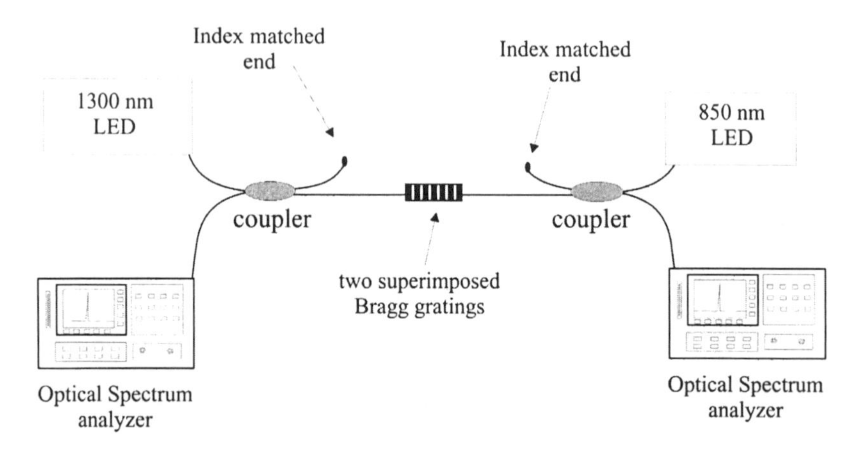

Figure 2.60. Figure Schematic diagram for measuring simultaneously strain and temperature using two superimposed Bragg gratings at 1300 nm and 850 nm

The elements of the Ψ matrix can be determined experimentally by separately measuring the Bragg wavelength changes with strain and temperature. Once Ψ is known, changes in both strain and temperature can be determined using the inverse of the above equation. The measured values of Ψ obtained in the work of Xu et al [186] (for the fiber used) were:

$$\Psi_{\varepsilon 1} = 0.96 \pm 6.5x10^{-3}\,\text{pm}/\mu\text{strain}$$
$$\Psi_{\varepsilon 2} = 0.59 \pm 3.4x10^{-3}\,\text{pm}/\mu\text{strain}$$
$$\Psi_{T1} = 8.72 \pm 7.7x10^{-2}\,\text{pm}/\,^{0}\text{C}$$
$$\Psi_{T2} = 6.30 \pm 3.7x10^{-2}\,\text{pm}/\,^{0}\text{C}$$

If the inverse matrix is used, the strain and temperature may be obtained from the two measured wavelength shifts.

2.6.6 Aerospace applications

The aerospace industry is a potentially important user of optical fibers, particularly as data links and through the use of optical fiber sensors. Although research projects have shown that optical fiber sensors can operate within prescribed limits for use in aircraft, they are still considered an immature technology. To date efforts are directed towards the sensor development for harsh environments unsuitable for conventional electro-mechanical sensors, taking advantage of radiation resistance and EMI immunity. Increases in sensor reliability, ease of installation and maintenance with little training and without special handling are demanded, ideally leading to the so-called “fit and forget” systems.

Sensing strategies for aerospace applications broadly follow the same conditions. The most important requirements are to have passive, low weight and ideally common sensors that may be multiplexed over optical links. By carefully defining sensor requirements it may be possible to specify a range of optical sensors, satisfying the majority of avionics applications that are either interchangeable or at least use common interrogation instrumentation. Currently, many sensor types perform similar functions while not being interchangeable. The Bragg grating sensor solves one of the major problems leading to drawbacks in optical fiber sensors: the lack of a standard demodulation approach, while maintaining a completely passive network. The largest class of sensors measures the position of flight control elements such as landing gear status, flap and rudder position and so forth. When taking into account high levels of system redundancy well in excess of 100 sensors are employed, therefore size and weight savings become critical. Fernando et al [187] have discussed a number of strategies for mounting optical fiber sensors, including Bragg gratings, in composites for aerospace and other applications.

2.6.7 Applications to civil engineering

There is growing concern over the state of civil infrastructure in both the US and Europe. It is essential that mechanical loading be measured for maintaining bridges, dams, tunnels, buildings and sport stadiums. By measuring the distributed strain in buildings one can predict the nature and grade of local loads, for example after an earthquake, whereas the mechanical health of bridges is increasingly under scrutiny, as old structures are often excessively loaded leading to a real possibility of increased structural failure rates. In fact, a 1996 US Department of Transportation survey estimates that 40% of all bridges in the United States are seriously deteriorated. There is concern with 50 year old railroad bridges in the US as regulatory limits on railcar loads are relaxed. The current inspection routine depends on periodic visual inspection. The use of modern optical-based sensors can lead to real time measurements, monitoring the formation and growth of defects and optical fibers sensors allow for data to be transmitted over long distances to a central monitoring location. The advantage of optical fibers is that they may either be attached to an existing structure or embedded into concrete decks and supports prior to pouring, thereby monitoring the curing cycle and the condition of the structure during its serviceable lifetime. One of the most important applications of Bragg gratings as sensors is for "smart structures" where the grating is embedded directly into the structure to monitor its strain distribution, however, for error-free, quasi-static strain measurement temperature compensation of thermal fluctuations is required. This could lead to structures that are self-monitoring or even self-scheduling of their maintenance and repair through the union of optical fiber sensors and artificial intelligence with material science and structural engineering. Several types of fiber optic sensor are capable of sensing structural strain, for example, the intrinsic and extrinsic fiber Fabry-Perot sensor. For example, Lee et al [188] have used a multiplexed array of 16 fiber Fabry-Perot sensors to monitor strain on the Union Pacific Bridge that crosses the Brazos River at Waco, Texas. The fiber sensors located at fatigue critical points for measuring dynamic loads induced by trains crossing the bridge, with the recorded data correlating well with those recovered by resistive strain gauges. Nevertheless, the general consensus is that fiber Bragg gratings are presently the most promising and widely used candidates for smart structures [189]. The instrumentation for multiplexing large grating sensor arrays can be the same, offering a potentially low cost solution for monitoring structural strain. As the wavelength shift with strain is linear and with zero offset, long-term measurements are possible and because the measurement can be interrupt-immune one can avoid perpetual monitoring of a structure, performing periodic measurements when necessary.

2.7 REFERENCES

1. Othonos, A and Kalli, K. (1999) *Fiber Bragg Gratings: Fundamentals and Applications in Telecommunications and Sensing*, Artech House.
2. Bennion, I., Williams, J. A. R., Zhang, L., Sugden, K. and Doran, N. J. (1996) UV-written in-fibre Bragg gratings. *Optical and Quantum Electronics*, **28**, 93-135.
3. Othonos, A. (1997) Fiber Bragg gratings. *Review of Scientific Instruments*, **68**, 4309-41.
4. Hill, K. O., Fujii, Y., Johnson, D. C. and Kawasaki, B. S. (1978) Photosensitivity in optical fiber waveguides: Application to reflection filter fabrication. *Applied Physics Letters*, **32**, 647-9.
5. Kawasaki, B. S., Hill, K. O., Johnson, D. C. and Fujii, Y. (1978) Narrow-band Bragg reflectors in optical fibers. *Optics Letters*, **3**, 66-8.
6. Lam, D. K. W. and Garside, B. K. (1981) Characterization of single-mode optical fiber filters. *Applied Optics*, **20**, 440-5.
7. Stone, J. (1987) Photorefractivity in GeO_2-doped silica fibers. *Journal of Applied Physics*, **62**, 4371-4.
8. Meltz, G., Morey, W. W. and Glenn, W. H. (1989) Formation of Bragg gratings in optical fibers by a transverse holographic method. *Optics Letters*, **14**, 823-5.
9. Hill, K. O., Bilodeau, F., Malo, B. and Johnson, D. C. (1991) Birefringent photosensitivity in monomode optical fiber: Application to the external writing of rocking filters. *Electronics Letters*, **27**, 1548-50.
10. Lemaire, P. J., Atkins, R. M., Mizrahi, V. and Reed, W. A. (1993) High-pressure H_2 loading as a technique for achieving ultrahigh UV photosensitivity and thermal sensitivity in GeO_2 doped optical fibres. *Electronics Letters*, **29**, 1191-3.
11. Hill, K. O., Malo, B., Bilodeau, F., Johnson, D. C. and Albert, J. (1993) Bragg gratings fabricated in monomode photosensitive optical fiber by UV exposure thorough a phase-mask. *Applied Physics Letters*, **62**, 1035-7.
12. Hosono, H., Abe, Y., Kinser, D. L., Weeks, R. A., Muta, K. and Kawazoe, H. (1992) Nature and origin of the 5-eV band in SiO_2: GeO_2 glasses. *Physical Review B*, **46**, 11445-51.
13. Nishii, J., Fukumi, K., Yamanaka, H., Kawamura, K., Hosono, H. and Kawazoe, H. (1995) Photochemical reactions in GeO_2-SiO_2 glasses induced by ultraviolet irradiation: Comparison between Hg lamp and excimer laser. *Physical Review B*, **52**, 1661-5.
14. Russell, P. St. J., Hand, D. P., Chow, Y. T. and Poyntz-Wright, L. J. (1991) Optically-induced creation, transformation and organisation of defects and colour-centres in optical fibres. *International Workshop on Photoinduced Self-Organization Effects in Optical Fiber*, Quebec City, Quebec, May 10-11, Proceedings SPIE, **1516**, 47-54.
15. Albert, J., Malo, B., Hill, K. O., Bilodeau, F., Johnson, D. C. and Theriault, S. (1995) Comparison of one-photon and two-photon effects in the photosensitivity of germanium-doped silica optical fibers exposed to intense ArF excimer laser pulses. *Applied Physics Letters*, **67**, 3529-31.
16. Hosono, H., Kawazeo, H. and Nishii, J. (1996) Defect formation in SiO_2: GeO_2 glasses studied by irradiation with excimer laser light. *Physical Review B*, **52**, R11921-3.

17. Hand, D. P. and Russell, P. St. J. (1990) Photoinduced refractive-index changes in germanosilicate fibers. *Optics Letters*, **15**, 102-4.
18. Limberger, H. G., Fonjallaz, P.-Y., Salathe, R. P. and Cochet, F. (1996) Compaction- and photoelastic-induced index changes in fiber Bragg gratings. *Applied Physics Letters*, **68**, 3069-71.
19. Poumellec, B., Niay, P., Douay, M. and Bayon, J. F. (1996) The UV-induced refractive index grating in Ge: SiO_2 preforms: additional CW experiments and the microscopic origin of the change in index. *Journal of Physics D: Applied Physics*, **29**, 1842-56.
20. Hill, K. O., Malo, B., Bilodeau, F., Johnson, D. C., Morse, J. F., Kilian, A., Reinhart, L. and Oh, K. (1991) Photosensitivity in Eu^{2+}: Al_2O_3 doped core fiber: preliminary results and applications to mode converters. *Conference on Optical Fiber Communication*, Technical Digest Series (Optical Society of America, Washington, DC), **14**, 14-7.
21. Broer, M. M., Cone, R. L. and Simpson, J. R. (1991) Ultraviolet-induced distributed-feeback gratings in Ce^{3+}-doped silica optical fibers. *Optics Letters*, **16**, 1391-3.
22. Bilodeau, F., Johnson, D. C., Malo, B., Vineberg, K. A., Hill, K. O., Morse, T. F., Kilian, A. and Reinhart, L. (1990) Ultraviolet-light photosensitivity in Er^{3+}-Ge-doped optical fiber. *Optics Letters*, **15**, 1138-40.
23. Williams, D. L., Ainslie, B. J., Armitage, J. R., Kashyap, R. and Campbell, R. (1993) Enhanced UV photosensitivity in boron codoped germanosilicate fibres. *Electronics Letters*, **29**, 45-7.
24. Niay, P., Bernage, P., Taunay, T., Xie, W. X., Boj, S., Delevaque, E., Poignant, H. and Monerie, M. (1994) Fabrication of Bragg gratings in fluorozirconate fibers and application to fiber lasers. *Conference on Laser and Electro-Optics*, Technical Digest Series (Optical Society of America, Washington, DC), **8**, Paper CPD 91/21.
25. Poumellec, B. and Kherbouche, F. (1996) The photorefractive Bragg gratings in the fibers for telecommunications. *Journal of Physics* III France, **6**, 1595-624.
26. Niay, P., Bernage, P., Legoubin, S., Douay, M., Xie, W. X., Bayon, J. F., Georges, T,. Monerie, M. and Poumellec, B. (1994) Behavior of spectral transmissions of Bragg gratings written in germania-doped fibers: Writing and erasing experiments using pulsed or CW UV exposure. *Optics Communications*, **113**, 176-92.
27. Poumellec, B., Guenot, P., Riant, I., Sansonetti, P., Niay, P., Bernage, P. and Bayon, J. F. (1995) UV induced densification during Bragg grating inscription in Ge: SiO_2 preforms. *Optical Materials*, **4**, 441-9.
28. Patrick, H. and Gilbert, S. L. (1993) Growth of Bragg gratings produced by continuous-wave ultraviolet light in optical fiber. *Optics Letters*, **18**, 1484-6.
29. Atkins, R. M., Mizrahi, V. and Erdogan, T. (1993) 248nm induced vacuum UV spectral changes in optical fibre preform cores: support for a colour centre model of photosensitivity. *Electronics Letters*, **29**, 385-7.
30. Mizrahi, V. and Atkins, R. M. (1992) Constant fluorescence during phase grating formation and defect band bleaching in optical fibres under 5.1eV laser exposure. *Electronics Letters*, **28**, 2210-1.
31. Anderson, D. Z., Mizrahi, V., Erdogan, T. and White, A. E. (1993) Production of in-fibre gratings using a diffractive optical element. *Electronics Letters*, **29**, 566-8.
32. Niay, P., Bernage, P., Douay, M., Taunay, T., Xie, W. X., Martenelli, G., Bayon, J. F., Poignant, H. and Delevaque, E. (1995) Bragg grating photoinscription within various types of fibers and glasses. *Proceedings of Topical Meeting on Photosensitivity and Quadratic Nonlinearity in Glass Waveguides: Fundamentals and Applications*, Technical Digest Series (Optical Society of America, Washington, DC), **22**, Portland, OR, Paper SuA1, 66-9.

33. Riant, I. and Haller, F. (1997) Study of the photosensitivity at 193nm and comparison with photosensitivity at 240nm influence of fiber tension: Type IIA aging. *IEEE Journal of Lightwave Technology*, **15**, 1464-9.
34. Xie, W. X., Niay, P., Bernage, P., Douay, M., Bayon, J. F., Georges, T., Monerie, M. and Poumellec, B. (1993) Experimental evidence of two types of photorefractive effects occurring during photoinscriptions of Bragg gratings written within germanosilicate fibres. *Optics Communications*, **104**, 185-95.
35. Archambault, J. L., Reekie, L. and Russell, P. St. J. (1993) 100% reflectivity Bragg reflectors produced in optical fibres by single excimer laser pulses. *Electronics Letters*, **29**, 453-5.
36. Erdogan, T., Mizrahi, V., Lemaire, P. J. and Monroe, D. (1994) Decay of ultraviolet-induced fiber Bragg gratings. *Journal of Applied Physics*, **76**, 73-80.
37. Dong, L. and Liu, W. F. (1997) Thermal decay of fiber Bragg gratings of positive and negative index changes formed at 193nm in a boron-codoped germanosilicate fiber. *Applied Optics*, **36**, 8222-6.
38. Douay, M., Xie, W. X., Taunay, T., Bernage, P., Niay, P., Cordier, P., Poumellec, B., Dong, L., Bayhon, J. F., Poignant, H. and Delevaque, E. (1997) Densification involved in the UV-based photosensitivity of silica glasses and optical fibers. *IEEE Journal of Lightwave Technology*, **15**, 1329-42.
39. Parent, M., Bures, J., Lacroix, S. and Lapierre, J. (1985) Proprietes de polarisation des reflecteurs de Bragg induits par photosensibilite dans les fibres optiques monomodes. *Applied Optics*, **24**, 354-7.
40. Meltz, G. and Morey, W. W. (1991) Bragg grating formation and germanosilicate fiber photosensitivity. *International Workshop on Photoinduced Self-Organization Effects in Optical Fiber*, Quebec City, Quebec, May 10-11, Proceedings SPIE, **1516**, 185-99.
41. Meyer, T., Nicati, P. A., Robert, P. A., Varelas, D., Limberger, H. G. and Salathe, R. P. (1996) Reversibility of photoinduced birefringence in ultralow-birefringence fibers. *Optics Letters*, **21**, 1661-3.
42. Russell, P. St. J. and Hand, D. P. (1990) Rocking filter formation in photosensitive high birefringence optical fibres. *Electronics Letters*, **52**, 1846-8.
43. Ouellette, F., Gagnon, D. and Poirier, M. (1991) Permanent photoinduced birefringence in a Ge-doped fiber. *Applied Physics Letters*, **58**, 1813-5.
44. Lauzon, J., Gagnon, D., LaRochelle, S., Blouin, A. and Ouellette, F. (1992) Dynamic polarization coupling in elliptical-core photosensitive optical fiber. *Optics Letters*, **17**, 1664-6.
45. Mizrahi, V., DiGiovanni, D. J. D., Atkins, R. M., Grubb, S. G., Park, Y. K. and Delavaux, J. M. P. (1993) Stable single-mode erbium fiber-grating laser for digital communications. *IEEE Journal of Lightwave Technology*, 11, 2021-5.
46. Kaiser, P. (1974) Drawing-induced coloration in vitreous silica fibers. *Journal of the Optical Society of America*, **64**, 475-81.
47. Friebele, E. J., Askins, C. G., Gingerich, M. E. and Long, K. J. (1984) Optical fiber waveguides in radiation environments II. *Nuclear Instruments and Methods in Physics Research B*, **1**, 355-69
48. Osterberg, U. and Margulis, W. (1986) Dye laser pumped by Nd:YAG laser pulses frequency doubled in a glass optical fiber. *Optics Letters*, **11**, 516-8.
49. Stolen, R. H. and Tom, H. W. K. (1987) Self-organized phase-matched harmonic generation in optical fibers. *Optics Letters*, **12**, 585-7.

50. Anderson, D. Z. (1989) Efficient second-harmonic generation in glass fibers: The possible role of photo-induced charge redistibution. *Nonlinear Properties of Materials*, SPIE Proceedings, **1148**, 186-96.
51. Osterberg, U. and Margulis, W. (1987) Experimental studies on efficient frequency doubling in glass optical fibers. *Optics Letters*, **12**, 57-9.
52. Anoikin, E. V., Dianov, E. M., Kazansky, P. G. and Stepanov, D. Yu. (1990) Photoinduced second-harmonic generation in gamma-ray-irradiated optical fibers. *Optics Letters*, **15**, 834-5.
53. Weeks, R. A. (1956) *Journal of Applied Physics*, **27**, 1376-81.
54. Neustruev, V. B. (1994) Colour centres in germanosilicate glass and optical fibres. *Journal of Physics, Condensed Matter*, **6**, 6901-36.
55. Jackson, J. M., Wells, M. E., Kordas, G., Kinser, D. L., Weeks, R. A. and Magruder III, R. H. (1985) Preparation effects on the UV optical properties of GeO_2 glasses. *Journal of Applied Physics*, **58**, 2308-11.
56. Yuen, M. J. (1982) Ultraviolet absorption studies of germanium silicate glasses. *Applied Optics*, **21**, 136-40.
57. Friebele, E. J., Griscom, D. L. and Sigel Jr, G. H. (1974) Defect centers in a germanium-doped silica-core optical fiber. *Journal of Applied Physics*, **45**, 3424-8.
58. Tsai, T. E., Griscom, D. L. and Friebele, E. J. (1987) On the structure of Ge-associated defect centers in irradiated high purity GeO_2 and Ge-doped SiO_2 glasses. *Diffusion and Defect Data*, **53-54**, 469-76.
59. Tsai, T. E., Griscom, D. L., Friebele, E. J. and Fleming, J. W. (1987) Radiation induced defect centers in high-purity GeO_2 glasses. *Journal of Applied Physics*, **62**, 2264-8.
60. Friebele, E. J. and Griscom, D. L. (1986) Color centers in glass optical fiber waveguides. *Materials Research Society Symposium Proceedings*, **61**, 319-31.
61. Tsai, T. E. and Griscom, D. L. (1991) Defect centers and photoinduced self-organization in Ge-doped silica core fiber. *International Workshop on Photoinduced Self-Organization Effects in Optical Fiber*, Quebec City, Quebec, May 10-11, Proceedings SPIE, **1516**, 14-28.
62. Dianov, E. M., Golant, K. M., Khrapko, R. R., Kurkov, A. S., Leconte, B., Douay, M., Bernage, P. and Niay, P. (1997) Grating formation in a germanium free silicon oxynitride fibre. *Electronics Letters*, **33**, 236-8.
63. Albert, J., Malo, B., Bilodeau, F., Johnson, D. C., Hill, K. O., Hibino, Y. and Kawachi, M. (1994) Photosensitivity in Ge-doped silica optical waveguides and fibers with 193-nm light from an ArF excimer laser. *Optics Letters*, **19**, 387-9.
64. Noguchi, K., Shibata, N., Uesugi, N. and Negishi, Y. (1985) Loss increase for optical fibers exposed to hydrogen atmosphere. *Journal of Lightwave Technology*, **LT-3**, 236-42.
65. Atkins, R. M., Lemaire, P. J., Erdogan, T. and Mizrahi, V. (1993) Mechanisms of enhanced UV photosensitivity via hydrogen loading in germanosilicate glasses. *Electronics Letters*, **29**, 1234-5.
66. Awazu, K., Kawazoe, H. and Yamane, M. (1990) Simultaneous generation of optical absorption abnds at 5.14 and 0.452 eV in $9SiO_2$: GeO_2 glasses heated under an H_2 atmosphere. *Journal of Applied Physics*, **68**, 2713-8.
67. Mizrahi, V., Lemaire, P. J., Erdogan, T., Reed, W. A., DiGiovanni, D. J. and Atkins, R. M. (1993) Ultraviolet laser fabrication of ultrastrong optical fiber gratings and of germania-doped channel waveguides. *Applied Physics Letters*, **63**, 1727-9.

68. Bilodeau, F., Malo, B., Albert, J., Johnson, D. C., Hill, K. O., Hibino, Y., Abe, M. and Kawachi, M. (1993) Photosensitization of optical fiber and silica-on-silicon/silica waveguides. *Optics Letters*, **18**, 953-5.
69. Camlibel, I., Pinnow, D. A. and Dabby, F. W. (1975) Optical aging characteristics of borosilicte clad fused silica core fiber optical waveguides. *Applied Physics Letters*, **26**, 185-7.
70. Dyer, P. E., Farley, R. J., Giedl, R., Byron, K. C. and Reid, D. (1994) High reflectivity fibre gratings produced by incubated damage using a 193nm ArF laser. *Electronics Letters*, **30**, 860-2.
71. Dong, L., Liu, W. F. and Reekie, L. (1996) Negative-index gratings formed by a 193-nm excimer laser. *Optics Letters*, **21**, 2032-4.
72. Starodubov, D. S., Grubsky, V., Feinberg, J., Kobrin, B. and Juma, S. (1997) Bragg grating fabrication in germanosilicate fibers by use of near-UV light: a new pathway for refractive-index changes. *Optics Letters*, **22**, 1086-8.
73. Watanabe, Y., Nishii, J., Moriwaki, H., Furuhashi, G., Hosono, H. and Kawazoe, H. (1997) Permanent refractive-index changes in pure GeO_2 glass slabs induced by irradiation with below-gap light. *Conference on Bragg Gratings, Photosensitivity, and Poling in Glass Fibers and Waveguides: Applications and Fundamentals*, Technical Digest Series (Optical Society of America, Washington, DC), **17**, 77-9.
74. Atkins, R. M. and Espindola, R. P. (1997) Photosensitivity and grating writing in hydrogen loaded germanosilicate core optical fibers at 325 and 351nm. *Applied Physics Letters*, **70**, 1068-9.
75. Herman, P. R., Beckley, K. and Ness, S. (1997) 157-nm photosensitivity in germanosilicate waveguides. *Conference on Bragg Gratings, Photosensitivity, and Poling in Glass Fibers and Waveguides: Applications and Fundamentals*, Technical Digest Series (Optical Society of America, Washington, DC), **17**, 159-61.
76. Atkins, R. M. and Mizrahi, V. (1992) Observations of changes in UV absorption bands of singlemode germanosilicate core optical fibres on writing and thermally erasing refractive index gratings. *Electronics Letters*, **28**, 1743-4.
77. Psaila, D. C., Martijn de Sterke, C. and Ouellette, F. (1996) Fabrication of rocking filters at 193nm. *Optics Letters*, **21**, 1550-2.
78. Dianov, E. M., Starodubov, D. S. and Frolov, A. A. (1996) UV argon laser induced luminescence changes in germanosilicate fibre preforms. *Electronics Letters*, **32**, 246-7.
79. Atkins, G. R., Wang, Z. H., McKenzie, D. R., Sceats, M. G., Poole, S. B. and Simmons, H. W. (1993) Control of defects in optical fibers - a study using cathodoluminescence spectroscopy. *IEEE Journal of Lightwave Technology*, **11**, 1793-801.
80. Svalgaard, M. (1995) Dynamics of ultraviolet induced luminescence and fiber Bragg grating formation in high fluence regime. *Proceedings of Topical Meeting on Photosensitivity and Quadratic Nonlinearity in Glass Waveguides: Fundamentals and Applications*, Technical Digest Series (Optical Society of America, Washington, DC), **22**, Portland, OR, Paper SuB18, 160-4.
81. Dianov, E. M., Starodubov, D. S., Vasiliev, S. A., Frolov, A. A. and Medvedkov, O. I. (1997) Refractive-index gratings written by near-ultraviolet radiation. *Optics Letters*, **22**, 221-3.
82. Williams, D. L., Davey, S. T., Kashyap, R., Armitage, J. R. and Ainslie, B. J. (1992) Direct observation of UV induced bleaching of 240nm absorption band in photosensitive germanosilicate glass fibres. *Electronics Letters*, **28**, 369-71.

83. Simmons, K. D., LaRochelle, S., Mizrahi, V., Stegeman, G. I. and Griscom, D. L. (1991) Correlation of defect centers with wavelength-dependent photosensitive response in germania-doped silica optical fibers. *Optics Letters*, **16**, 141-3.
84. Tsai, T. E., Friebele, E. J. and Griscom, D. L. (1993) Thermal stability of photoinduced gratings and paramagnetic centers in Ge- and Ge/P-doped silica optical fibers. *Optics Letters*, **18**, 935-7.
85. Bernadin, J. P. and Lawandy, N. M. (1990) Dynamics of the formation of Bragg gratings in germanosilicate optical fibers. *Optics Communications*, **79**, 194-9.
86. Chiang, K. S., Sceats, M. G. and Wong, D. (1993) Ultraviolet photolytic-induced changes in optical fibers: the thermal expansion coefficient. *Optics Letters*, **18**, 965-7.
87. Hosono, H., Mizuguchi, M., Kawazoe, H. and Nishii, J. (1996) Correlation between GeE' centers and optical absorption bands in SiO_2:GeO_2 glasses. *Japanese Journal of Applied Physics*, **35**, L234-6.
88. Malo, B., Albert, J., Johnson, D. C., Bilodeau, F. and Hill, K. O. (1992) Elimination of photoinduced absorption in Ge-doped silica fibres by annealing of ultraviolet colour centres. *Electronics Letters*, **28**, 1598-9.
89. Cordier, P., Doukhan, J. C., Fertein, E., Bernage, P., Niay, P., Bayon, J. F. and Georges, T. (1994) TEM characterization of structural changes in glass associated to Bragg grating inscription in a germanosilicate optical fiber preform. *Optics Communications*, **111**, 269-75.
90. Sceats, M. G., Atkins, G. R. and Poole, S. B. (1993) Photo-induced index changes in optical fibers. *Annual Reviews in Material Science*, **23**, 381-410.
91. Tsai, T. E., Williams, G. M. and Friebele, E. J. (1997) Index structure of fiber Bragg gratings in Ge-SiO_2 fibers. *Optics Letters*, **22**, 224-6.
92. Dong, L., Archambault, J. L., Reekie, L., Russell, P. St. J. and Payne, D. N. (1995) Photoinduced absorption changes in germanosilicate preforms: evidence for the color-center model of photosensitivity. *Applied Optics*, **34**, 3436-40.
93. Tsai, T. E., Friebele, E. J., Griscom, D. L. and Saifi, M. (1990) Defect centers induced by harmonic wavelength of 1.06μm light in Ge/P-doped fibers. *Electro-Optics and NonLinear Optics*, in Ceramic Transactions, **14**, 127-36.
94. Fiori, C. and Devine, R. A. B. (1986) Evidence for a wide continuum of polymorphs in a-SiO_2. *Physical Review B*, **33**, 2972-4.
95. Fiori, C. and Devine, R. A. B. (1986) Ultraviolet irradiation induced compaction and photoetching in amorphous thermal SiO_2. *Materials Research Society Symposium Proceedings*, **61**, 187-95.
96. Bazylenko, M. V., Moss, D. and Canning, J. (1998) Complex photosensitivity observed in germanosilica planar waveguides. *Optics Letters*, **23**, 697-9.
97. Grimsditch, M. (1984) Polymorphism in amorphous SiO_2. *Physical Review Letters*, **52**, 2379-81.
98. O'Reilly, E. P. and Robertson, J. (1983) Theory of defects in vitreous silicon dioxide. *Physical Review B*, 3780-95.
99. Fiori, C. and Devine, R. A. B. (1985) High resolution ultraviolet photoablation of SiO_x films. *Applied Physics Letters*, **47**, 361-2.
100. Poumellec, B., Riant, I., Niay, P., Bernage, P. and Bayon, J. F. (1995) UV induced densification during Bragg grating inscription in Ge:SiO_2 preforms: Interferometric microscopy investigations. *Optical Materials*, **4**, 404-9.
101. Fonjallaz, P. Y., Limberger, H. G., Salathe, R. P., Cochet, F. and Leuenberger, B. (1995) Tension increase correlated to refractive-index change in fibers containing UV-written Bragg gratings. *Optics Letters*, **20**, 1346-8.

102. Attard, A. E. (1992) Fermi level shift in $Bi_{12}SiO_{20}$ vis photon-induced trap level occupation. *Journal of Applied Physics*, **71**, 933-7.
103. Payne, F. P. (1989) Photorefractive gratings in single-mode optical fibres. *Electronics Letters*, **25**, 498-9.
104. Lawandy, N. M. (1989) Light induced transport and delocalization in transparent amorphous systems. *Optics Communications*, **74**, 180-4.
105. Miotello, A. and Kelly, R. (1992) Laser irradiation effects in Si^+-implanted SiO_2. *Nuclear Instruments and Methods in Physics Research B*, **65**, 217-22.
106. Russell, P. St. J., Archambault, J. L. and Reekie, L. (1993) Fibre gratings. *Physics World*, October, 41-6.
107. Hill, K. O. and Meltz, G. (1997) Fiber Bragg grating technology fundamentals and overview. *IEEE Journal of Lightwave Technology*, **15**, 1263-76.
108. Delevaque, E., Boj, S., Bayon, J. F., Poignant, H., Le Mellot, J. M., Monerie, M., Niay, P. and Bernage, P. (1995) Optical fiber design for strong gratings photoimprinting with radiation mode suppression. *Conference on Optical Fiber Communication*, Postdeadline paper PD5.
109. Albert, J., Hill, K. O., Malo, B., Theriault, S., Bilodeau, F., Johnson D. C. and Erickson, L. E. (1995) Apodisation of the spectral response of fibre Bragg gratings using a phase mask with variable diffraction efficiency. *Electronics Letters*, **31**, 222-3.
110. Malo, B., Theriault, S., Johnson, D. C., Bilodeau, F., Albert, J. and Hill, K. O. (1995) Apodised in-fibre Bragg grating reflectors photoimprinted using a phase mask. *Electronics Letters*, **31**, 223-5.
111. Kersey, A. D., Davis, M. A., Patrick, J., LeBlanc, M., Koo, K. P., Askins, C. G., Putnam, M. A. and Friebele, E. J. (1997) Fiber grating sensors. *IEEE Journal of Lightwave Technology*, **15**, 1442-63.
112. Ball, G. A., Morey, W. W. and Glenn, W. H. (1991) Standing-wave monomode erbium fiber laser. *IEEE Photonics Technology Letters*, **3**, 613-5.
113. Hillmer, H. et al (1994) Novel tunable semiconductor lasers using continuously chirped distributed feedback gratings with ultrahigh spatial precision. *Applied Physics Letters*, **65**, 2130-2.
114. Ball, B. A. and Morey, W. W. (1992) Continuously tunable single-mode erbium fiber laser. *Optics Letters*, **17**, 420-2.
115. Othonos, A., Alavie, A. T., Melle, S., Karr, S. E. and Measures, R. M. (1993) Fiber Bragg grating laser sensor. *Optical Engineering*, **32**, 2841-6.
116. Ball, G. A., Morey, W. W. and Cheo, P. K. (1994) Fiber laser source/analyzer for Bragg grating sensor array interrogation. *IEEE Journal of Lightwave Technology*, **12**, 700-3.
117. Alavie, A. T., Karr, S. E., Othonos, A. and Measures, R. M. (1993) A multiplexed Bragg grating fiber laser sensor system. *IEEE Photonics Technology Letters*, **5**, 1112-4.
118. Meltz, G. and Morey, W. W. (1991) *Conference on Optical Fiber Communciation*, Paper TuM2.
119. Kashyap, R., Wyatt, R. and McKee, P. F. (1993) Wavelength flattened saturated erbium amplifier using multiple side-tap Bragg gratings. *Electronics Letters*, **29**, 1025-6.
120. Ouellette, F. (1987) Dispersion cancellation using linearly chirped Bragg grating filters in optical waveguides. *Optics Letters*, **12**, 847-9.
121. Byron, K. C., Sugden, K., Bircheno, T. and Bennion, I. (1993) Fabrication of chirped Bragg gratings in photosensitive fibre. *Electronics Letters*, **29**, 1659-60.
122. Sugden, K., Bennion, I., Molony, A. and Copner, N. J. (1994) Chirped gratings produced in photosensitive optical fibres by fibre deformation during exposure. *Electronics Letters*, **30**, 440-2.

123. Kashyap, R., Mckee, P. F., Campbell, R. J. and Williams, D. L. (1994) Novel method of producing all fibre photoinduced chirped gratings. *Electronics Letters*, **30**, 996-7.
124. Russell, P. St. J. and Ulrich, R. (1985) Grating fiber-coupler as a high-resolution spectrometer. *Optics Letter*, **10**, 291-3.
125. Askins, C. G., Tsai, T. E., Williams, G. M., Puttnam, M. A., Bashkasnsky, M. and Friebele, E. J. (1992) Fiber Bragg reflectors prepared by a single excimer pulse. *Optics Letters*, **17**, 833-5.
126. Dong, L., Archambault, J. L., Reekie, L., Russell, P. St. J. and Payne, D. N. (1993) Single pulse Bragg gratings written during fibre drawing. *Electronics Letters*, **29**, 1577-8.
127. Othonos, A., Lee, X. and Measures, R. M. (1994) Superimposed multiple Bragg gratings. *Electronics Letters*, **30**, 1972-3.
128. Eggleton, B. J., Krug, P. A., Poladian, L. and Ouellette, F. (1994) Long periodic superstructure Bragg gratings in optical fibres. *Electronics Letters*, **30**, 1620-2.
129. Morey, W. W., Ball, G. A. and Meltz, G. (1994) Photoinduced Bragg gratings in optical fibers. *Optics and Photonics News*, Optical Society of America, February, 8-14.
130. Agrawal, G. P. and Dutta, N. K. (1993) *Semiconductor Lasers*, Van Nostrand Reinhold, New York, 1993, Chapter 7.
131. Haus, H. A. and Shank, C. V. (1976) *IEEE Journal of Quantum Electronics*, **QE-12**, 352.
132. Alferness, R. C. et al (1986) Narrowband grating resonator filters in InGaAsP/InP waveguides. *Applied Physics Letters*, **49**, 125-7.
133. Kashyap, R., McKee, P. F. and Armes, D. (1994) UV written reflection grating structures in photosensitive optical fibres using phase-shifted phase masks. *Electronics Letters*, **30**, 1977-8.
134. Canning, J. and Sceats, M. G. (1994) Pi-phase-shifted periodic distributed structures in optical fibres by UV post-processing. *Electronics Letters*, **30**, 1344-5.
135. Uttamchandani, D. and Othonos, A. (1996) Phase shifted Bragg gratings formed in optical fibres by post-fabrication thermal processing. *Optics Communications*, **127**, 200-4.
136. Dockney, M. L., James, J. W. and Tatam, R. P. (1996) Fiber Bragg grating fabricated using a wavelength tunable source and a phase-mask based interferometer. *Measurement Science and Technology*, **7**, 445.
137. Kashyap, R., Armitage, J. R., Wyatt, R., Davey, S. T. and Williams, D. L. (1990) All-fiber narrow band reflection grating at 1500 nm. *Electronics Letters*. **26**, 730-2.
138. Eggleton, B. J., Krug, P. A. and Poladian, L. (1994) Experimental demonstration of compression of dispersed optical pulses by reflection from self-chirped optical fiber Bragg gratings. *Optics Letters*, **19**, 877-80.
139. Limberger, H. G., Fonjallaz, P. Y., Lambelet, P., Zimmer, Ch., Salathe, R. P. and Gilgen, H. H. (1993) Photosensitivity and Self-Organization in Optical Fibers and Waveguides. *SPIE*, **2044**, Photosensitivity and Self-Organization in Optical Fibers and Waveguides, 272.
140. Othonos, A. and Lee, X. (1995) Narrow linewidth excimer laser for inscribing Bragg gratings in optical fibers. *Review of Scientific Instruments*, **66**, 3112-5.
141. Cannon, J. and Lee, S. (1994) Fiberoptic Product News. *Laser Focus World*, **2**, 50-1.
142. Othonos, A. and Lee, X. (1995) Novel and improved methods of writing Bragg gratings with phase-masks. *IEEE Photonics Technology Letters*, **7**, 1183-5.

143. Dyer, P. E., Farley, R. J. and Giedl, R. (1996) Analysis and application of a 0/1 order Talbot interferometer for 193 nm laser grating formation. *Optics Communications*, **129**, 98-108.
144. Kashyap, R., Armitage, J. R., Campbell, R. J., Williams, D. L., Maxwell, G. D., Ainslie, B. J. and Millar, C. A. (1993) Light-sensitive optical fibers and planar waveguides. *BT Technol. J.*, **11**, 150-60.
145. Malo, B., Hill, K. O., Bilodeau, F., Johnson, D. C., Albert, J. (1993) Point-by-point fabrication of micro-Bragg gratings in photosensitive fiber using single excimer pulse refractive index modification techniques. *Electronics Letters*, **29**, 1668-9.
146. Hill, K. O., Malo, B., Vineberg, K. A., Bilodeau, F., Johnson, D. C. and Skinner, I. (1990) Efficient mode-conversion in telecommunication fiber using externally written gratings. *Electronics Letters*, **26**, 1270-2.
147. Mihailov, S. and Gower, M. (1994) Recording of efficient high-order Bragg reflectors in optical fibers by mask image projection and single pulse exposure with an excimer laser. *Electronics Letters*, **30**, 707-8.
148. Malo, B., Johnson, D. C., Bilodeau, F., Albert, J. and Hill, K. O. (1993) Single-excimer-pulse writing of fiber gratings by use of a zero-order mulled phase-mask: grating spectral response and visualization of index perturbations. *Optics Letters*, **18**, 1277-9.
149. Martin, J. and Ouellette, F. (1994) Novel writing technique of long and highly reflective in-fiber gratings. *Electronics Letters*, **30**, 811-2.
150. Hill, K. O., Bilodeau, F., Malo, B., Kitagawa, T., Theriault, S., Johnson, C., Albert, J. and Takiguchi, K. (1994) Aperiodic in-fiber Bragg gratings for optical fiber dispersion compensation. *Proc. OFC'94*, PF-77 postdeadline paper.
151. Loh, H. W., Cole, M. J., Zervas, M. N., Barcelos, S. and Laming, R. I. (1995) Complex grating structures with uniform phase-masks based on the moving fiber-scanning beam technique. *Optics Letters*, **20**, 2051.
152. Yariv, A. (1973) Coupled-mode theory for guided-wave optics. *IEEE Journal of Quantum Electronics*, **QE-9**, 919-33.
153. Yamada, M. and Sakuda, K. (1987) Analysis of almost-periodic distributed feedback slab waveguide via a fundamental matrix approach. *Applied Optics*, **26**, 3474-8.
154. Hill, K. O., Theriault, S., Malo, B., Bilodeau, F., Kitagawa, T., Johnson, D. C., Albert, J., Takiguchi, K., Kataoka, T. and Hagimoto, K. (1994) Chirped in-fiber Bragg grating disperion compensators: Linearization of the dispersion characteristic and demonstration of dispersion compensation in a 100 km, 10 Gbit/s optical fiber link. *Electonics Letters*, **30**, 1755-6.
155. Mizrahi, V. and Sipe, J. E. (1993) Optical properties of photosensitive fiber phase gratings. *IEEE Journal of Lightwave Technology*, **11**, 1513-7.
156. Kashyap, R., Swanton, A. and Armes, D. J. (1996) Simple technique for apodising chirped and unchirped fibre Bragg gratings. *Electronics Letters*, **32**, 1226-8.
157. Bird, D. M., Armitage, J. R., Kashyap, R., Fatah, R. M. A. and Cameron, K. H. (1991) Narrow line semiconductor laser using fiber grating. *Electronics Letters*, **27**, 1115-6.
158. Morton, P. A., Mizrahi, V., Andrekson, P. A., Tanbun-Ek, T., Logan, R. A., Lemaire, P., Coblentz, D. L., Sergent, A. M., Wecht, K. W. and Sciortino Jr., P. F. (1993) Mode-locked Hybrid soliton pulse source with extremely wide operating frequency range. *IEEE Photonics Technology Letters*, **5**, 28-31.
159. Reekie, L., Mears, R. J., Poole, S. B. and Payne, D. N. (1986) Tunable single-mode fiber laser. *Journal of Lightwave Technology*, **LT4**, 956-7.
160. Ball, G. A., Morey, W. W. and Waters, J. P. (1990) Nd^{3+} fiber laser utilizing intra-core Bragg reflectors. *Electronics Letters*, **26**, 1829-30.

161. Ball, G. A. and Glenn, W. H. (1992) Design of a single-mode linear-cavity erbium fiber laser utilizing Bragg reflector. *Journal of Lightwave Technology*, **10**, 1338-43.
162. Ball, G. A., Glenn, W. H., Morey, W. W. and Cheo, P. K. (1993) Modeling of short, single-frequency fiber laser in high-gain fiber. *IEEE Photonics Technology Letters*, **5**, 649-51.
163. Ball, G. A., Morey, W.W. and Cheo, P. K. (1993) Single- and multi-point fiber-laser sensors. *IEEE Photonics Technology Letters*, **5**, 267-70.
164. Othonos, A., Lee, X. and Tsai, D. P. (1996) Spectrally broadband Bragg grating mirror for an erbium-doped fiber laser. *Optical Engineering*, **35**, 1088-92.
165. Zyskind, J. L., Mizrahi, V., DiGiovanni, D. J. and Sulhoff, J. W. (1992) Short single frequency erbium-doped fiber laser. *Electronics Letters*, **28**, 1385-6.
166. Zyskind, J. L., Sulhoff, J. W., Magill, P. D., Reichmann, K. C., Mizrahi, V. and DiGiovanni, D. J. (1993) Transmission at 2.5 Gbits/s over 654 km using an erbium-doped fiber grating laser source. *Electronics Letters*, **29**, 1105-6.
167. Kringlebotn, J. T., Archambault, J. L., Reekie, L., Townsend, J. E. and Payne, D. N. (1994) Highly efficient low-noise grating-feedback Er^{3+}:Yb^{3+} co-doped fiber laser. *Electronics. Letters*, **30**, 972-3.
168. Grubb, G. S. (1995) High-power 1.48 μm cascaded Raman laser in germanosilicate fibers in *OFC'95, the Conference on Optical Communication*, **8**, Technical Digest Series, Postconference Edition, 41-2.
169. Hand, D. P. and Russell, P. St J. (1989) *7th International Conference on Integrated Optics and Optical Fiber Communication (IOOC'89)*, Kobe, Technical Digest, 64.
170. Hill, K. O., Johnson, D. C., Bilodeau, F. and Faucher, S. (1987) Narrow-bandwidth optical waveguide transmission filters: A new design concept and applications to optical fiber communications. *Electronics Letters*, **23**, 464-5.
171. Bilodeau, F., Hill, K.O., Malo, B., Johnson, D. C. and Albert, J. (1994) High-return-loss narrowband all-fiber bandpass Bragg transmission filter. *IEEE Photonics Technology Letters*, **6**, 80-2.
172. Johnson, D. C., Hill, K. O., Bilodeau, F. and Faucher, S. (1987) New design concept for a narrowband wavelength-selective optical tap and combiner. *Electronics Letters*, **23**, 668-9.
173. Fielding, A., Cullen, T. J., Rourke, H. N. (1994) Compact all-fiber wavelength drop and insert filter. *Electronics Letters*, **30**, 2160-1.
174. Reid, D. C., Ragdale, C. M., Bennion, I., Robbins, D. J., Buus, J. and Stewart, W. J. (1990) Phase-shifted Moiré grating fiber resonators. *Electronics Letters*, **26**, 10-1.
175. Legoubin, S., Fertein, E., Douay, M., Bernage, P., Niay, P., Bayon, F. and Georges, T. (1991) Formation of Moiré grating in core of germanosilicate fiber by transverse holographic double exposure method. *Electronics Letters*, **27**, 1945-6.
176. Zhang, L., Sugden, K., Bennion, I. and Molony, A. (1995) Wide-stopband chirped fiber moiré grating transmission filters. *Electronics Letters*, **31**, 477-9.
177. Morey, W. W. (1991) Tuneable narrow-line bandpass filter using fiber gratings in *Proc. Conference on Optical Fiber Communications*, *OFC'91*, San Diego, California PDP 20, 96.
178. Williams, J. A. R., Bennion, I., Sugden, K. and Doran, N. J. (1994) Fiber dispersion compensation using a chirped in fiber Bragg grating. *Electronics Letters*, **30**, 985-7.
179. Melle, S. M., Liu, K. and Measures, R. M. (1992) A passive wavelength demodulation system for guided-wave Bragg grating sensors. *IEEE Photonics Technology Letters*, **4**, 516-8.

180. Kersey, A. D., Berkoff, T. A. and Morey, W. W. (1993) Multiplexed fiber Bragg grating strain-sensor system with a fiber Fabry-Perot wavelength filter. *Optics Letters*, **18**, 1370-2.
181. Geiger, H., Xu, M. G., Eaton, N. C. and Dakin, J. P. (1995) Electronic tracking system for multiplexed fibre grating sensors. *Electronics Letters*, **31**, 1006-7.
182. Jackson, D. A., Lobo Ribeiro, A. B., Reekie, L. and Archambault, J. L. (1993) Simple multiplexing scheme for fiber-optic grating sensor network. *Optics Letters*, **18**, 1192-4.
183. Kersey, A. D., Berkoff, T. A. and W Morey,. W. (1992) High-resolution fibre-grating based strain sensor with interferometric wavelength-shift detection. *Electronics Letters*, **28**, 236-8.
184. Morey, W. W., Meltz, G. and Weiss, J. M. (1992) Evaluation of a fibre Bragg grating hydrostatic pressure sensor. *Proceedings of the Optical Fiber Sensors Conference (OFS-8)*, Monterey, USA, Postdeadline paper PD-4.4.
185. Xu, M. G., Archambault, J. L., Reekie, L. and Dakin, J. P. (1994) Thermally-compensated bending gauge using surface mounted fibre gratings. *International Journal of Optoelectronics*, **9**, 281-3.
186. Xu, M. G., Archambault, J. L., Reekie, L. and Dakin, J. P. (1994) Discrimination between strain and temperature effects using dual-wavelength fibre grating sensors. *Electronics Letters*, **30**, 1085-7.
187. Fernando, G. F., Crosby, P. A. and Liu, T. (1999) The application of optical fiber sensors in advanced fiber reinforced composites. Parts, I-III. *Optical Fiber Sensor Technology*, **3**, Eds: Grattan, K. T. V. and Meggitt, B. T., Kluwer Academic Publishers, London, 25-129.
188. Lee, W., Lee, J., Henderson, C., Taylor, H. F., James, R., Lee, C. E., Swenson, V., Gibler, W. N., Atkins, R. A. and Gemeiner, W. G. (1997) Railroad bridge instrumentation with fiber optic sensors. *Proceedings of the Optical Fiber Sensors Conference (OFS-12)*, Williamsburg, USA, 412-5.
189. Merzbacher, C. I., Kersey, A. D. and Friebele, E. J. (1999) Fiber optic sensors in concrete structures: a review. *Optical Fiber Sensor Technology*, **3**, Eds: Grattan, K. T. V. and Meggitt, B. T., Kluwer Academic Publishers, London, 1-24.

3

Nonlinear Optics and Optical Fibers

A. J. Rogers

3.1 GENERAL INTRODUCTION

In the various discussions concerning the propagation of light in material media such as silica optical fibers, in the previous chapters we have been dealing with linear processes. By this we mean that a light beam of a certain optical frequency which enters a given medium will leave the medium with the same frequency, although the amplitude and phase of the wave will, in general, be altered.

The fundamental physical reason for this linearity lies in the way in which the wave propagates in time, t, through a material medium. The effect of the electric field of the optical wave on the medium is to set the electrons of the atoms (of which the medium is composed) into forced oscillation; these oscillating electrons then radiate secondary wavelets (since all accelerating electrons radiate) and the secondary wavelets combine with each other and with the original (primary) wave, to form a resultant wave. Now the important point here is that all the forced electrons oscillate at the same frequency, ω, (but differing phase, in general) as the primary, driving wave, and thus we have the sum of waves all of the same frequency, but with different amplitudes (a_1 and a_2) and phases (φ_1 and φ_2)

If two such sinusoids are added together, the combined effect is given by:

$$A_T = a_1 \sin(\omega t + \varphi_1) + a_2 \sin(\omega t + \varphi_2)$$

and we have, from simple trigonometry:

$$A_T = a_T \sin(\omega t + \varphi_T)$$

where

K.T.V. Grattan and B.T. Meggitt (eds.), Optical Fiber Sensor Technology, 189–240.

$$a_T^2 = a_1^2 + a_2^2 + 2a_1a_2 \cos(\varphi_1 - \varphi_2)$$

and

$$\tan \varphi_T = \frac{a_1 \sin \varphi_1 + a_2 \sin \varphi_2}{a_1 \cos \varphi_1 + a_2 \cos \varphi_2}$$

In other words, the resultant is a sinusoid of the same frequency but of different amplitude and phase. It follows, then, that no matter how many more such waves are added, the resultant will always be a wave of the same frequency, i.e.,

$$A_T = \sum_{n=0}^{N} a_n \sin(\omega t + \varphi_n) = \alpha \sin(\omega t + \beta)$$

where α and β are expressible in terms of a_n and φ_n.

It follows, further, that if there are two primary input waves, each will have the effect described above independently of the other, for each of the driving forces will act independently and the two will add to produce a vector resultant. We call this the 'principle of superposition' for linear systems since the resultant effect of the two (or more) actions is just the sum of the effects of each one acting on its own. This has to be the case whilst the displacements of the electrons from their equilibrium positions in the atoms vary linearly with the force of the optical electric fields. Thus, if we pass into a medium, along the same path, two light waves, of angular frequencies ω_1 and ω_2, emerging from the medium will be two light waves (and only two) with those same frequencies, but with different amplitudes and phases from the input waves.

Suppose now, however, that the displacement of the electrons is not linear with the driving force and for example, that the displacement is so large that the electron is coming close to the point of breaking free from the atom altogether. We are now in a nonlinear regime. Strange things happen here. For example, a given optical frequency input into the medium may give rise to waves of several different frequencies at the output. Two frequencies ω_1 and ω_2 passing in may lead to, among others, sum and difference frequencies $\omega_1 \pm \omega_2$ coming out.

The fundamental reason for this is that the driving sinusoid has caused the atomic electrons to oscillate non-sinusoidally (fig. 3.1). A knowledge of Fourier analysis tells us that any periodic non-sinusoidal function contains, in addition to the fundamental component, components at harmonic frequencies, i.e., integral multiples of the fundamental frequency.

This is a fascinating regime. All kinds of interesting new optical phenomena occur here. As might be expected, some are desirable, some are not. Some are valuable in new applications, some just comprise sources of noise. But to use them to advantage, and to minimize their effects when they are a nuisance, we must, of course, understand them better. This we shall now try to do.

3.2 NONLINEAR OPTICS AND OPTICAL FIBERS

Let us begin by summarizing the conditions which give rise to optical nonlinearity.

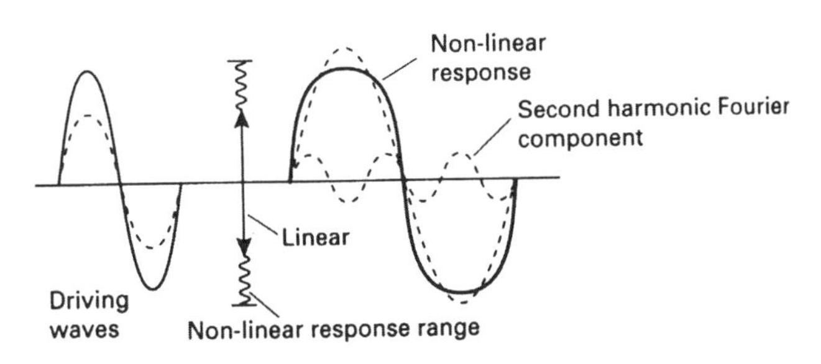

Figure 3.1. Nonlinear response to a sinusoidal drive

In a semiclassical description of light propagation in dielectric media, the optical electric field drives the atomic/molecular oscillators of which the material is composed, and these oscillators become secondary radiators of the field; the primary and secondary fields then combine vectorially to form the resultant wave. The phase of this wave (being different from that of its primary) determines a velocity of light different from that of free space, and its amplitude determines a scattering/absorption coefficient for the material.

Nonlinear behavior occurs when the secondary oscillators are driven beyond the linear response; as a result, the oscillations become non-sinusoidal. Fourier theory dictates that, under these conditions, frequencies other than that of the primary wave will be generated (fig. 3.1).

The fields necessary to do this depend upon the structure of the material, since it is this which dictates the allowable range of sinusoidal oscillation at given frequencies. Clearly, it is easier to generate large amplitudes of oscillation when the optical frequencies are close to natural resonances, and one expects (and obtains) enhanced nonlinearity there. The electric field required to produce nonlinearity in a material therefore varies widely, from $\sim 10^6$ Vm^{-1} up to $\sim 10^{11}$ Vm^{-1}, the latter being comparable with the atomic electric field. Even the lower of these figures, however, corresponds to an optical intensity of $\sim 10^9$ Wm^{-2}, which is only achievable practically with laser sources. It is for this reason that the study of nonlinear optics only really began with the invention of the laser, in 1960.

The magnitude of any given nonlinear effect will depend upon the optical intensity, the optical path over which the intensity can be maintained, and the size of the coefficient which characterizes the effect.

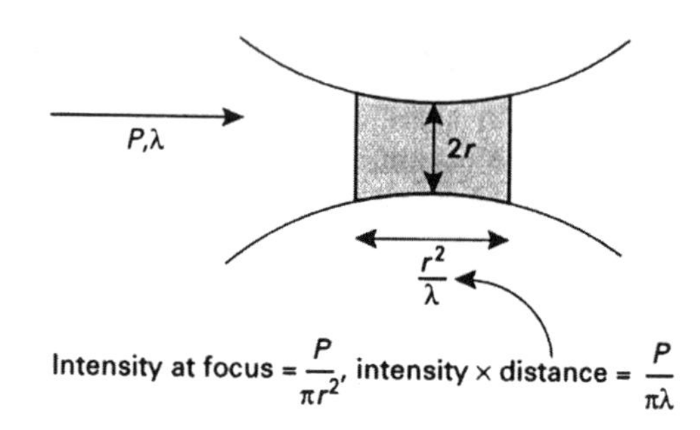

(a) The Rayleigh distance for free-space focusing

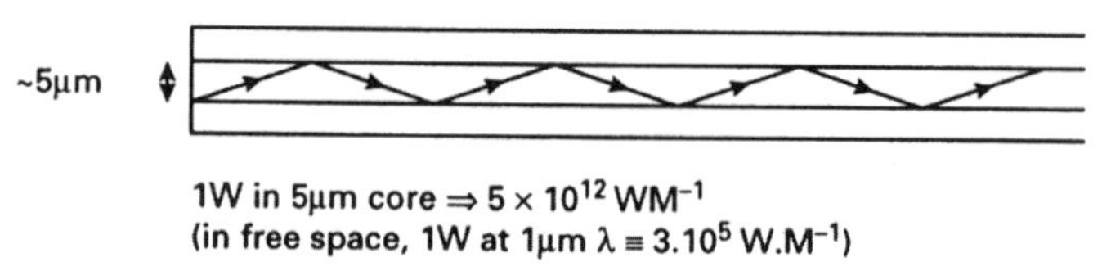

(b) Non-linear facility in optical fibres

Figure 3.2. The intensity-distance product for non-linearity

In bulk media, the magnitude of any nonlinearity is limited by diffraction effects. For a beam of power,P watts, and wavelength, λ, focused to a spot of radius, r, the intensity, $P/\pi r^2$, can be maintained (to within a factor of ~2) over a distance $\sim r^2/\lambda$ (the Rayleigh distance), beyond which diffraction will rapidly reduce it. Hence the product of intensity and distance is $\sim P/\pi\lambda$, independent of r, and of the propagation length (fig. 3.2(a)).

However, in an optical fiber the waveguiding properties, in a small diameter core, serve to maintain a high intensity over lengths of up to several kilometers (fig. 3.2(b)). This simple fact allows magnitudes of nonlinearities, in fibers, which are many orders greater than in bulk materials. Further, for maximum overall effect, the various components effects per elemental propagation distance must add coherently over the total path. This implies a requirement for phase coherence throughout the path which, in turn, implies a single propagation mode: monomode rather than multimode fibers must, in general be used.

3.3 THE FORMALISM OF NONLINEAR OPTICS

We assume a linear relationship between the electric polarization (*P*) of a medium and the electric field (*E*) of an optical wave propagating in it, by taking:

$$\chi = \frac{P}{E}$$

(for convenience the constant ε_0 has been absorbed into χ; this only implies a change of units) where χ is the volume susceptibility of the medium, and is assumed constant. The underlying assumption for this is that the separation of atomic positive and negative charges is proportional to the imposed field, leading to a dipole moment per unit volume (*P*) which is proportional to the field.

However, it is clear that the linearity of this relationship cannot persist for ever-increasing strengths of field. Any resonant physical system must eventually be torn apart by a sufficiently strong perturbing force and, well before such a catastrophe occurs, we expect the separation of oscillating components to vary nonlinearly with the force. In the case of an atomic system under the influence of the electric field of an optical wave, we can allow for this nonlinear behavior by writing the electric polarization of the medium in the more general form:

$$P(E) = \chi_1 E + \chi_2 E^2 + \chi_3 E^3 + \cdots \chi_j E^j + \cdots \tag{3.1}$$

The value of χ_j (often written $\chi^{(j)}$) decreases rapidly with increasing j for most materials. Also the importance of the j-th term, compared with the first, varies as $(\chi_j/\chi_1)E^{(j-1)}$, and so depends strongly on E. In practice, only the first three terms are of any great importance, and then only for laser-like intensities, with their large electric fields. It is not until one is dealing with power densities of $\sim 10^9$ Wm^{-2}, and fields $\sim 10^6$ Vm^{-1}, that $\chi_2 E^2$ becomes comparable with $\chi_1 E$.

Let us now consider the refractive index of the medium. We remember that the relative permittivity of a medium is given by:

$$\varepsilon = 1 + \chi, \quad n^2 = \varepsilon$$

Hence

$$n = (1 + \chi)^{1/2} = (1 + \frac{P}{E})^{1/2}$$

i.e.,

$$n = (1 + \chi_1 + \chi_2 E + \cdots \chi_j E^{j-1} + \cdots)^{1/2} \qquad (3.2)$$

Hence we note that the refractive index has become dependent on E. The optical wave, in this nonlinear regime, is altering its own propagation conditions as it travels. This is a central feature of nonlinear optics.

3.4 SECOND HARMONIC GENERATION (SHG) AND PHASE MATCHING

Probably the most straightforward consequence of nonlinear optical behavior in a medium is that of the generation of the second harmonic of a fundamental optical frequency. To appreciate this mathematically, let us assume that the electric polarization of an optical medium is quite satisfactorily described by the first two terms of equation (3.1), i.e.:

$$P(E) = \chi_1 E + \chi_2 E^2 \qquad (3.3)$$

Before proceeding, there is a quite important point to make about equation (3.3).

Let us consider the effect of a change in sign of E. The two values of the field, $\pm$E, will correspond to two values of P:

$$P(+E) = \chi_1 E + \chi_2 E^2$$
$$P(-E) = -\chi_1 E + \chi_2 E^2$$

These two values clearly have different absolute magnitudes. Now if a medium is isotropic (as is the amorphous silica of which optical fiber is made) there can be no directionality in the medium and thus the matter of the sign of E, i.e., whether the electric field points up or down, cannot be of any physical relevance and cannot possibly have any measurable physical effect. In particular, it cannot possibly affect the value of the electric polarization (which is, of course, readily measurable). We should expect that changing the sign of E will merely change the sign of P, but that the magnitude of P will be exactly the same: the electrons will be displaced by the same amount in the opposite direction, all directions being equivalent. Clearly this can only be so if $\chi_2 = 0$. The same argument extended to higher order terms evidently leads us to the conclusion that all even-order terms must be zero for amorphous (isotropic) materials, i.e., $\chi_{2m} = 0$. This is a point to remember. The corollary of this argument is, of course, that in order to retain any even order terms the medium must exhibit some anisotropy. It must, for example, have a crystalline structure without a center of symmetry. It follows that equation (3.3) refers to such a medium.

Suppose now that we represent the electric field of an optical wave entering such a crystalline medium by:

$$E = E_0 \cos \omega t$$

Then substituting into equation (3.3) we find

$$P(E) = \chi_1 E_0 \cos \omega t + \frac{1}{2}\chi_2 E_0^2 + \frac{1}{2}\chi_2 E_0^2 \cos 2\omega t$$

The last term, the second harmonic term at twice the original frequency, is clearly in evidence. Fundamentally, it is due to the fact that it is easier to polarize the medium in one direction than in the opposite direction, as a result of the crystal asymmetry. A kind of 'rectification' occurs.

Now the propagation of the wave through the crystal is the result of adding the original wave to the secondary wavelets from the oscillating dipoles which it induces. These oscillating dipoles are represented by P. Thus $\partial^2 P / \partial t^2$ leads to electromagnetic (e.m.) waves, since radiated power is proportional to the acceleration of charges, and waves at all of *P's* frequencies will propagate through the crystal.

Suppose now that an attempt is made to generate a second harmonic over a length, L, of crystal. At each point along the path of the input wave a second harmonic component will be generated. But, since the crystal medium will almost certainly be dispersive, the fundamental and second harmonic components will travel at different velocities. Hence the successive portions of second harmonic component generated by the fundamental will not, in general, be in phase with each other, and thus will not interfere constructively. Hence, the efficiency of the generation will depend upon the velocity difference between the waves.

A rigorous treatment of this process requires a manipulation involving Maxwell's equations, but a semi-analytical treatment which retains a firm grasp of the physics will be given here.

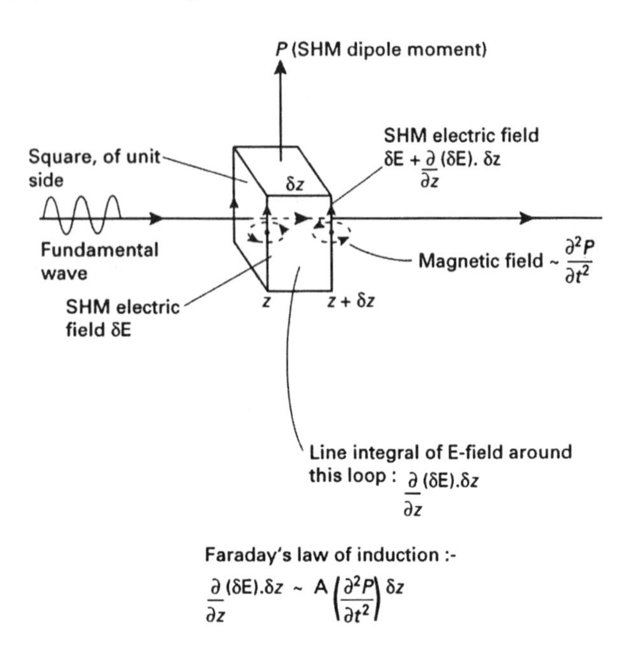

Figure 3.3. Infinitesimals for second harmonic generation along the path of the fundamental

Suppose that the amplitude of the fundamental (driving) wave between distances z and z + dz along the optical path in the crystal is $e\cos(\omega t - kz)$. Then from equation (3.1) , there will be a component of electric polarization (dipole moment per unit volume) of the form: $\chi_2 e^2 \cos^2(\omega t - kz)$ giving a time-varying second harmonic term $\chi_2 e^2 \cos 2(\omega t - kz)/2$, as before. Consider, then, a slab, in the medium, of unit cross-section, and thickness *dz* (fig. 3.3).

For this slab the dipole moment will be:

$$P = \frac{1}{2}\chi_2 e^2 \cos 2(\omega t - kz)dz \tag{3.4}$$

Now a time-varying dipole moment represents a movement of charge, and therefore an electric current. This current will create a magnetic field. A time-varying magnetic field (second derivative, $\partial^2 P/\partial t^2$, of the dipole moment) will generate a voltage around any loop through which it threads (Faraday's law of electromagnetic induction). From fig. 3.3 it can be seen that, if δE is the elemental component of the second harmonic electric field generated by the changing dipole moment in the thin slab, then this voltage is proportional to $\partial(\delta E)/\partial z$. Hence, we have

$$\frac{\partial(\delta E)}{\partial z} = A\frac{\partial^2 P}{\partial t^2}$$

where A is a constant.

Hence, from (4):

$$\frac{\partial(\delta E)}{\partial z} = -A2\omega^2\chi_2 e^2 \cos 2(\omega t - kz)dz$$

Integrating this w.r.t. z gives:

$$\delta E = A\frac{\omega^2}{k}\chi_2 e^2 \sin 2(\omega t - kz)dz$$

and, with $\omega / k = c$, we have:

$$\delta E = Ac\omega\chi_2 e^2 \sin 2(\omega t - kz)dz$$

as the element of the second harmonic electric field generated by the slab between z and z + dz. But the second harmonic component now propagates with wavenumber k_s, say (since the crystal will have a different refractive index at frequency 2ω, compared with that at ω), so when this component emerges from the crystal after a further distance (L - z), it will have become:

$$\delta E_L = Ac\omega\chi_2 e^2 \sin\left[2\omega t - 2kz - k_s(L - z)\right]dz$$

Hence the total electric field amplitude generated over the length, L, of crystal will be, on emergence:

$$E_L(2\omega) = \int_0^L Ac\omega\chi_2 e^2 \sin\left[2\omega t - 2kz - k_s(L - z)\right]dz$$

Performing this integration gives:

$$E_L(2\omega) = Ac\chi_2 e^2 L\omega \frac{\sin(k - \frac{1}{2}k_s)L}{(k - \frac{1}{2}k_s)L} \sin[2\omega t - (2k + k_s)L]$$

The intensity of the emerging second harmonic will be proportional to the square of amplitude of this, i.e.:

$$I_L(2\omega) = B\chi_2^2 e^4 L^2 \omega^2 \left[\frac{\sin(k - \frac{1}{2}k_s)L}{(k - \frac{1}{2}k_s)L} \right]^2$$

where B is another constant. Now the intensity of the fundamental wave is proportional to e^2, so the intensity of the second harmonic is proportional to the square of the intensity of the fundamental, i.e.:

$$I_L(2\omega) = B'\chi_2^2 I_L^2(\omega) L^2 \omega^2 \left[\frac{\sin(k - \frac{1}{2}k_s)L}{(k - \frac{1}{2}k_s)L} \right]^2 \qquad (3.5)$$

where B' is yet another constant. From this we can define an efficiency η_{SHG} for the second harmonic generation process as:

$$\eta_{SHG} = \frac{I_L(2\omega)}{I_L(\omega)}$$

Note that η_{SHG} varies as the square of the fundamental frequency and of the length of the crystal; note also that it increases linearly with the power of the fundamental.

From equation (3.5) it is clear that, for maximum intensity, we require that the sinc^2 function has its maximum value, i.e., that:

$$k_s = 2k_f$$

This is the *phase matching condition* for second harmonic generation. Now the velocities of the fundamental and the second harmonic are given by:

$$c_f = \frac{\omega}{k_f}, \qquad c_s = \frac{2\omega}{k_s}$$

These are equal when $k_s = 2k_f$, so the phase-matching condition is equivalent to a requirement that the two velocities are equal. This is to be expected, since it means that the fundamental generates, at each point in the material, second harmonic components which will interfere constructively. We encounter the same argument when dealing with coupling between polarization eigenmodes in a 'hi-bi' fiber.

The phase-match condition usually can be satisfied by choosing the optical path to lie in a particular direction within the crystal. It has already been noted that the material must be anisotropic for second harmonic generation to occur; it will also, therefore, exhibit birefringence. One way of solving the phase-matching problem, therefore, is to arrange that the velocity difference resulting from birefringence is cancelled by that resulting from material dispersion. In a crystal with normal dispersion the refractive index of both the eigenmodes (i.e. both the ordinary and extraordinary rays) increases with frequency. Suppose we consider the specific example of quartz, which is a positive uniaxial crystal. This means that the principal refractive index for the extraordinary ray is greater than that for the ordinary ray, i.e.,

$$n_e > n_0$$

Since quartz is also normally dispersive, it follows that:

$$n_e^{(2\omega)} > n_0^{(\omega)}$$
$$n_0^{(2\omega)} > n_0^{(\omega)}$$

Hence the index ellipsoids for the two frequencies are as shown in fig. 3.4(a). Now it will be remembered that the refractive indices for the 'o' and 'e' rays for any given direction in the crystal are given by the major and minor axes of the ellipse in which the plane is normal to the direction, and passing through the center of the index ellipsoid, intersects the surface of the ellipsoid. The geometry (fig. 3.4(a)) thus makes it clear that a direction can be found[1] for which

$$n_0^{(2\omega)}(\vartheta_m) = n_e^{(\omega)}(\vartheta_m)$$

so SHG phase matching occurs provided that

$$n_0^{(2\omega)} < n_e^{(\omega)}$$

The above is indeed true for quartz over the optical range. Simple trigonometry allows ϑ_m to be determined in terms of the principal refractive indices as:

$$\sin^2 \vartheta_m = \frac{\left(n_e^{(\omega)}\right)^{-2} - \left(n_e^{(2\omega)}\right)^{-2}}{\left(n_0^{(2\omega)}\right)^{-2} - \left(n_e^{(2\omega)}\right)^{-2}}$$

Hence ϑ_m is the angle at which phase matching occurs. It also follows from this that, for second harmonic generation in this case, the wave at the fundamental frequency must be launched at angle ϑ_m with respect to the crystal axis and *must have the 'extraordinary' polarization*; and that the second harmonic component *will appear in the same direction and will have the 'ordinary' polarization*, i.e., the two waves are collinear and have orthogonal linear polarizations. Clearly, other crystal-direction and polarization arrangements also are possible in other crystals.

The required conditions can be satisfied in many crystals but quartz is an especially good one owing to its physical robustness, its ready obtainability with good optical quality and its high optical power-handling capacity.

Provided that the input light propagates along the chosen axis, the conversion efficiency ($\omega \rightarrow 2\omega$) is a maximum compared with any other path (per unit length) through the crystal. Care must be taken, however, to minimize the divergence of the beam (so that most of the energy travels in the chosen direction) and to ensure that the temperature remains constant (since the birefringence of the crystal will be temperature dependent).

The particle picture of the second harmonic generation process is viewed as an annihilation of two photons at the fundamental frequency, and the creation of one photon at the second harmonic frequency. This pair of processes is necessary in order to conserve energy i.e.,

$$2h\nu_f = h(2\nu_f) = h\nu_s$$

The phase-matching condition is then equivalent to conservation of momentum. The momentum of a photon of wave number k is given by:

$$p = \frac{h}{2\pi} k$$

and thus conservation requires that:

$$k_s = 2k_f$$

as in the wave treatment.

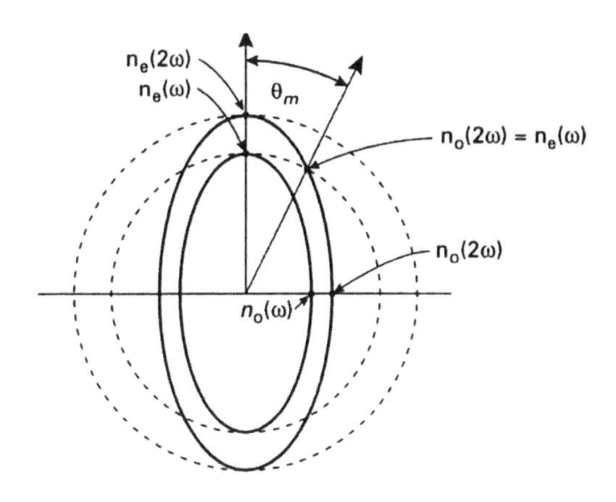

(a) Phase matching with the birefringence index ellipsoids

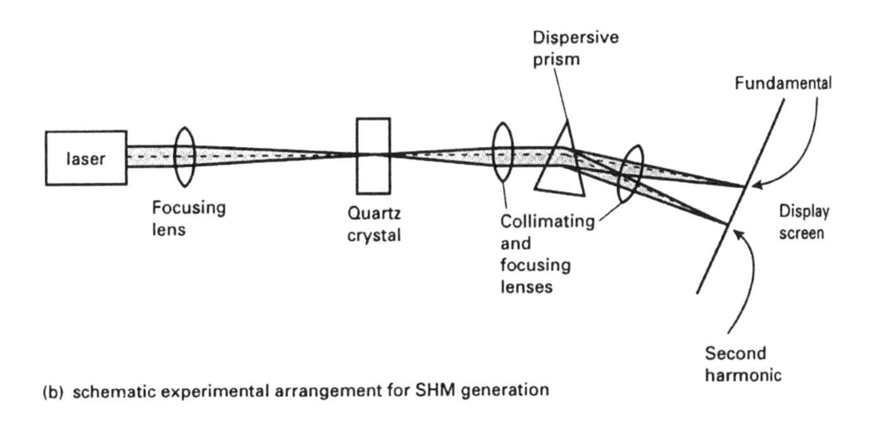

(b) schematic experimental arrangement for SHM generation

Figure 3.4. Conditions for second harmonic generation in quartz

Quantum processes which have no need to dispose of excess momentum are again the most probable, and thus this represents the condition for maximum conversion efficiency in the particle picture.

The primary practical importance of second harmonic generation is that it allows laser light to be produced at the higher frequencies, into the blue and ultraviolet, where conditions are not intrinsically favorable for laser action. In this context we note again, from equation (3.5), that the efficiency of the generation increases as the square of the fundamental frequency, which is of assistance in producing these higher frequencies.

3.5 OPTICAL MIXING

Optical mixing is a process closely related to second harmonic generation. If, instead of propagating just one laser wave through the same nonlinear crystal, we superimpose two (at different optical frequencies) simultaneously along the same direction, then we shall generate sum and difference frequencies, i.e.,

$$E = E_1 \cos \omega_1 t + E_2 \cos \omega_2 t$$

and thus again using equation (3.3):

$$P(E) = \chi_1 (E_1 \cos \omega_1 t + E_2 \cos \omega_2 t) + \chi_2 (E_1 \cos \omega_1 t + E_2 \cos \omega_2 t)^2$$

This expression for P(E) is seen to contain the term

$$2\chi_2 E_1 E_2 \cos \omega_1 t \cos \omega_2 t = \chi_2 E_1 E_2 \cos(\omega_1 + \omega_2)t + \chi_2 E_1 E_2 \cos(\omega_1 - \omega_2)t$$

giving the required sum and difference frequency terms. Again, for efficient generation of these components, we must ensure that they are phase matched. For example, to generate the sum frequency efficiently, we require that:

$$k_1 + k_2 = k_{(1+2)}$$

which is equivalent to

$$\omega_1 n_1 + \omega_2 n_2 = (\omega_1 + \omega_2) n_{(1+2)}$$

where the n represent the refractive indices at the suffix frequencies. The condition again is satisfied by choosing an appropriate direction relative to the crystal axes.

This mixing process is particularly useful in the reverse sense. If a suitable crystal is placed in a Fabry-Perot cavity which possesses a resonance ω_1, say, and is 'pumped' by laser radiation at $\omega_{(1+2)}$, then the latter generates both ω_1 and ω_2. This process is called parametric oscillation: ω_1 is called the signal frequency and ω_2 the idler frequency. It is a useful method for 'down conversion' of an optical frequency, i.e., conversion from a higher to a lower value.

The importance of phase matching in nonlinear optics cannot be overstressed. If waves at frequencies different from the fundamental are to be generated efficiently they must be produced with the correct relative phase to allow constructive interference, and this, as we have seen, means that velocities must be equal to allow phase matching to occur. This feature dominates the practical application of nonlinear optics.

3.6 INTENSITY-DEPENDENT REFRACTIVE INDEX

It was noted in section 3.4, all the even-order terms in the expression (3.1) for the non-linear susceptibility (χ) are zero for an amorphous (i.e. isotropic) medium. This means, of course, that, in an optical fiber, made from amorphous silica, we can expect that $\chi_{(2m)} = 0$, so it will not be possible to generate a second harmonic according to the principles outlined in section 3.4. (However, second harmonic generation has been observed in fibers[2] for reasons which took some time to understand) It is possible to generate a third harmonic, however, since to a good approximation the electric polarization in the fiber can be expressed by:

$$P(E) = \chi_1 E + \chi_3 E^3 \tag{3.6}$$

Clearly, though, if we wish to generate the third harmonic efficiently we must again phase match it with the fundamental, and this means that somehow we must arrange for the two relevant velocities to be equal, i.e., $c_\omega = c_{3\omega}$. This is very difficult to achieve in practice, although it has been done.

There is, however, a more important application of equation (3.6) in amorphous media. We know that the effective refractive index in this case can be written:

$$n_e = (1 + \chi_1 + \chi_3 E^2)^{1/2}$$

and, if $\chi_1, \chi_3 E^2 \ll 1$,

$$n_e \approx 1 + \frac{1}{2}\chi_1 + \frac{1}{2}\chi_3 E^2$$

Hence:

$$n_e = n_0 + \frac{1}{2}\chi_3 E^2 \tag{3.7a}$$

where n_0 is the 'normal', linear refractive index of the medium. But we know that the intensity (power/unit area) of the light is proportional to E^2, so that we can write:

$$n_e = n_0 + n_2 I \tag{3.7b}$$

where n_2 is a constant for the medium. Equation (3.7b) is very important and has a number of practical consequences. We can see immediately that it means that the refractive index of the medium depends upon the intensity of the propagating light: the light is influencing its own velocity as it travels.

In order to fix ideas to some extent, let us consider some numbers for silica. For amorphous silica $n_2 \sim 3.2 \times 10^{-20}$ m^2W^{-1}, which means that a 1% change in refractive index (readily observable) will occur for an intensity $\sim 5\times10^{17}$ Wm^{-2}. For a fiber with a core diameter $\sim$ 5 μm, this requires an optical power level of 10 MW. Peak power levels of this magnitude are readily obtainable, for short durations, with modern lasers.

It is interesting to note that this phenomenon is another aspect of the electro-optic effect. Clearly the refractive index of the medium is being altered by an electric field. This will now be considered in more detail.

3.6.1 Optical Kerr effect

The normal electro-optic Kerr effect is an effect whereby an electric field imposed on a medium induces a linear birefringence with its slow axis parallel with the field (fig. 3.5(a)). The value of the induced birefringence is proportional to the square of the electric field. In the optical Kerr effect the electric field involved is that of an optical wave, and thus the birefringence probed by one wave may be that produced by another (fig. 3.5(b)).

The phase difference introduced by an electric field, E, over an optical path, L, is given by:

$$\Delta\varphi = \frac{2\pi}{\lambda}\Delta n L$$

where $\Delta n = KE^2$, K being the Kerr constant.

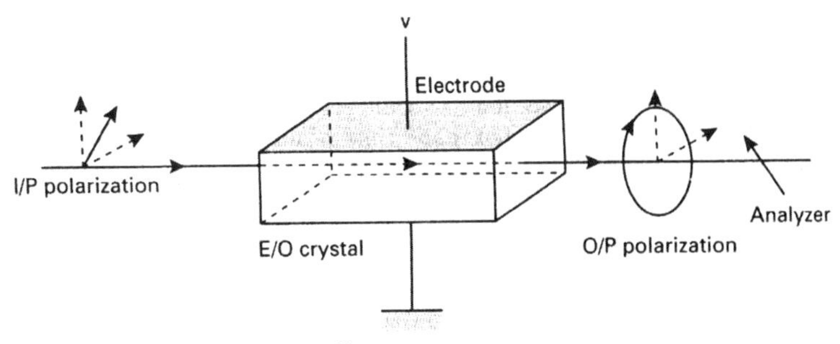

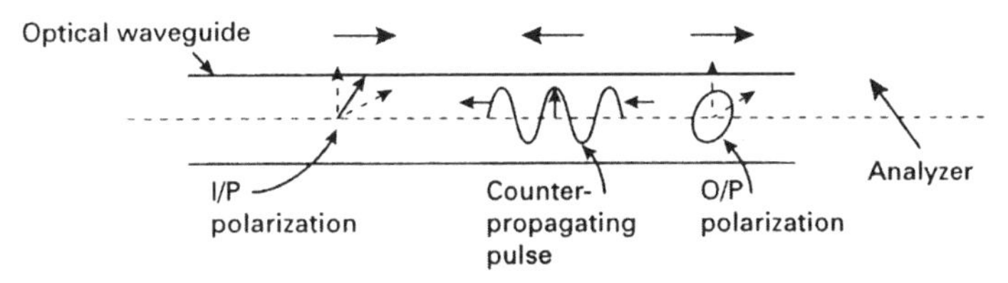

Figure 3.5. 'Normal' and 'optical' Kerr effects

Now from (7a) and (7b) we have:

$$\Delta n = n_2 I = \frac{1}{2}\chi_3 E^2 = KE^2 \tag{3.8}$$

From elementary electromagnetics we have:

$$I = c\varepsilon E^2$$

Hence we have, from (8):

$$K = n_2 c\varepsilon = \frac{1}{2}\chi_3$$

showing that the electro-optic effect, whether the result of an optical or an external electric field, is a nonlinear phenomenon, depending on χ_3. Using similar arguments it can easily be shown that the electro-optic Pockels effect also is a nonlinear effect, depending on χ_2. (Remember that the Pockels effect can only occur in anisotropic media, so that χ_2 will be non-zero).

The optical Kerr effect has several other interesting consequences. One of these is self-phase modulation, which is the next topic for consideration.

3.6.2 Self-phase modulation (SPM)

The fact that refractive index can be dependent on optical intensity clearly has implications for the phase of the wave propagating in nonlinear medium. We have:

$$\varphi = \frac{2\pi}{\lambda} nL$$

Hence for $n = n_0 + n_2 I$,

$$\varphi = \frac{2\pi L}{\lambda}(n_0 + n_2 I)$$

Suppose now that the intensity is a time-dependent function $I(t)$. It follows that φ also will be time dependent, and, since $\omega = \frac{d\varphi}{dt}$, the frequency spectrum will be changed by this effect, which is known as *self-phase modulation* (SPM).

In a dispersive medium, a change in the spectrum of a temporally varying function (e.g. a pulse) will change the shape of the function. For example, pulse broadening or pulse compression can be obtained under appropriate circumstances. To see this, consider a Gaussian pulse (fig. 3.6(a)). The Gaussian shape modulates an optical carrier of frequency ω_0, say, and the new instantaneous frequency becomes

$$\omega' = \omega_0 + \frac{d\varphi}{dt}$$

If the pulse is propagating in the z direction:

$$\varphi = -\frac{2\pi z}{\lambda}(n_0 + n_2 I) \tag{3.9a}$$

and we have

$$\omega' = \omega_0 - \frac{2\pi z}{\lambda} n_2 \frac{dI}{dt} \tag{3.9b}$$

At the leading edge of the pulse $dI/dt > 0$, hence

$$\omega' = \omega - \omega_I(t)$$

At the trailing edge

$$\frac{dI}{dt} < 0$$

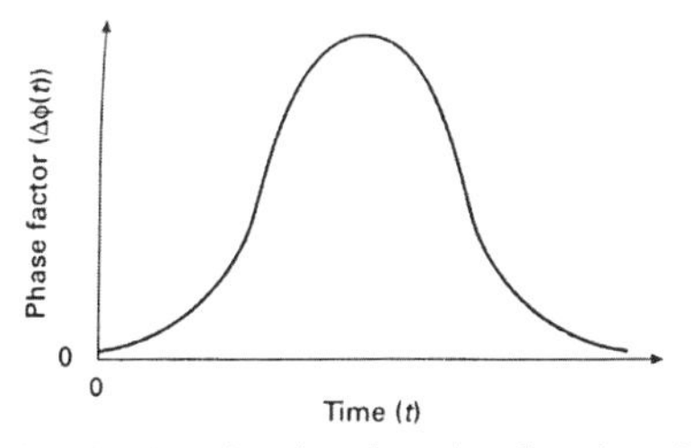

(a) Intensity-dependent phase factor for a Gaussian pulse

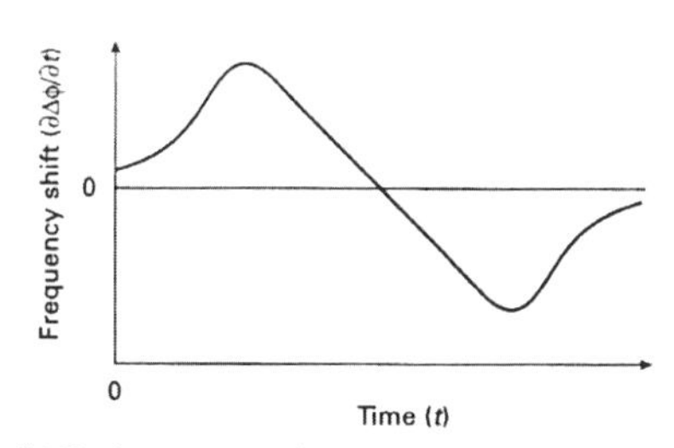

(b) The instantaneous frequency shift for (a)

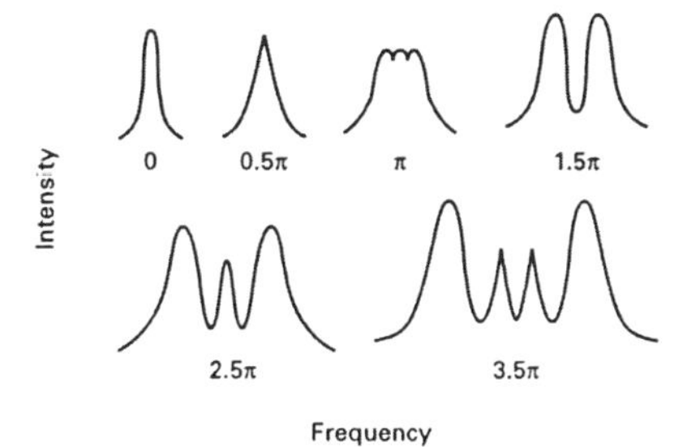

(c) Frequency spectra for (a) designated by maximum phase shift, at peak

Figure 3.6. Self-phase modulation for a Gaussian pulse[6]

and

$$\omega' = \omega + \omega_I(t)$$

Hence the pulse is now 'chirped', i.e., the frequency varies across the pulse. Fig.3.6(b) shows an example of this effect.

Suppose, for example, a pulse from a mode-locked argon laser, of initial width 180 ps, is passed down 100 m of optical fiber. As a result of self-phase modulation the frequency spectrum is changed by the propagation. Fig.3.6(c) shows how the spectrum varies as the initial peak power of the pulse is varied. The peak power will lead to a peak phase change, according to equation (3.9a) and this phase change is shown for each of the spectra. It can be seen that the initial spectrum ($\Delta\varphi = 0$) is due just to the modulation of the optical sinusoid (the Fourier spectrum of a Gaussian pulse) and, as the value of $\Delta\varphi$ increases, the first effect is a broadening. At $\Delta\varphi = 1.5\pi$ the spectrum has split into two clear peaks, corresponding to the frequency shifts at the back and front edges of the pulse. The spectra then develop multiple peaks.

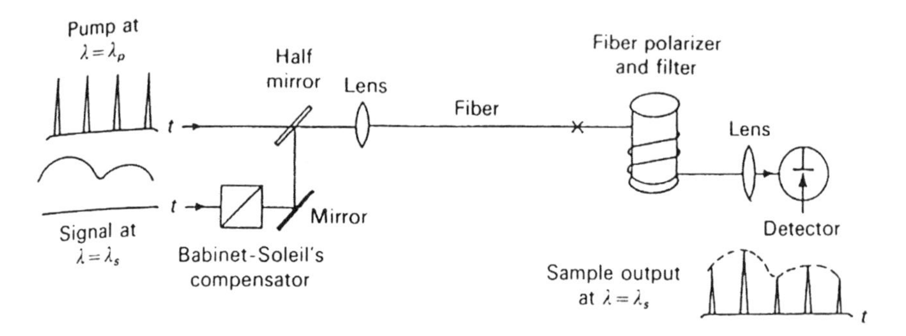

Figure 3.7. Optical sampling by means of a fiber optic Kerr shutter. (After Kitayama et al[16])

It is important to realize that this does not necessarily change the shape of the pulse envelope, just the optical frequency within it. However, if the medium through which the pulse is passing is dispersive, the pulse shape will change. This is an interesting possibility and it will be considered further in section 3.9.

An argon laser power of 1 W was sufficient to induce a phase shift of π (at He-Ne wavelength 633 nm) over a fiber length of 50 m[14,15]. The most important feature of the optical Kerr effect in optical fiber is its very high speed (<1 ps), which has implications for digital sampling (fig. 3.7)[16]. It also means that distributed sensing may be implemented, with its aid, with very high spatial resolution, using counter-propagating beams (fig. 3.8). (Light, in a fiber, travels only 0.2 mm in 1ps). However, the optical Kerr effect in fibers is relatively insensitive to external fields, including temperature (fig. 3.8)[17]. An unwanted consequence of the optical Kerr effect is that of the so-called 'nonreciprocity' in the optical-fiber gyroscope (it is actually not a true nonreciprocity) (fig. 3.9). In the optical-fiber gyro, two counter-propagating CW beams interfere after passing around a fiber coil. It transpires that, via n_2, the phase shifts for the two waves, A_+ and A_-, are given by (fig. 3.9):

$$\Delta\phi_+ = 2\pi n_2 (|A_+|^2 + 2|A_-|^2)\frac{L}{\lambda}$$

$$\Delta\phi_- = 2\pi n_2 (|A_-|^2 + 2|A_+|^2)\frac{L}{\lambda}$$

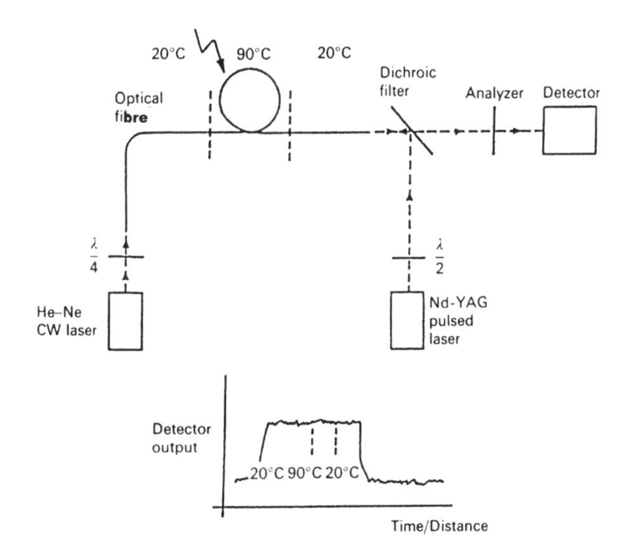

Figure 3.8. Distributed sensing with the optical Kerr effect: (a) experimental set-up; (b) results. (After Dakin[17])

Hence the effect, on a given wave, of the other one, is twice that of the wave on itself. This means that $\Delta\phi_+ = \Delta\phi_-$ only if $A_+ = A_-$. In fact, a difference of only 1 μW leads to a phase discrepancy of ~10^{-6}radian, which is equivalent to a rotation of ~0.2 °h^{-1}. The devices seek to measure ~0.01 °h^{-1}. The problem may be overcome by square-wave amplitude modulation of the waves[18]. Since the cross-effect is now present for only half the time, the factors of 2 in the expressions for $\Delta\phi$ are reduced to unity. Another solution is to use a broadband source. The Gaussian spectrum of a superluminescent diode also reduces the effect to zero.

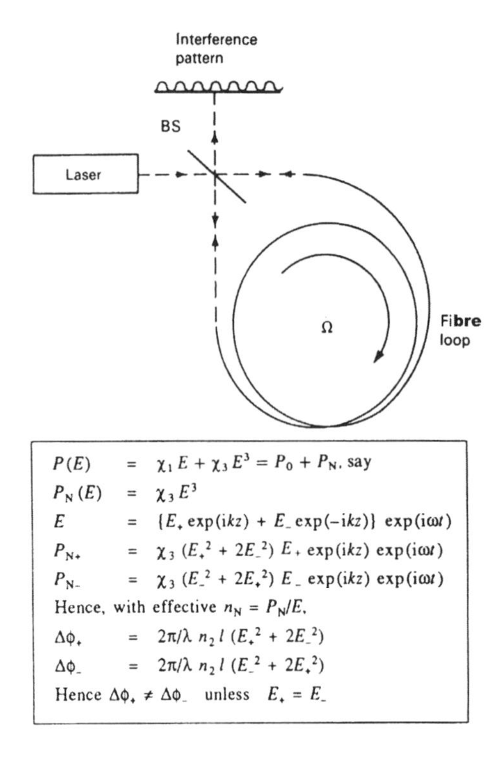

Figure 3.9. Optical-fiber gyroscope characteristics

3.7 FOUR-PHOTON MIXING (FPM)

In sections 3.4 and 3.5, we saw how two photons of certain frequencies could be 'mixed' to generate photons at different frequencies in the processes of second harmonic generation, and of sum and difference frequency generation. These processes were 'mediated' (as we say) by χ_2. In section 3.6 we saw how the electro-optic effect in amorphous media (Kerr effect) was mediated by χ_3. Can χ_3 be used to generate new frequencies? If it can, it would be very convenient if it could do so in amorphous silica because then it could be used with the high intensity, long path lengths associated with optical fibers, and efficient generation could be expected. The two-photon mixing processes considered in the preceding sections relied upon the mixing of two fields, and thus on two photons, in the squared-field second term of equation (3.1): $\chi_2 E^2$. If χ_3 is to be used via the third term, $\chi_3 E^3$, we naturally expect, therefore, that three photons will be involved.

Let us consider three optical waves of frequencies ω_p, ω_s and ω_a and further suppose (for reasons which will soon become clear) that these are related by:

$$2\omega_p = \omega_s + \omega_a \tag{3.10}$$

(Note that four photons are involved here).

It is clear that, under this condition, the term $\chi_3 E^2_{\omega_p} E_{\omega_s}$ will generate a frequency:

$$2\omega_p - \omega_s = \omega_a$$

Hence ω_s and ω_a are continuously generated by each other, with the assistance of ω_p and χ_3. However, as we know very well, this can only take place efficiently if there is phase matching, i.e., if the velocities of ω_p, ω_s and ω_a are all the same.

Interestingly, this can be achieved using high-linear-birefringence fiber. Remember that the two linearly polarized eigenmodes in such a fiber have different velocities. Also remember that there is material and waveguide dispersion to take into account. The result is that, provided that the 'pump' frequency (ω_p) is chosen correctly in relation to the dispersion characteristic and launched as a linearly polarized wave with its polarization direction aligned with the fiber slow axis, then, as in the case of second harmonic generation in a crystal (section 3.4), the combination of the velocity difference in the other polarization eigenmode (fast axis) and the dispersion in the fiber (material and waveguide) can allow two other frequencies, ω_s and ω_a, to have the same velocity in the fast mode as ω_p has in the slow mode, and also to satisfy (10). ω_s is called the Stokes frequency and ω_a the anti-Stokes frequency when $\omega_a > \omega_s$. This process clearly involves four photons (ω_p, ω_p ω_s, ω_a) and hence the name *four-photon mixing* (FPM). (It is sometimes also referred to as 'three-wave mixing', for obvious reasons.)

The process is analogous to that known as parametric down-conversion in microwaves, where it is used to produce a down-converted frequency (ω_s) known as the 'signal' and an (unwanted) up-converted frequency (ω_a) known as the 'idler'. An optical four-photon mixing frequency spectrum generated in hi-bi fiber is shown in fig. 3.10.

Four-photon mixing has a number of uses. An especially valuable one is that of an optical amplifier. If a pump is injected at ω_p, it will provide gain for the signals injected (in the orthogonal polarization of course) at ω_s or ω_a. The gain can be controlled by injecting signals at ω_s and ω_a simultaneously, and then varying their relative phase. The pump will provide more gain to the component which is the more closely phase matched (fig. 3.11).

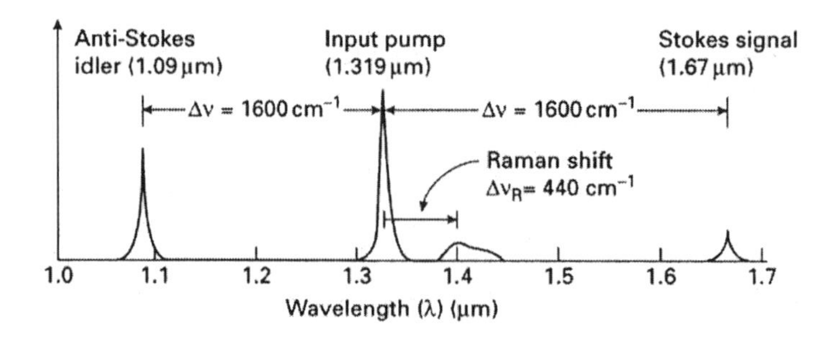

Figure 3.10. Four-photon mixing spectrum in hi-bi fiber[7]

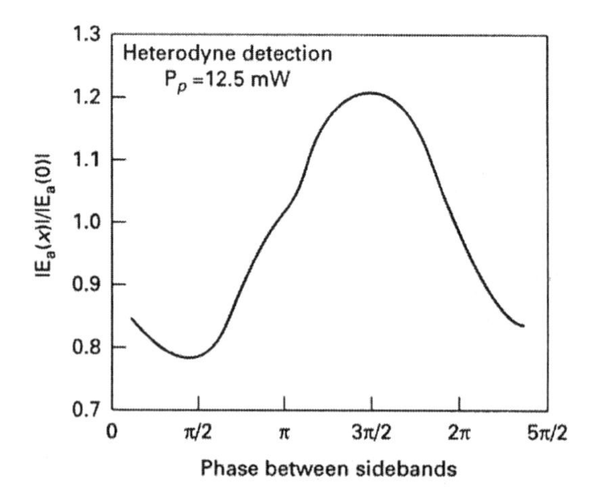

Figure 3.11. Dependence of parametric gain on phase matching in four-photon mixing[8]

Another useful application is that of determining which is the 'fast' and which the 'slow' axis of a hi-bi fiber. Only when the pump is injected into the slow axis will FPM occur. This determination is surprisingly difficult by any other method. By measuring accurately the frequencies ω_s and ω_a, variations in birefringence can be tracked, implying possibilities for use in optical-fiber sensing of any external influences which affect the birefringence (e.g. temperature, stress).

Any field which alters the birefringence of a fiber can be sensed by observing the sideband frequencies generated as a function of time, by a propagating high peak power pump pulse (fig. 3.12). The value of the frequency at a particular time is relatively small, however, and large peak powers would be required. An advantage is the large separation between sideband and pump frequencies; this eases detection problems.

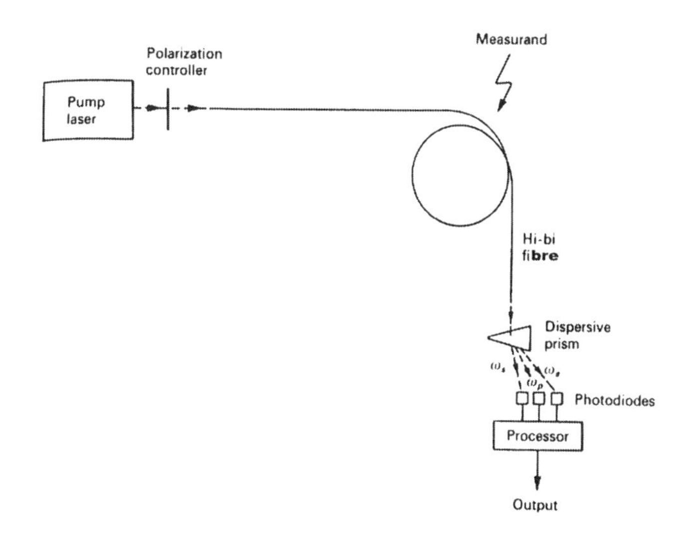

Figure 3.12. Optical-fiber sensor using four-photon mixing

Finally the effects of FPM also can be unwanted. In optical-fiber telecommunications the generation of frequencies other than that of the input signal, via capricious birefringence effects, can lead to cross-talk in multichannel systems (e.g. wavelength-division-multiplexed (WDM) systems).

3.8 PARAMETRIC AND INELASTIC PROCESSES

In the discussion of non-linear optical processes so far we have considered effects due to the nonlinear susceptibility, χ. Now χ is a measure of the ease with which an imposed electric field can separate the centers of positive and negative electric charge, and this separation is almost entirely due to the movement of the electron charge distributions, since electrons are so much more mobile than the positive nuclei. Hence it follows that the nonlinear processes considered so far are the result of the near-instantaneous responses of atomic electrons to fields which push them beyond their linear displacements. These processes are referred to as 'parametric' nonlinear effects, since they effectively rely on the parameter χ.

However, there is another class of nonlinear optical effects in materials, the so-called 'inelastic' class, and it concerns the 'inelastic' scattering of primary, propagating radiation. The word inelastic refers to the fact that the optical energy is not simply redistributed into other optical waves via the 'mediation' effects of the atomic electrons but, in this case, is converted into other forms of energy: heat energy or acoustic energy, for example.

The two best-known inelastic scattering effects are the Raman effect and the Brillouin (pronounced *Breelooah*, after the famous Frenchman) effect. Classically, these two effects are broadly explicable in terms of Doppler shifts. When light is incident upon a moving atom or molecule, the light which is scattered will be Doppler shifted in frequency. If the scatterer is moving away from the incident light, either as a result of bulk movement of the material or as a result of electron oscillation within the molecule, then the frequency of the scattered light will be downshifted; if it is moving towards the scatterer, it will be upshifted. Downshifted, or lower frequency, scattered light is called Stokes radiation; upshifted light is called anti-Stokes radiation. (Hence the designations for ω_s and ω_a in the case of FPM, previous section.)

It should be emphasized, however, that only one photon in $\sim 10^6$ takes part in a frequency-shift Stokes or anti-Stokes process. The vast majority simply re-radiate at the same frequency to give rise to Rayleigh scattering.

When the frequency shift is due to motion resulting from molecular vibration or rotations, the phenomenon is referred to as the Raman effect; when it is due to bulk motions of large numbers of molecules, as when a sound wave is passing through the material, it is called the Brillouin effect.

Of course all such motions are quantized at the molecular level. Transitions can only take place between discrete energy levels, and scattering occurs between photons and photons in Raman scattering, and between photons and phonons (quantized units of acoustic energy) in Brillouin scattering.

We shall deal firstly with the Raman effect.

3.8.1 Raman scattering

When an intense laser beam of angular frequency ω_L is incident upon a material the radiation scattered from the medium contains frequencies higher and lower than ω_L. As ω_L varied, the spectrum of frequencies moves along with ω_L (fig. 3.13). It is, in other words, the difference between ω_L and the spectrum of frequencies which the medium scatters which is characteristic of the medium.

These difference frequencies are just the vibrational and rotational modes of the material molecular structure, and thus Raman spectroscopy is a powerful means by which this structure can be examined. For a given (quantified) vibrational frequency ω_v we have:

$$\omega_s = \omega_L - \omega_v \tag{3.11a}$$

where ω_s is the Stokes (downshifted) frequency and

$$\omega_a = \omega_L + \omega_v \tag{3.11b}$$

where ω_a is the anti-Stokes (upshifted) frequency.

It is often useful to begin with a classical (i.e., non-quantum) explanation of a physical effect (if possible) since this provides our inadequate thought processes with 'pictures' which make us feel more comfortable, but also, and probably more importantly, gives us a better idea of which other physical quantities might influence the effect. The classical explanation of the Raman effect resides in the notion of a variable susceptibility for molecules. The normal definition of susceptibility is:

$$\chi = \frac{P}{E}$$

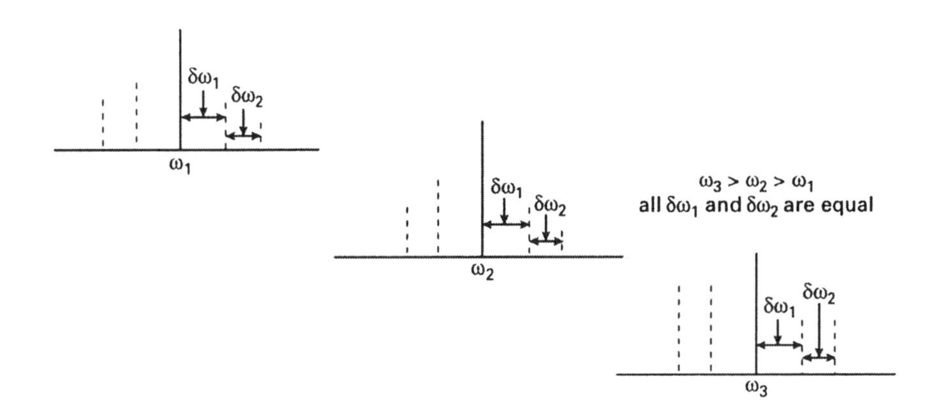

Figure 3.13. Raman spectra at different pump frequencies

(where the ε_0 is again absorbed into χ for convenience). However, we know from equation (3.1) that *P*/*E*, which is a measure of the ease with which positive and negative charges are separated by an electric field, is not a constant, but can be expressed as a power series in the electric field. The fundamental reason for this is, of course, that the force needed further to separate the charges depends, to some extent, on the actual separation. Clearly, if the simple case of positive and negative point charges $+q$ and $-q$ is considered, the force between them is given by:

$$F = \frac{Cq^2}{r^2}$$

(where C is a constant), so the force required for further separation must be greater than this, and hence this force will vary as $1/r^2$.

In the classical picture of a vibrating molecule the distance between the centers of positive and negative electric charge varies sinusoidally, with small amplitude, about a mean value, so that we can expect the volume susceptibility, χ, also to vary sinusoidally, with the same frequency, according to:

$$\chi = \chi_0 + \chi_0' \sin \omega_m t$$

say, with $\chi_0' << \chi_0$

If now, with this molecular oscillation occurring, an optical electric field at frequency ω_e, of the form:

$$E = E_0 \sin \omega_e t$$

is incident upon the molecule, the electric polarization will become:

$$P = (\chi_0 + \chi_0' \sin \omega_m t) E_0 \sin \omega_e t$$

i.e.,

$$P = \chi_0 E_0 \sin \omega_e t + \frac{1}{2} \chi_0' E_0 \cos(\omega_e - \omega_m) - \frac{1}{2} \chi_0' E_0 \cos(\omega_e + \omega_m)$$

The first term in this expression represents normal Rayleigh scattering from the molecule. It is the dominant term, and leads to scattered radiation at the same frequency as the incident optical wave. The other two, much smaller terms give downshifted and upshifted components of P. We know from section 3.4 that a time-varying electric polarization generates radiation at the frequency of variation, so the shifted P-terms lead to the Raman radiation at frequencies ω_e-ω_m (Stokes) and ω_e+ω_m (anti-Stokes).

This simple, classical picture is useful quantitatively in indicating that the scattered radiation will have an intensity proportional to the intensity of the incident optical wave, and to the square of the differential susceptibility, χ_0'.

However, for a proper quantification of the Raman effect it is necessary to return to the quantum description, and we begin this by considering some of the consequences of equations (3.11a) and (3.11b). In (3.11a) a laser photon has interacted with the atom to give rise to a lower frequency Stokes photon. To conserve energy we must have:

$$h\omega_L = h\omega_s + h\omega_v$$

Hence the emission of the Stokes photon must be accompanied by the excitation of the molecule to an excited vibrational state. But how can this be if $h\omega_L / 2\pi$ does not correspond to a molecular transition? To answer this we have to invoke the notion of *virtual* energy levels. A molecule can exist in a virtual level which differs from a real level of the system by an energy $\Delta\varepsilon$, but only for a time $\Delta\tau$, where $\Delta\varepsilon\cdot\Delta\tau \sim h$, the quantum constant.

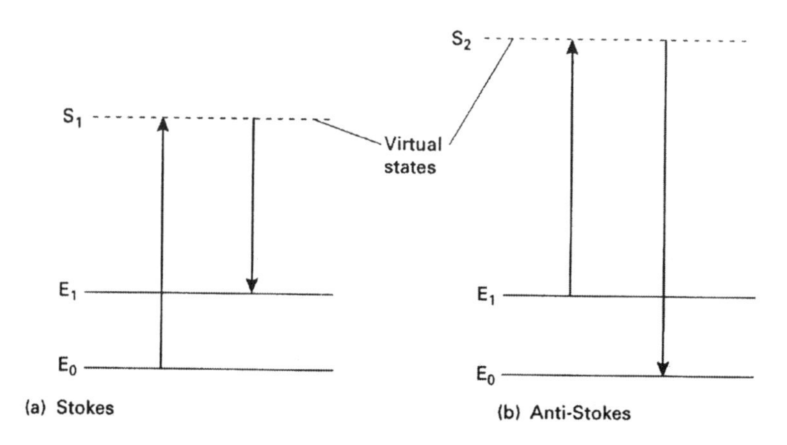

Figure 3.14. Raman energy-level transitions

This is another manifestation of the quantum uncertainty principle. In this case the emission can be explained by means of the diagram shown in fig. 3.14(a). The laser photon raises the molecule from the ground state (E_o) to a short-lived virtual state (S_1) which then decays to an excited state. The anti-Stokes case is shown in fig. 3.14(b). The laser photon now raises an already-excited molecule to an even higher virtual state (S_2) which then decays to the ground state. We now have a satisfactory explanation of the phenomenon in quantum terms.

Of course, in thermal equilibrium, the ratio of the number of molecules in the excited state, E_1, to the number in the ground state, E_0, is given by the Boltzmann factor:

$$\frac{n_1}{n_0} = \exp(-\frac{E_1 - E_0}{kT}) = \exp(-\frac{h\omega_v}{2\pi kT}) \qquad (3.12)$$

and will be very much less than unity. Consequently, the anti-Stokes radiation will be at a much lower level than the Stokes radiation. This is partly offset by the larger scattering efficiency at the higher frequency: the efficiency increases as the fourth power of the frequency. Hence the ratio of anti-Stokes to Stokes radiation levels when the medium is in thermal equilibrium is given by:

$$\frac{I_a}{I_s} = \frac{\omega_a^4}{\omega_s^4} \exp(-\frac{h\omega_v}{2\pi kT}) \tag{3.13}$$

where

$$\omega_v = \omega_L - \omega_s = \omega_a - \omega_L$$

The same phase-matching conditions apply as always. For efficient generation of the Stokes radiation, for example, we must have the velocity of the laser beam equal to that of the Stokes-frequency component and this must be arranged, if dealing with a crystal medium, by choice of a suitable direction, as before.

3.8.2 Stimulated Raman scattering (SRS)

The Raman scattering process we have been considering up to now can be called 'spontaneous' scattering, since it depends, for the generation of Stokes and anti-Stokes radiation, upon the spontaneous decay of the molecule from its virtual states down to the lower states. This will be the case when the relative density of photons at the Stokes and anti-Stokes frequencies is close to the equilibrium levels given by equation (3.12), since one way of regarding a spontaneous decay is that it is 'stimulated' to decay by the photons present within the correct energy range in equilibrium. (And even for an excited, isolated atom there are 'vacuum fluctuations' which perform the same task: it is never possible to say that energy within a given volume, and frequency range, is zero, for this would mean that the energy was known exactly, and there will always be uncertainty of knowledge ΔE, where $\Delta E \cdot \Delta\tau \sim h$).

Suppose now that there are many Stokes photons, caused by a particularly intense laser pumping beam. Then these Stokes photons will stimulate other Stokes transitions from the virtual level to the excited level; this will cause an increase in the excited level population which will then increase the anti-Stokes radiation, which will itself become stimulated. The whole system becomes self-driving and self-sustaining. This is stimulated

Raman scattering (SRS). It occurs quite readily in optical fibers, for all the reasons already explained. Hence, we should now study the Raman effect in optical fibers.

3.8.3 The Raman effect in optical fibers

Optical fibers, as we know, allow long, high intensity, optical path interaction lengths, and thus we would expect the Raman effect, and especially the stimulated Raman effect, to be relatively easy to observe within them. This is indeed the case.

The fiber is made from fused silica, an amorphous medium. Consequently, there is a large variety of vibrational and rotational frequencies among all the varying-strength and varying-orientation chemical bondings. Hence there is a broad spectrum of Raman radiation (fig. 3.15).

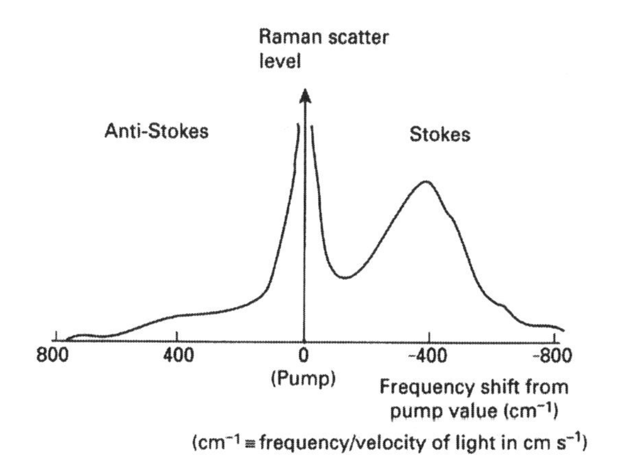

Figure 3.15. Raman spectrum for silica

The fact of the broad spectrum eases the phase-matching restrictions. This is because each of the scatterings occurs with a large measure of phase independence (since the structure is random) and thus the resultant intensity is simply the sum of the separate scattered intensities from the individual centers and this is true for any direction. Hence the Raman scattering process in amorphous silica can quite readily be observed in multimode fibers (in addition to single-mode fibers) since there is no phase-matching condition to satisfy and therefore no requirement for phase coherence from the pump wave: the random contributions from all the modes can add incoherently just as can the contributions from the random scattering centers. The extra advantage of being able to use a multimode fiber is that it is easier to launch into it large levels of pump power, and therefore to generate high levels of Raman radiation.

When the pump laser signal is very intense, as has been noted, it is possible to produce stimulated Raman radiation in the fiber. In this case the Stokes photons are so plentiful that new Stokes photons are more likely to be generated from other Stokes photons, by stimulated emission, than by spontaneous decay following pump excitation. In this case, then, the rate of increase of Stokes radiation is proportional to both the pump laser photon density n_L and the Stokes photon density, n_s. In fact, taking the fiber axis as Oz, we can express the rate at which the Stokes photon density increases as the pump propagates along Oz by a differential equation of the form:

$$\frac{dn_s}{dz} = An_L(n_s + 1) \tag{3.14}$$

where A is a constant for the material, and the '1' is necessary in the third factor (n_s+1) in order to allow the process to begin at the front end, when n_s=0, from the stimulation. The single photon (per unit volume) which the '1' represents will be generated spontaneously.

Of course, it is very quickly the case that $n_s \gg 1$, so that equation (3.14) becomes, for most of the fiber

$$\frac{dn_s}{dz} = An_L n_s$$

the solution of which is:

$$n_s = n_s(z_0)\exp(An_L z) \tag{3.15a}$$

where $n_s(z_0)$ is the value of the Stokes density at that value of z (i.e., z_0) where $n_s \gg 1$ first can be regarded as valid.

Equation (3.15a) is more conveniently expressed in the form:

$$I_s(z) = I_s(z_0)\exp(gI_L z) \tag{3.15b}$$

where I_s is now the Stokes intensity, I_L is the pump intensity and g is called the 'Raman gain'. Since g must be positive (from the structure of the exponent in (15a)), it is clear that the Stokes photon density rises exponentially with distance along the fiber axis. Further, if a beam of radiation at the Stokes frequency is injected into the fiber at the same time as the pump, then this light acts to cause stimulated Stokes radiation and hence is amplified by the pump. In this case:

$$I_s(z) = I_s(0)\exp(gI_L z)$$

where $I_s(0)$ is now the intensity of the injected signal. Hence we have a Raman amplifier. Moreover, since there are no phase conditions to satisfy, pump and signal beams can even propagate in opposite directions, allowing easy separation of the two components.

Finally, if the Stokes radiation is allowed to build up sufficiently, over a long length of fiber, it can itself act as a pump which generates second-order Stokes radiation, at a frequency now given by $2\nu_v$ lower than the original pump. This second-order radiation can then generate a third-order Raman signal, etc. To date, five orders of Stokes radiation have been observed in an optical fiber (fig. 3.16).

Such Stokes Raman sources are very useful multiple-laser-line sources which have been used (among other things) to measure monomode fiber dispersion characteristics.

3.8.4 Practical applications of the Raman effect

It will be useful to summarize some of the uses and consequences of the Raman effect (especially the effect in optical fibers) and, at the same time, to fix ideas by providing some numbers.

Spontaneous Raman scattering will always occur to some extent when an intense optical beam is passed through a material. It provides valuable information on the molecular structure of the material. The spontaneous Raman effect is also used in distributed sensing.

Stimulated emission will occur when there is more chance of a given virtual Stokes excited state being stimulated to decay, than of decaying spontaneously. Stimulated emission will be the dominant propagation when its intensity exceeds that of the pump. For a typical monomode fiber this occurs for a laser pump power ~5 W. This means that the power-handling capacity of fibers is quite severely limited by Raman processes. Above ~ 5 W, the propagation breaks up into a number (at least three) of frequency components. There are also implications for cross-talk in multichannel optical-fiber telecommunications.

The Raman spectrum shown in fig. 3.15 has a spectral width ~40 nm, emphasizing the lack of coherence and the broad gain bandwidth for use in optical amplification. Remember also that it means that amplification can occur in both forward and backward directions with respect to the pump propagation. The broad bandwidth is a consequence of the large variety of rotational-vibrational energy transitions in an amorphous material. The Raman cross-section, and thus the gain, can be enhanced by the use of suitable dopants in the fiber: GeO_2 is a well-known one, and it may be remembered that this dopant is also used to increase the core refractive index in elliptically-cored hi-bi fibers. These latter fibers are thus very useful for Raman applications.

Gains of up to 45 dB have been obtained with fiber Raman lasers. The maximum gain is obtained when both pump and signal have the same (linear) polarization, thus indicating another advantage for the use of elliptically-cored, hi-bi fibers.

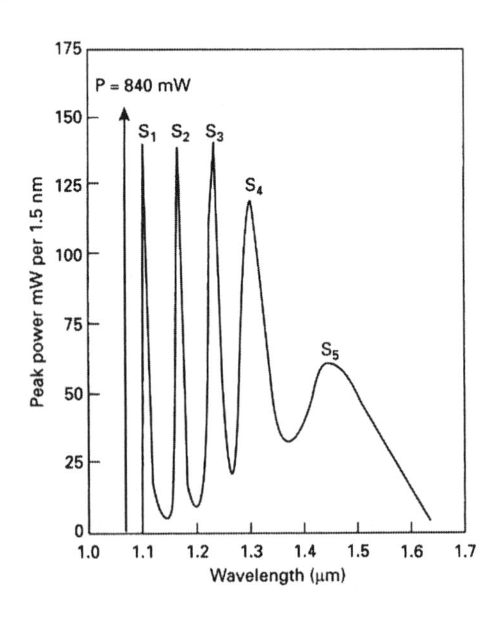

Figure 3.16. Spectrum of multiple-order Stokes emission[9]

By increasing the fiber input pump power to ~1 kW, up to five orders of Stokes radiation can be generated, as explained in the preceding section. Higher orders than five are broadened into a quasi-continuum by the effects of self-phase modulation and four-photon mixing discussed in sections 3.6.2 and 3.7. By placing the fiber within a Fabry-Perot cavity, as for the fiber laser, a fiber Raman laser can be constructed. By tuning the cavity length, this laser can be tuned over ~30 nm of the Raman spectral width (fig. 3.17). This tunability is extremely useful, in a source which is so readily compatible with other optical fibers, for a range of diagnostic procedures in optical fiber technology.

It is possible to measure temperature distributively along an optical fiber, using the Raman effect[19]. As has been stated, the ratio of anti-Stokes to Stokes levels depends only on absolute temperature, so that the time-resolved Raman back-scatter from a pulse propagating along a fiber will reveal the distribution of temperature along it (fig. 3.18). Commercial systems based on this idea now exist [20]. Such systems can measure temperature to ±2°C with spatial resolutions of a few meters over lengths of several kilometers, in times ~1min. Fig.3.19 shows another application of the Raman effect to distributed sensing. In this case the monomode fiber is Raman pumped by a pulse from a tunable Nd-YAG-pumped dye laser, and the populated Stokes levels are stimulated to emit by a CW He-Ne probe beam counter-propagating down the fiber. The probe beam thus receives gain from the pump. The amount of gain received at any point in the fiber depends upon the relative polarization states of pump and probe beam at that point; this, in turn, depends upon the distribution of the fiber polarization characteristics, which can be modified by a stress field [21]. Thus the time dependence of the emergent probe beam power level provides the information necessary to determine the distribution of the stress along the length of the fiber. Figure (3.19(b)) shows the effect brought about by applying a localized stress to one section of fiber.

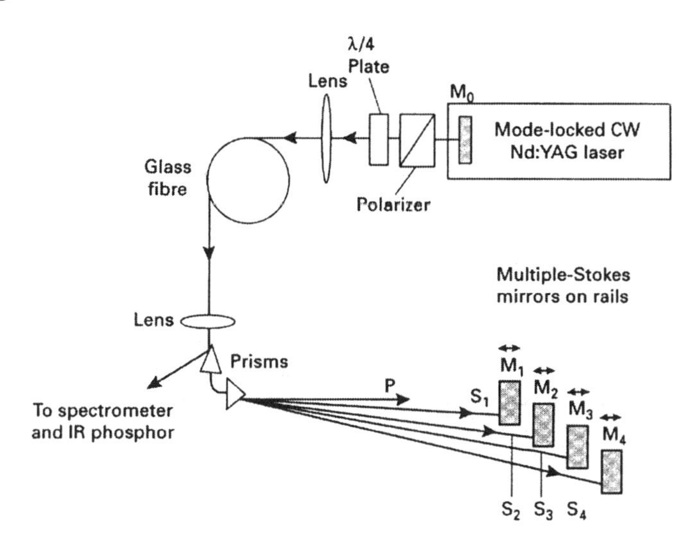

Figure 3.17. Schematic for a fiber-Raman laser[10]

3.8.5 Brillouin scattering

Brillouin scattering is the result of an interaction between a pump laser and the bulk motions of the medium in which it is propagating. The bulk motions effectively comprise an acoustic wave in the medium. The phenomenon can thus be regarded essentially as a Bragg-type scattering from the propagating acoustic wave in the medium.

In quantum mechanical terms the effect can be explained in just the same way as was the Raman effect. It is essentially the same effect, the only difference being that the excitation energy is not now due to molecular vibration/rotation but to the bulk motion. The bulk motion must be quantized, and the relevant quanta are called phonons in this case. The rest of the explanation is the same.

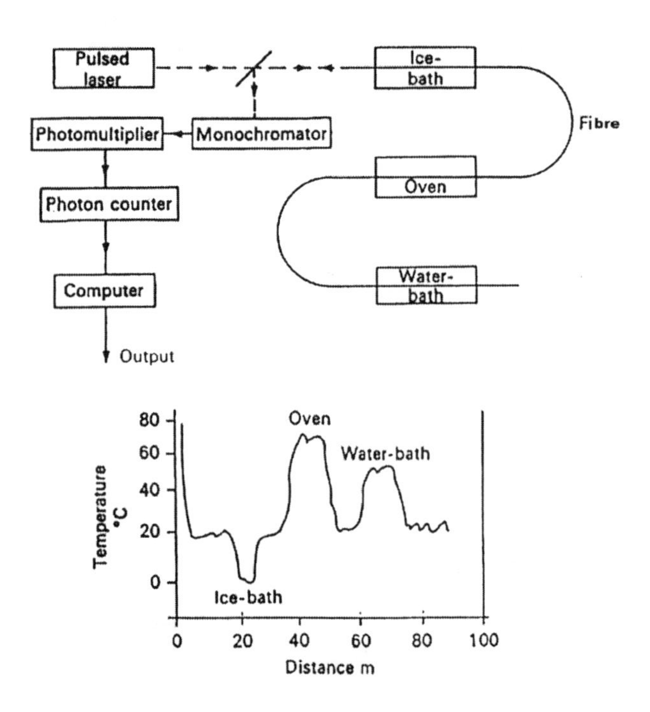

Figure 3.18. Distributed Raman temperature measurement: (a) experimental arrangement; (b) results

However, we are in a different regime and the values of the physical quantities are very different. The strength of the interaction is much greater, the bandwidth is much narrower and, since the medium moves, acoustically, as a coherent whole, phase-matching conditions are now important, even in amorphous media.

In the case of Brillouin scattering, the various phenomena are probably, initially, best understood in terms of the classical interaction between the optical pump (laser) and the acoustic wave in the medium. This is the approach we shall adopt.

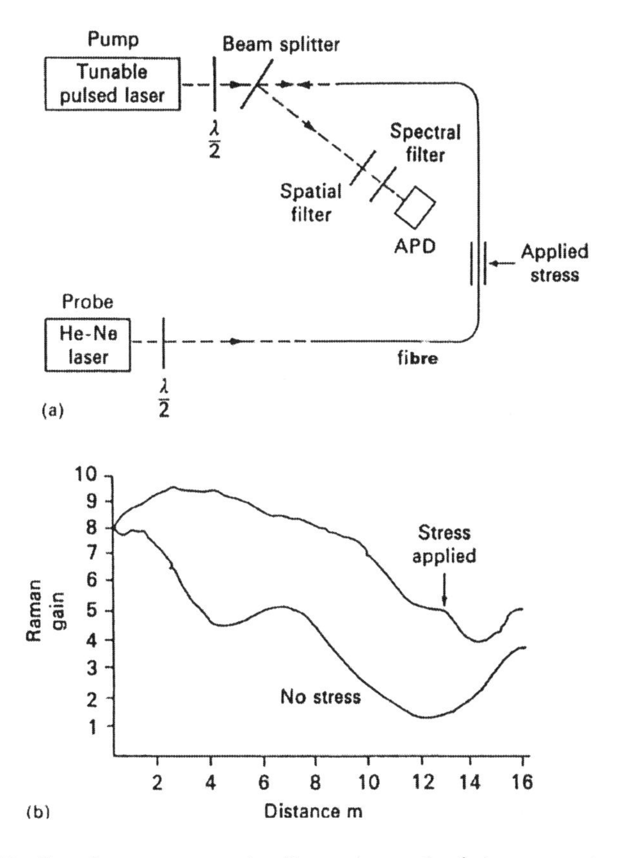

Figure 3.19. Distributed stress sensor using Raman interaction between counter-propagating pulse and wave: (a) experimental arrangement; (b) results

If an acoustic wave propagates in a medium, the variations in pressure give rise to variations in strain which, via the strain-optic effect, give rise to corresponding variations in refractive index in the medium. Sound waves will always be present in a medium at a temperature above absolute zero, since the molecules are in thermal motion, and the consequent dynamic interaction will couple energy into the natural vibrational modes of the structure. Hence a propagating optical wave (a pump laser) will be scattered from these refractive index variations (*photon-phonon* scattering). These effects will comprise spontaneous scattering and, since the acoustic waves are of low energy, this will be at a very low level.

As the power of the pump laser is increased, however, some of its power may be backscattered from an acoustic wave travelling along the same path, either forwards or backwards. The velocity of the acoustic wave will, of course, have a definite value, characteristic of the medium.

Now since the laser radiation is being backscattered from what is, essentially, a moving Bragg diffraction grating, there will be a Doppler shifting of the backscattered optical radiation to either above or below the pump frequency (fig. 3.20)

This Doppler-shifted wave now interferes with the forward-propagating laser radiation to produce an optical standing wave in the medium which, owing to the Doppler frequency difference, moves through the medium, at just the acoustic velocity (this will be proved, analytically, shortly).

The standing wave so produced will consist of large electric fields at the anti-nodes and small fields at the nodes (fig. 3.20).

Now whenever an electric field is applied to a medium, there will be a consequent mechanical strain on the medium. This is a result of the fact that the field will perturb the inter-molecular forces which hold the medium together, and will thus cause the medium to expand or contract. This phenomenon is known as electrostriction, and, as might be expected, its magnitude varies enormously from material to material.

The result of electrostriction in this case is to generate an acoustic wave in sympathy with the optical standing wave. Hence the backscattered wave has generated a moving acoustic diffraction grating from which further backscattering can occur. Then the pump wave and the Doppler-shifted scattered waves combine to produce diffraction gratings which move forwards and backwards at the acoustic velocity; each of the three-wave interactions is stable; the forward acoustic wave producing the Stokes backscattered signal, and the backward wave the anti-Stokes signal. The complete self-sustaining system comprises the stimulated Brillouin scattering (SBS) phenomenon. As always, for an understanding of the phenomenon sufficient to be able to use it, it is necessary to quantify the above ideas.

The Doppler frequency shift from an acoustic wave moving at velocity, v, is given by:

$$\frac{\delta\omega}{\omega_p} = \frac{2v}{c} \tag{3.16}$$

where $\delta\omega$ is the angular frequency shift, ω_p the angular frequency of the optical pump and c the velocity of light in the medium.

For the sum of forward (pump) and backward (scattered) optical waves at frequencies ω_p and ω_B, respectively, we may write

$$S = E_p \cos(\omega_p t - k_p z) + E_B \cos(\omega_B t + k_B z)$$

which gives, on manipulation:

$$S = (E_p - E_B)\cos(\omega_p t - k_p z) +$$
$$E_B[\cos(\omega_p t - k_p z) + \cos(\omega_B t + k_B z)]$$

i.e.

$$S = (E_p - E_B)\cos(\omega_p t - k_p z) +$$
$$2E_B \cos\frac{1}{2}[(\omega_p + \omega_B)t + (k_B - k_p)z]\cos\frac{1}{2}[(\omega_p - \omega_B)t + (k_B + k_p)z]$$

This expression represents a wave (first term) travelling in direction *Oz*, plus a standing wave (second term) whose amplitude is varying as

$$\cos\frac{1}{2}[(\omega_p - \omega_B)t + (k_B + k_p)z]$$

This comprises an envelope which moves with velocity

$$v_e = \frac{\omega_p - \omega_B}{k_p + k_B}$$

but $k_p = \omega_p / c$ and $k_B = \omega_B / c$, so that

$$k_p + k_B = \frac{\omega_p + \omega_B}{c} = \frac{2\omega_p + \delta\omega}{c} \approx \frac{2\omega_p}{c}$$

since $\delta\omega/\omega_p$ is very small.

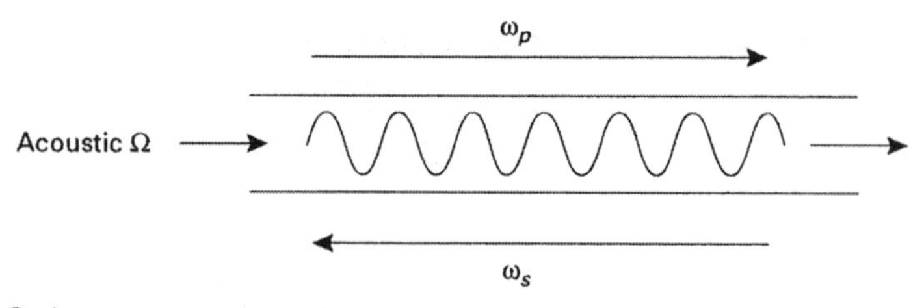

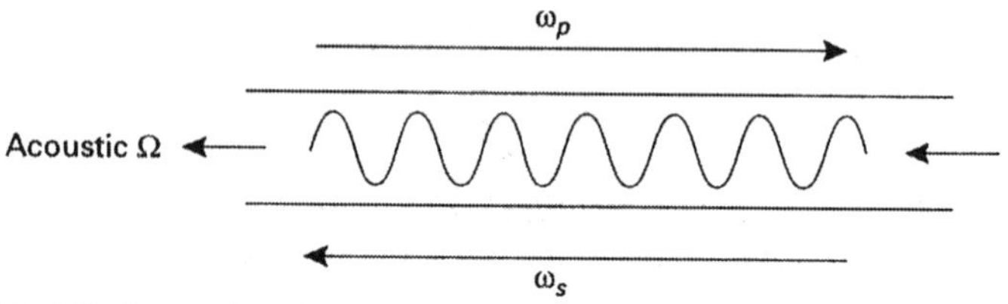

Figure 3.20. Basic Brillouin scatter processes in optical fiber

Thus, $k_p + k_B \approx 2k_p$ and

$$v_e = \frac{\delta\omega}{2k_p} = v$$

(from (16)), and hence the standing wave moves at the acoustic velocity.

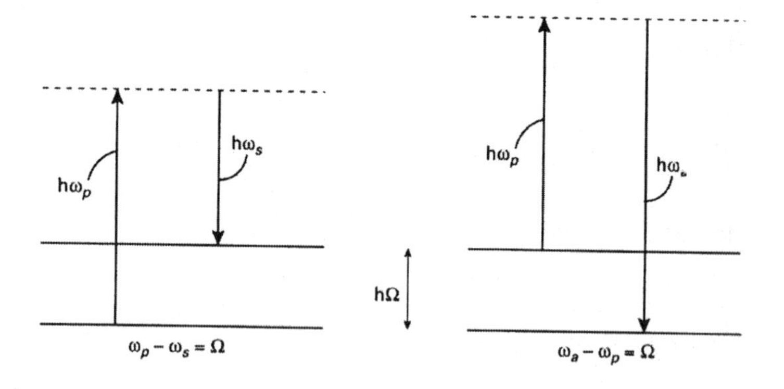

Figure 3.21. Frequency relations for Brillouin scattering

Further, the standing wave will have a distance between successive anti-nodes of $\lambda_p / 2$, the same, to first order, for both directions of propagation, again because $\delta\omega/\omega_p \ll 1$. Hence this (i.e. $\lambda_p/2$) will be the acoustic wavelength.

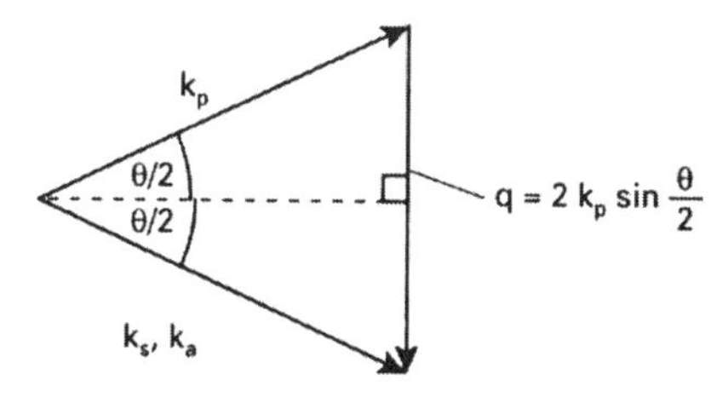

Figure 3.22. Momentum conservation for Brillouin scattering

Let us now look at the matching conditions. Energy conservation requires that (fig. 3.21):

$$\delta\omega = \omega_p - \omega_s = \omega_a - \omega_p = \Omega$$

where Ω is the acoustic angular frequency and the subscripts p, a, s refer to the pump, anti-Stokes and Stokes components, respectively. Momentum conservation requires that:

$$q = k_p - k_s = k_a - k_p$$

where q is the acoustic wavenumber. If ϑ is the angle between the pump propagation direction and that of either of the two scattered waves, we have (fig. 3.22) with $k_a \approx k_s \approx k_p$:

$$q = 2k_p \sin\frac{1}{2}\vartheta$$

Now for an optical fiber, to first order, we have only two possible values of ϑ, i.e., 0 or π. If ϑ=0, q=0 and there can be no acoustic wave. If $\vartheta = \pi$, q = $2k_p$ and the acoustic wave has a wavelength $\lambda_p/2$, which is the case we considered qualitatively. Thus the acoustic wavelength is half the optical pump wavelength. The magnitude of the Stokes or anti-Stokes frequency shift will be given by:

$$\delta f = \frac{\Omega}{2\pi} = \frac{2\mathrm{v}}{\lambda_p} = \frac{2n\mathrm{v}}{\lambda_0}$$

where λ_0 is the free-space optical wavelength, and n is the refractive index of the fiber medium. (This result also could have been obtained from the expression the Doppler shift.) For silica at $\lambda_0 = 1.5$ μm we find that $\delta f \sim$ 11.5 GHz. Hence the frequency shift in the Brillouin effect is some two orders of magnitude smaller than for the Raman effect.

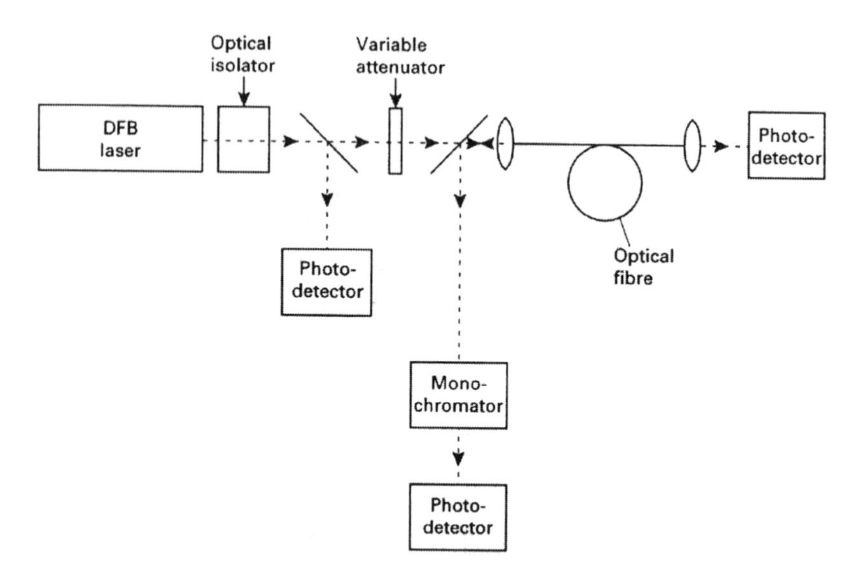

Figure 3.23. Schematic arrangement for observing stimulated Brillouin scatter in an optical fiber[3]

An important feature of Brillouin backscatter in fibers is the optical bandwidth with which it is associated. This bandwidth is determined by the decay time of the acoustic wave, which is about 5ns at these frequencies. Hence for coherent scatter from an acoustic wave, the optical wave also must itself remain coherent for ~5 ns which implies an optical bandwidth ~100 MHz, for optimum Brillouin effect. This is a narrow bandwidth, even for laser sources, so in general only a small fraction of the light from a given source will suffer Brillouin backscatter. This explains why Raman scattering usually dominates: it is able to utilize a much greater fraction of light emitted from conventional laser sources. However, for low dispersion and, especially, for coherent optical communications systems very small optical bandwidths are used, much less than 100 MHz, and Brillouin scattering then becomes the limiting factor on the level of transmitted power. A threshold for SBS has been measured[3] at only 5 mW for a 13.6 km length of fiber, with a 1.6 MHz optical bandwidth from a distributed feedback (DFB) laser (fig. 3.23). SBS can be suppressed by applying a phase modulation to the source light. This effectively destroys the phase coherence necessary for constructive backscatter, and thus allows higher power levels to be transmitted. This has considerable advantages for coherent optical communications systems[3] (figs. 3.24(a) and (b)).

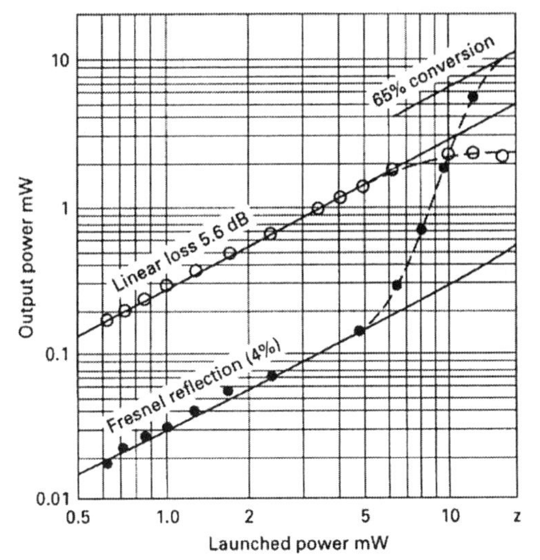

(a) Transmitted and reflected power in SBS: ○ Transmitted
● Reflected

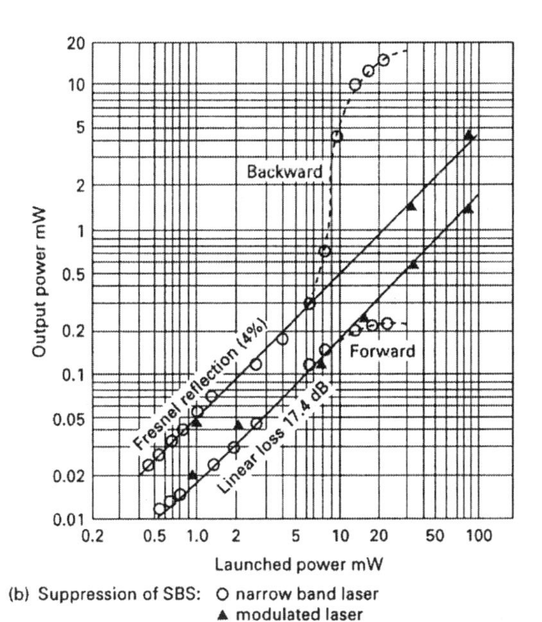

(b) Suppression of SBS: ○ narrow band laser
▲ modulated laser

Figure 3.24. Power dependence and suppression of SBS in an optical fiber[3]

Brillouin amplifiers and oscillators (fig. 3.25) can be constructed much as for the Raman effect. The significant advantages which they possess over those for the Raman effect are that the frequency differences are much smaller (~10 GHz) and they thus allow electronic, rather than optical, filtering and detection techniques, and that the bandwidth of the oscillation and amplification can be very narrow (< 25 MHz). This latter advantage renders Brillouin amplifiers very suitable for amplification of the optical carrier, as opposed to the modulation sidebands, in coherent optical communications systems.

3.9 SOLITONS

No account of nonlinear optics even at this introductory level would be complete without a brief discussion of optical solitons. A soliton is a 'solitary wave', a wave pulse which propagates, even in the face of group velocity dispersion (GVD) in the medium, over long distances without change of form. It thus has enormous potential for application to long-distance optical-fiber digital communications.

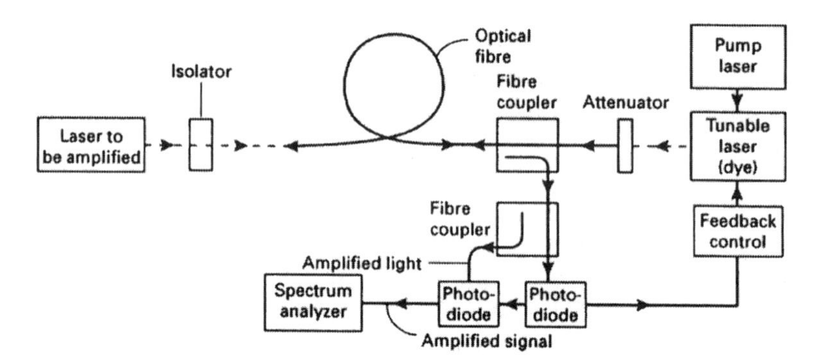

Figure 3.25. Schematic for a stimulated Brillouin amplifier

The soliton is not limited to optics. It comprises a particular set of solutions of the nonlinear Schrödinger wave equation and is a possibility whenever nonlinear wave motion occurs in a dispersive medium. It was first observed in 1834 (before any theory was worked out and certainly long before Schrödinger formulated his wave equation) as a large amplitude water wave propagating along the Union canal which connects Edinburgh and Glasgow, in Scotland. John Scott Russell, a Scottish civil engineer, was exercising his horse alongside the canal in the summer of 1834 when he noticed a wave pulse which had been generated by the sudden halting of a horse-drawn barge. This wave traveled without change of shape or size, and Russell followed it on his horse for two kilometers, noting how puzzlingly stable it was. This turned out to be a water-wave soliton, but solitons were subsequently observed in many other branches of physics, wherever wave motion can occur, in fact, because all restorative systems are capable of being drive into nonlinearity.

The detailed mathematical analysis of this phenomenon is complex, but the basic ideas are relatively straightforward; moreover they follow from ideas which are well known in other contexts: the phenomenon of group velocity dispersion (GVD) where different frequencies already present in an optical pulse will travel at different group velocities in a material medium; and self-phase modulation (SPM) which has been discussed in section 3.6.2 of this chapter. These two ideas must be put together in order to understand solitons.

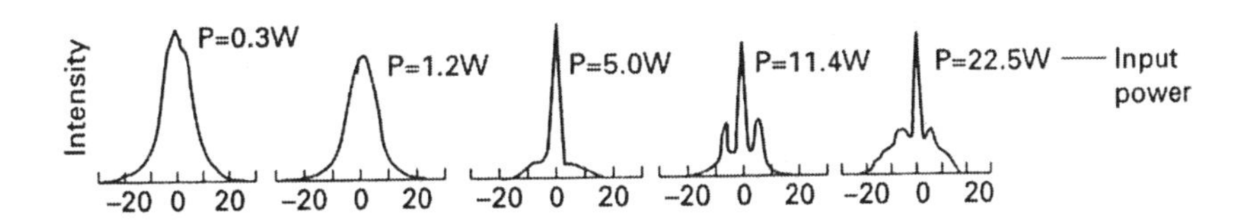

Figure 3.26. Measured solitons emerging from an optical fiber[11]

Let us, as in section 3.6.2, take an optical wave pulse with a Gaussian intensity envelope, and pass it into a dispersive medium. Since the source of the pulse will have a non-zero spectral width, the GVD will act on this pulse to broaden it: the positive GVD will cause the lower frequencies to arrive at the output first, and the negative GVD the higher frequencies first. Clearly, in both cases the pulse is broadened. If self-phase modulation (SPM) also is present, however, we know that, from the consequences of equation (3.9b), higher frequencies are produced at the trailing edge of the pulse, while lower ones are produced at the leading edge. If this effect happens in the presence of negative GVD, then the trailing edge of the pulse will tend to catch up with leading edge: the pulse will be compressed. It is easy to see, that, under certain circumstances, this compressive effect might exactly balance the spreading effect due to the source spectral width, and the pulse width can then remain constant throughout the propagation.

This balance can indeed be struck and the result is a soliton. Fig.3.27 illustrates this process.

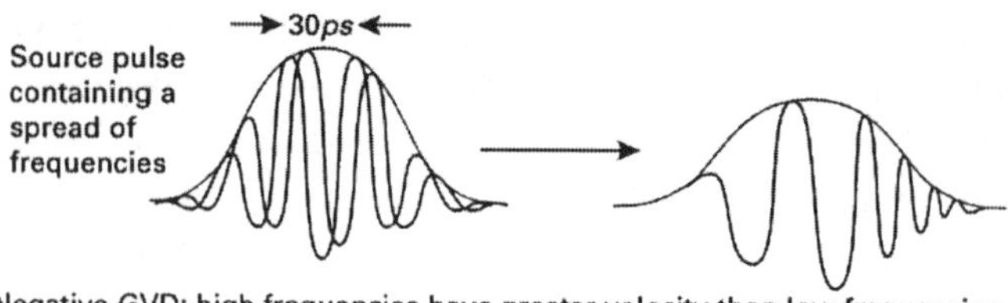

(a) Negative GVD: high frequencies have greater velocity than low frequencies and the pulse is broadened

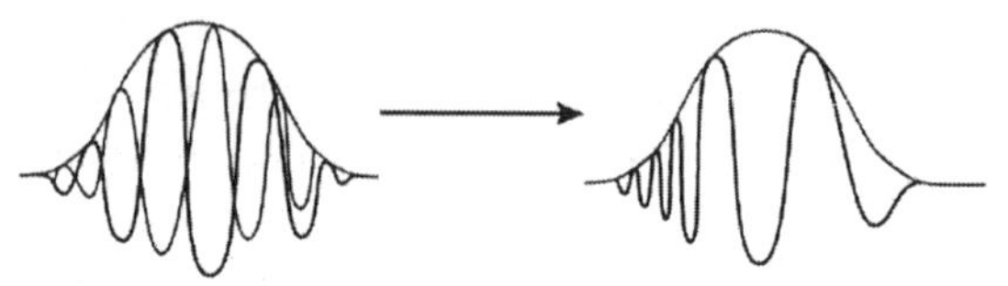

(b) Self phase modulation: high frequencies are 'chirped' to the trailing edge of the pulse

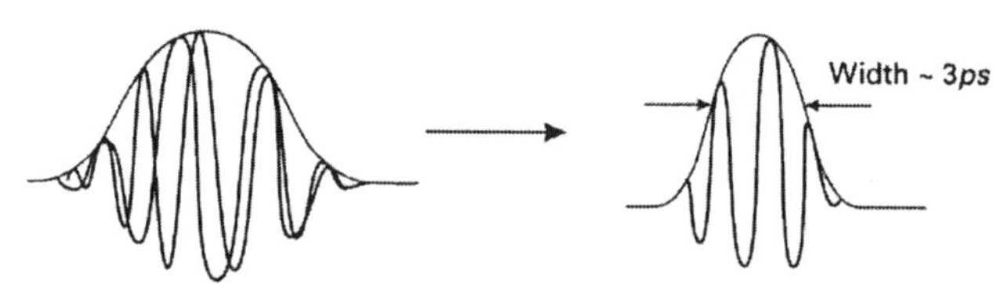

(c) Self phase modulation together with negative GVD: the effects in (a) and (b) balance to generate a stable 'soliton'

Figure 3.27. Essentials of soliton formation

Solitons have been observed in optical fibers (fig. 3.26) and could lead to optical communications systems of phenomenal bandwidth and distance products, perhaps as high as 10,000 GHz km. However, the theory shows solitons to be unstable in lossy media. In addition they tend to attract each other when closer than ~10 pulse widths apart. These features clearly limit their advantages in the communications area. Their potential remains considerable, however, and the research into them undoubtedly will continue.

An interesting corollary to the discussion of the effects which give rise to solitons involves the interaction of a Gaussian pulse with a positive GVD. Fig.3.28 shows the effect, on the spectrum of this pulse, of SPM with a *positive* GVD medium. It is seen there that the result is a much extended region of linear chirp (fig. 3.28(c)) where the frequency varies linearly from the front edge to the back edge of the pulse. If, after this has been done, the pulse is then passed into a purely *negative* GVD medium (e.g. another fiber), the pulse can be very strongly compressed. Pulse widths as small as 8fs (8×10^{-15}s) have been produced, using such a method, at a wavelength of 620 nm. Such pulses contain only about four optical cycles.

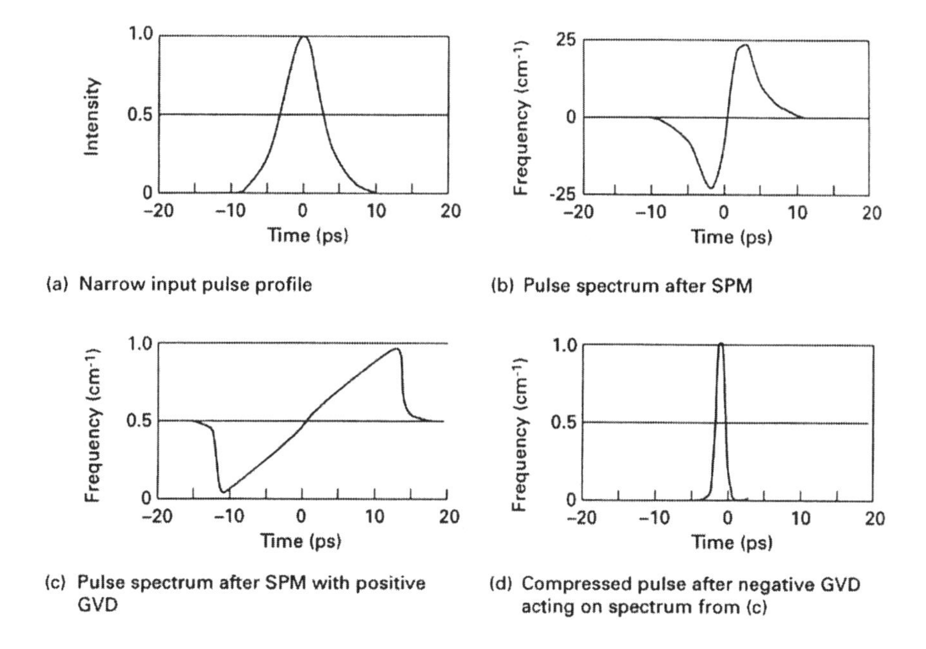

Figure 3.28. Pulse compression using SPM with positive GVD followed by negative GVD

Pulses such as these can be used in research to study, for example, very fast molecular processes, single-atom chemical reactions and very fast switching phenomena.

3.10 PHOTOSENSITIVITY

It is possible for light to bring about changes in a fiber refractive index which persist long after the optical stimulus has ceased. Sometimes the changes are, in fact, permanent (or semi-permanent).

The first such phenomenon was observed in 1978[4]. In this, argon laser light was launched into ~50cm of a Ge-doped monomode fiber, and both the backscattered and transmitted beams were carefully monitored (fig. 3.29). The reflected light level was observed to rise quite steeply over a period of a few minutes, with a corresponding fall in the transmitted level. Investigation revealed that a standing-wave interference pattern had been set up in the fiber, via the light Fresnel-reflected (initially) from the fiber end. This intensity pattern had written, over a period of minutes, a corresponding periodic refractive index variation in the fiber, which then acted to Bragg-reflect the light. Clearly, as the reflection increased in strength, so also did the 'visibility' of the interference pattern: positive feedback was present.

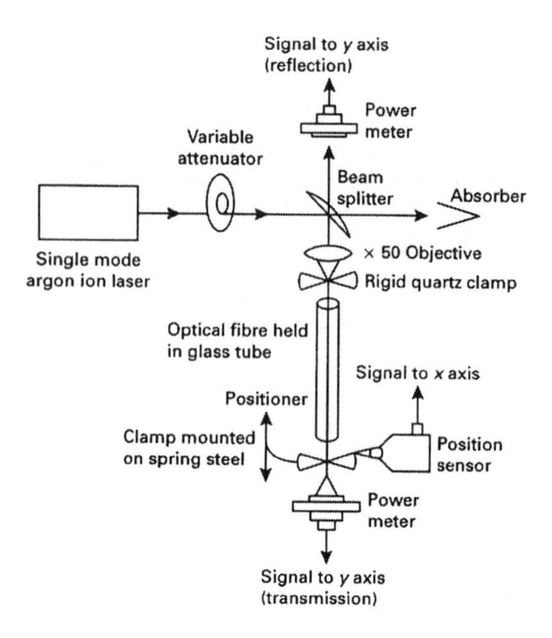

Figure 3.29. Schematic arrangement for recording reflection gratings in an optical fiber[2]

Up to 90% of the light could be backreflected in this way (fig. 3.30). The 'grating' persists when the source light is switched off. Such 'fiber filters' have many potential uses. First, they are spectrally very narrow (~300MHz over 30cm of fiber); secondly, they are tunable over a limited range (e.g., by stretching the fiber or changing its temperature). Interesting polarization signatures may also be written into the fiber using this technique.

A different kind of application involves the dispersion characteristic in the region of the fabricated grating absorption line. This can allow a negative GVD to be generated at a convenient wavelength, so as to provide the possibility of pulse compression without an external delay line. It can also provide the means by which the GVD accumulated over a long optical fiber communication path can be compensated, with consequent annulment of the pulse distortion caused by the GVD, and thus offering a greatly increased communications bandwidth.

Photosensitivity is also thought to be the cause of second harmonic generation(SHG), which has been observed[5] in single-mode fiber exposed to pulsed 1.06 μm radiation from a Nd-YAG laser. SHG should not be possible in an amorphous fiber (as discussed in section 3.6, $\chi_2 = 0$). However, Fig.3.31 shows how the SHG component grew to ~3% of the input power over a 10 hour period. The exact mechanism for this generation is not (at the time of writing) fully understood, although it is thought to be due to a 'poling' (i.e., ordering) of the silica molecules by the optical electric field.

Neither, in fact, is the general nature of the 'writing' process well understood. It is believed to be due to the effect of the germanium atom on the silica lattice in producing lattice 'defects', or 'dangling bonds', which become traps for electrons (F-centers). The refractive index depends upon the occupancy of these traps, and the occupancy depends upon the input light intensity; photons can eject the electrons from their traps. However, there remain some puzzling features, and a fully quantified understanding has yet to be worked out.

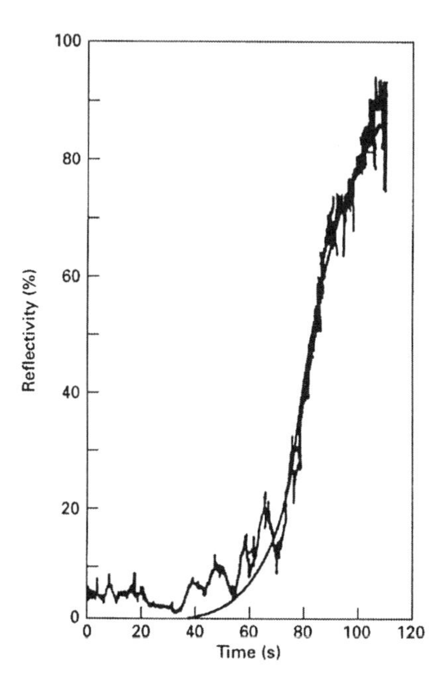

Figure 3.30. Growth of reflectivity with time during formation of a photosensitive grating in an optical fiber[13]

The photosensitivity of optical fibers is very important as the basic mechanism by which Bragg and other grating phenomena can be created in optical fibers [22].

3.11 CONCLUSIONS

We have seen in this chapter that nonlinear optics has its advantages and disadvantages. When it is properly under control it can be enormously useful; but on other occasions it can intrude, disturb and degrade.

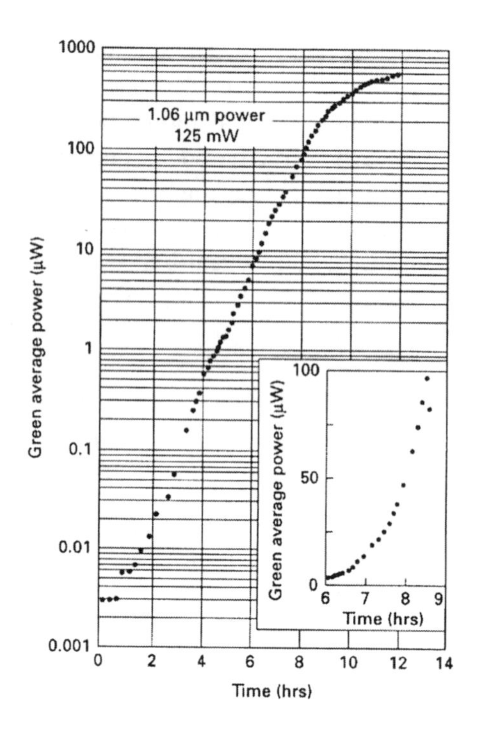

Figure 3.31. Growth of second harmonic component in a monomode optical fiber exposed to laser radiation at 1.06 μm wavelength[5]

The processes by which light waves produce light waves of other frequencies require very high optical electric fields and thus high peak intensities. It was for this reason that nonlinear optics only became a serious subject with the advent of the laser. Optical fibers provide a convenient means by which peak intensities can be maintained over relatively long distances, and are thus very useful media for the study and control of nonlinear optical effects.

We must also remember that in order to cause one optical frequency component to generate another, the second must be generated in phase with itself along the generation path: phase matching is an important feature of such processes.

We have noted the difference between parametric effects (optical interactions are 'mediated' by an optical material) and 'inelastic' effects (the material joins in enthusiastically). And we have seen how the effects which occur when electrons are stretched beyond the comfortable sinusoidal oscillations in their atoms or molecules can yield useful extra optical waves (Raman, Brillouin) and can influence their own propagation conditions (optical Kerr effect, SPM), and how when balancing two well-known effects (SPM, GVD) we can generate the amazing distortion-free solitons. Finally we can also use light to alter, permanently or semi-permanently, the optical properties of a medium, and thus provide the means whereby a new class of optical components, especially fiber components, can be fabricated.

There is a wealth of potential here. The exploration of possibilities for non-linear optics, especially in regard to new, natural or synthetic optical materials (e.g. organics, high T_c superconductors, etc.), has not even really begun. The prospects, for example, for new storage media, fast switching of light by light, and three-dimensional television, which will be opened up in the future by such materials. are intriguing, and it could well be that nonlinear optical technology soon will become a powerful subject in its own right.

3.12 REFERENCES

1. Nye, J. F. (1976) *Physical Properties of Crystals,* Clarendon Press, Oxford, Chap.13.
2. Fujii, Y., Kawasaki, B. S., Hill, K. O. and Johnson, D. C. (1980) Sum-frequency light generation in optical fibres, *Opt. Lett.,* **5**, 48.
3. Cotter, D. (1982) Suppression of stimulated Brillouin scattering during transmission of high power narrowband laser light in monomode fibre. *Electron. Lett.,* **18**, 638.
4. Hill, K. O., Fujii, Y., Johnson, D. C. and Kawasaki, B. S. (1978) Photosensitivity in optical-fibre waveguides, *Appl. Phys. Lett.,* **32**, 647.
5. Osterberg, U., Margulis, W. (1986) Dye laser pumped by Nd-YAG laser pulses frequency doubled in a glass optical fibre, *Opt. Lett.,***11**, 516.
6. Stolen, R. H. and Lin, C. (1978) Self-phase modulation in optical fibres, *Phys. Rev. A,* **17**, 1448-52.
7. Lin, C. et al (1981) Phase matching in the minimum chromatic dispersion region of SM Fibres for stimulated FPM, *Opt. Lett.,* **6**, 493.
8. Bar-Joseph, I. et al (1986) Parametric interaction of a modulated wave in an SM fibre, *Opt. Lett.,* **11**, 534.
9. Cohen, L. G. and Lin, C. (1978) A universal fibre-optic measurement system based on a Near IR fibre Raman laser', *IEEE J. Quant. Elect.,* **QE-14**, 855.
10. Lin, C. and French, W. G. (1979) A near IR fibre Raman oscillator, *Appl. Phys. Lett.,* **34**, 10.
11. Mollenauer, L. F., Stolen, R. H. and Gordon, J. P. (1980) Experimental observation of picosecond narrowing and solitons in optical fibres, *Phys. Rev. Lett.*, **45**, 1095.
12. Kawasaki, B. S., Hill, K.O., Johnson, D. C. and Fujii, Y. (1978) Narrow-band Bragg reflectors in optical fibres, *Opt. Lett.,* **3,** 66.
13. Bures, J., Lapierre, J. and Pascale, D. (1980) Photosensitivity in optical fibres: a model for growth of an interference filter', *Appl. Phys. Lett.,* **37,** 860.
14. Ayral, J. L., Pocholle, J. P., Raffy, J. and Papuchon, M. (1984) Optical Kerr coefficient measurement at 1.15μm in SM optical fibres. *Optics Commun.,* **49**, 405.
15. Dziedzic, J. M., Stolen, R. H. and Ashkin, A. (1981) Optical Kerr effect in long fibres. *Appl. Optics.,* **20**, 1403.
16. Kitayama, K., Kimura, Y., Okamato, K. and Seikoi, S. (1985) Optical sampling using an all-fibre optical shutter. *Appl. Phys. Lett.,* **46**, 623.

17. Dakin, J. P. (1987) Distributed fibre temperature sensor using the optical Kerr effect. *Proc. SPIE,* **798**, *Fibre Optic Sensors 11,* 149-156.
18. Bergh, R. A., Lefevre, H. C. and Shaw, A. J. (1982) Compensation of the optical Kerr effect in fibre-optic gyroscopes. *Optics Lett.,* **7**, 282.
19. Dakin, J. P., Pratt, D. J., Bibby, G. W. and Ross, J. N. (1985) Distributed anti-Stokes Raman thermometry, in *Proceedings 3rd International Conference on Optical Fibre Sensors,* San Diego, February, postdeadline paper.
20. Hartog, A. (2000) Distributed fiber optic sensors: principles and applications in *Optical Fiber Sensor Technology 5*, Eds Grattan, K. T. V. and Meggitt, B. T., 239-300.
21. Farries, M. C. and Rogers, A. J. (1984) Distributed sensing using stimulated Raman interaction in an optical fibre, in *Proceedings 2nd International Conference on Optical-Fibre Sensors,* Stuttgart, paper 4.5, pp. 121-32.
22. Othonos, A. (2000) Bragg gratings in optical fibers: fundamentals and applications in *Optical Fiber Sensor Technology 5*, Eds Grattan, K. T. V. and Meggitt, B. T., 79-186.

3.13 FURTHER READING

23. Agrawal, G. P. (1989) *Nonlinear Fiber Optics*, Academic Press.
24. Boyd, R. W. (1992) *Nonlinear Optics*, Academic Press.
25. Guenther, R. D. (1990) *Modern Optics*, John Wiley and Sons, Chap.15

4

Distributed Fiber-Optic Sensors: Principles and Applications

A. Hartog

4.1 INTRODUCTION

Optical fiber sensors have been researched now for a number of years and a wide body of knowledge has been accumulated, as witnessed by the work reported in the other chapters in this book. Although much of the initial development of these sensors was technology-driven, the most successful examples of fiber sensors are those where one or more of the often-cited benefits of fiber sensors bring a fundamental advantage to a particular application. For example, the fiber gyroscope has been able to compete on cost with the laser gyroscope and yet retain some of the advantages of the latter, e.g. zero spool-up time and complete elimination of moving parts. More generally, certain industries have noted the benefits that all-dielectric sensors could bring, in particular the gas and electricity supply industries, where the removal of electrical sensors has significant and specific advantages. In both cases, these are industries where statutory requirements on safety and security of supply have forced a certain degree of caution in the introduction of new technology.

It is recognized, however, that eventually optical fiber sensors will have to be at least roughly price-competitive with electrical sensors in order to gain widespread acceptance. Single point sensors seldom utilize the very high intrinsic bandwidth of the optical fiber transmission medium and, where a large number of points is to be measured, many fibers must be used to connect the sensor heads and the terminal electronics, thus negating some of the advantages of the fiber transmission medium. There is therefore an opportunity to save cost by multiplexing a number of sensing elements onto a single fiber or fiber pair and thus to form a fiber-optic sensor network distributed in space. A similar trend exists in electrical sensors where

K.T.V. Grattan and B.T. Meggitt (eds.), Optical Fiber Sensor Technology, 241–301.

intelligent, networked, sensors are becoming available. In both cases, the benefits of networking are not simply the derived from sharing terminal equipment: the major cost saving is usually in simplified wiring and reduced installation costs.

One solution to this problem is to multiplex a number of single point optical fiber sensors onto a single fiber in order to transmit their outputs to the terminal equipment. In this case, the sensors can be identical to, or adapted from, some of the well-known single point devices. Techniques successfully used for multiplexing fiber-optic sensors include the use of unbalanced interferometers, wavelength multiplexing and time-of-flight discrimination.

This chapter is concerned with a class of optical fiber sensors which are used to monitor the measurand continuously along the sensing element and are able to provide a continuous reading of the measurand as a function of position along the sensor.

In contrast, traditional single *point* sensors provide a single output denoting the value of the measurand at the sensor (or sometimes the average over a given volume in space). Where it is necessary to measure the spatial distribution of the measurand, in that case with single point sensors, it is necessary to install as many sensors as may be required to achieve an adequate sampling of the distribution of interest and there is always the possibility that a relevant feature of the distribution falls between two sample points and is missed by the sensor array. Of course, the number of points sensed in such schemes is minimized on the grounds of cost and in practice the sampling is thus inadequate in many cases.

The optical fiber distributed sensor has no real equivalent in other sensing media: there are electrical linear heat detectors, but these give only an indication based on the temperature of the hottest point along the cable. The closest equivalents are techniques such as nuclear magnetic resonance imagery, ultra-sonic structure monitoring and weather radar.

Fully distributed sensors have been implemented almost invariably as intrinsic fiber sensors, i.e. here the fiber is the sensor as well as the transmission medium. The sensing principles and implementation are quite different from the methods typically used for single-point devices.

Distributed fiber-optic sensors allow very many points, (typically 10 000 in the case of the York DTS-800, released in 1996) to be measured simultaneously on a single optical fiber. This far exceeds the capacity of other schemes to multiplex fiber-optic sensor outputs. The equipment is thus shared between a large number of measured points and therefore the cost of the equipment per measured point can be very low. This is especially the case with the DTS-800, where up to six fibers can be addressed by a single instrument, increasing the number of measurement points that are accessible by a single instrument to 60 000. In many practical cases, then, the cost of the measurement is dominated by the cost of the fiber and its installation. Here again the distributed approach provides an advantage since only one fiber needs to be installed to monitor the temperature of many places.

In this chapter, we shall review the principles behind the various types of distributed optical fiber sensors (restricted largely to those sensors which are able to determine the spatial distribution of the measurand). Aspects of the engineering of such sensors, of the market requirements (as presently perceived) and examples and practical implementations (i.e. commercially available equipment or instruments which at least undergone field trials) will also be given.

4.2 CLASSIFICATION OF DISTRIBUTED OPTICAL FIBERS SENSORS

The resolution of the measurand at many points along an optical fiber requires a means of identifying that portion of the signal originating uniquely from a given section of fiber. All of the methods proposed to date rely on distinguishing sections of fiber by differences in propagation delay. This may be achieved either by a reflective method where the differences in the transit time from the launching end are used, or forward-propagation approaches, where differences in transit times between modes travelling in the same direction identify positions of mode conversion, where the mode conversion can be measurand-dependent. A further possibility is that of pulse-collision measurements, where two counter-propagating pulses interact only at the point along the fiber where they meet. The position of the collision is determined by the relative timing of the launching of the pulses. In each case, a wide range of modulation schemes, signal derivation and optical arrangements have been proposed.

Almost two decades of active research in the field has led to a plethora of proposed implementations. To help see some structure amongst the vast number of possibilities, it is helpful to separate the implementations under the following headings:

Network architecture: The time of arrival of the signal is normally used to determine the location of a point of interest. For distributed sensors the options are for the architecture to be transmissive or reflective. In multiplexed sensor arrays, the additional topologies allowed include series, parallel or star interconnections.

Origin of the detected signal: This depends on whether the light detected has been transmitted from one end of the fiber to the other, reflected off a well-defined interface, or undergone one of a variety of scattering mechanisms such as:

- Rayleigh scattering
- Inelastic spontaneous scattering (Raman or Brillouin scattering)
- Fluorescence
- Stimulated Raman or Brillouin scattering.
- Conversion into a different guiding channel.
- Reflection by a distributed artefact.

Encoding of the data: whether the modulation by the measurand affects the intensity of the signal, the derivative of intensity with respect to distance, the state of polarization, its frequency, or the timing of any returned pulses.

The interrogation method used: whether it is based on time domain reflectometry, on baseband frequency-domain reflectometry (whether with swept-or stepped frequencies), on direct frequency modulation of the optical carrier (e.g. using frequency modulated, continuous-wave (FMCW) techniques).

In this chapter, the subject is structured according to the methods used for spatial discrimination and to achieve sensitivity to the measurand, examples being given, where appropriate, of relevant results. We shall start with the reflectometric approaches.

4.3 PRINCIPLES OF OPERATION

4.3.1 Optical Time-Domain Reflectometry

The technique of Optical Time-Domain Reflectometry (OTDR)[1] is now used routinely in the optical communications industry for the evaluation of fibers, cables and installed links. An optical time-domain reflectometer typically consists of (figure 4.1) a laser source, a directional coupler connected to the fiber under test, an optical receiver, followed by further electronic circuitry used for data acquisition, signal averaging and processing, measurement control and display of the results.

A short, high-intensity, optical pulse is launched into the fiber and a measurement is made of its backscatter as a function of time. The signal consists of light scattered during the progress of the pulse down the fiber and re-captured by the waveguide in the return direction; it takes the following well-known [1] form as a function of the position z of the scattering element *du*:

$$P_s(z) = P_0 W v_g \alpha_s S \exp\left(\int_0^z -2\alpha(u) du \right) \tag{4.1}$$

where P_0, W and v_g are the power launched, the pulse width and the group velocity, respectively. Here α_s and α are the scattering and total loss coefficients, respectively; i.e. they represent the proportion of the pulse energy lost by scattering mechanisms (or all loss mechanisms) per unit length of fiber and are expressed in Neper/m. Both can be functions of position along the fiber.

S is the capture fraction, i.e. that proportion of the scattered light collected by the optical system. S is given approximately by the following relation between the numerical aperture (NA) of the fiber and its core index n_1, for both single-mode [2] and graded-index multimode fibers [3]:

$$S = 0.25\left(\frac{\mathrm{NA}}{n_1}\right)^2 \tag{4.2}$$

Of course, the measurement position z is determined by the time for the pulse to travel the distance from the launching end to z and return.

In order to use optical reflectometry in a distributed sensor, it is clear from Eq. (4.1) that the scattering loss α_s [4], the local capture fraction *S*, both of which affect the signal directly, or the local fiber attenuation α_s [5], which affects the derivative of the signal, must be functions of the quantity to be measured. Moreover, it must be possible to isolate the effect of any other environmental factors. Results reported to date have involved the measurement of temperature, strain, side pressure [6] and the detection of magnetic fields [7]. The basic OTDR technique has been shown to be capable of detecting, as a function of position along a fiber, changes several of the fundamental properties, including fiber diameter, numerical aperture and scattering loss in multimode and single mode fiber [8-11].

The following examples will give an indication of the signal levels which are present in a distributed optical fiber sensor. In all case a spatial resolution of 1m (i.e. the pulse width, *W*, is just below 10 ns) has been assumed.

Example 1: OTDR powered by a semiconductor laser operating at 850 nm on 50/125 graded-index fiber.

In this case, the power emitted by the source is limited by the available devices to a few Watts peak power (although higher power devices exist, their emitting aperture is significantly higher and the power which can be launched into the fiber is no higher). Allowing for the splitting network (50% transmission) and the launching efficiency from the broad contact, high divergence source, a launched power below 200 mW is usually achieved. The scattering loss in a graded-index fiber, where the index-raising dopant is GeO_2 is approximated by

$$\alpha_s = 0.7(1+66\Delta)\cdot\lambda^{-4}\cdot(2.3\times10^{-4}) \tag{4.3}$$

where Δ is the relative index difference between core and cladding (1% for this type of fiber, λ is the wavelength, expressed in μm and the final factor converts the scattering loss from dB/km to Neper/m. Thus the scattering loss at 0.85 μm is 2.22 dB/km, i.e. 5.12×10^{-4} Neper/m. The capture fraction, S, may evaluated from equation (4.2) as 0.469% (since for these fibers NA=0.20).

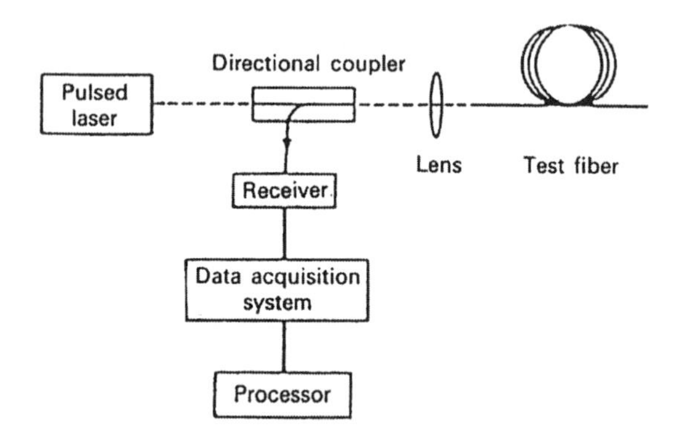

Figure 4.1. Functional diagram of an optical time-domain reflectometer

Thus the near end signal, from equation (4.1) evaluates to about 5 μW. After allowing for losses in the return optics, the optical signal arriving at the detector will be about 2 μW. For every km of fiber addressed, the signal weakens by, typically, 3 dB in both the return and forward directions. Thus for every km of sensing fiber, the signal is reduced by a factor of 4.

Example 2: Raman based distributed temperature sensor operating at 1064 nm.

The sources (e.g. Q-switched diode-pumped Nd:YAG or Nd:glass laser) available at this wavelength are sufficiently powerful that the energy launched is limited by the onset of non-linear effects in the fiber. Assuming the same 50/125 graded-index fiber design, the peak power which can be launched is of order 20 W – depending on the length of the fiber. The pulse energy, for a typical 7ns pulse duration is thus 14 nJ. The Rayleigh backscatter from the near-end amounts to some 1.3 μW, but after filtering losses, the anti-Stokes Raman component reaches the detector weakened by typically three orders of magnitude. The near end signal is thus of order 1 nW and after propagation through 10km of fiber in both directions, the signal available for the temperature measurement is a few pW.

4.3.2 Modulation of the Fiber Loss

According to (1), if the loss of the fiber varies with the measurand of interest, this should be detectable by reflectometry. Thus, all other properties (numerical aperture, scattering loss, fiber geometry) of the fiber being uniform, a change in the fiber loss will show up on the backscatter waveform as a change in the rate of decay of the exponential term in (1) or, on a logarithmic display, as a change of slope.

This has been demonstrated by inserting thin color-glass filters at selected positions in the fiber [12] and more recently [5] in fibers doped with rare-earth ions. In either case, the physical mechanism providing a sensitivity of loss to temperature is the shift in the edges of absorption bands with temperature: as the temperature of the material is increased, the absorption bands broaden towards longer wavelengths. If the wavelength of the probe pulse is arranged to lie on the long-wavelength edge of an absorption band, a significant change (which depends on the ion and host material used) is observed in the absorption loss as a function of temperature.

Other attenuation mechanisms, which are inescapable in a real system, may be separated from the effect of the temperature on loss by referencing the measurement to another wavelength at which the loss insensitive to temperature (i.e. a wavelength well separated from any absorption edges)[12].

Other means of causing the fiber loss to vary according to a measurand include microbending of the fiber, e.g. in response to external pressure. Work along these lines is currently under way in a number of groups (e.g. at the Virginia Polytechnic Institute [13]) notably for investigating the characteristics of composite materials for the aerospace industry. A commercial product was supplied by Ericsson AB (but now withdrawn from the market), in which a standard OTDR was used to interrogate a special cable in which the fiber was spiral-wound with a plastic thread and close proximity with a wax [14]. Upon heating above its melting point, the wax expands rapidly and forces the fiber against the plastic thread thus inducing localized microbending losses. A distributed temperature switch was thus provided, primarily for fire-detection applications; by changing the composition of the wax, a variety of switching temperatures could be selected.

A similar concept, in which the wax is replaced by a hydrogel compound (which swells when wet) has been the subject of research at Strathclyde University, Scotland for moisture detection [15].

Changes in the guidance of the fiber (i.e. its ability to guide light around tight bends, dictated by the design of the core) have also been used. A detailed proposal was made to this effect by Gottlieb and Brandt [16], where glasses of dissimilar thermo-optic coefficient were used for the core and cladding of a single-mode optical fiber. The index-difference (and therefore the field penetration into the cladding) is thus modulated by the temperature of the fiber. The cladding material is, in addition, selected to have a far higher attenuation than the core material. The overall loss of the fiber is thus modulated by its temperature. Similar principles could be applied to the measurement, e.g., of strain by selecting materials with different strain-optical coefficients. It must be noted however that the material constraints imposed by the need for large differential guidance may lead to fundamentally very lossy fibers, which would severely restrict the range of the sensor.

The main drawback of the loss-modulation approach is that the number of sensing points is limited by attenuation induced directly by the measurand: if the fiber is sensitive, its loss will sometimes be high, which will then leave little power to probe the following measurement point. In practice, for temperature sensing, approximately 10 hot-spots can be measured simultaneously, which could be sufficient in a number of applications. In other applications, it is desirable to use approaches which do not require the fiber loss to be high; this is the case when the scattering loss, the capture fraction or the polarization of the light is modulated.

Nevertheless, the loss-modulation approach is very appropriate where it is sufficient for the sensing system to detect the existence, and possibly the location of, a single point of markedly different value from the remainder of the fiber, such as in some fire detection applications, hot-spot monitoring or crack propagation. It has also been used successfully in cryogenic leak detection applications, e.g. in a device developed by British Gas [17] in collaboration with Pilkington and commercialized by the latter company. In this type of cryogenic leak detector, the principle used is again that of the change of guidance: a plastic-clad silica fiber is the sensing medium, the difference in thermo-optic coefficients between core and cladding being so large that the fiber losses all guidance below a certain temperature, where the refractive index *v.* temperature functions for the core and cladding cross-over. Note that this sensor, although it provides a very reliable indication of a leak anywhere along a fiber, does give the location of the leak or any other temperature information.

4.3.3 Polarization Effects

In single-mode fibers, the backscatter signal carries information on the evolution of the state of polarization to the scattering point and back [18-19], since the scattering process in silica fibers preserves, to a large (95%) degree, the state of polarization of the incident light. This approach (polarization OTDR, or POTDR) was first demonstrated in 1980 [18] and has been used [7] to detect magnetic fields via the Faraday effect. The Faraday effect is a non-reciprocal circular birefringence exhibited by an optical element when exposed to a longitudinal magnetic field; it is particularly useful in fiber-optic current sensors since the integral of the magnetic field experienced by a turn of fiber wrapped around a conductor is proportional to the current carried by that conductor. The method was never taken much beyond the initial laboratory demonstration stage owing to considerable difficulties in separating the information of interest from a number of spurious effects which mask the desired signal. In recent years, however, concern in the telecommunications industry for the effects of Polarization Mode Dispersion (PMD) has led to renewed interest in the technique and further theoretical developments which might serve to advance its application to distributed sensors.

POTDR has the merit that it can be used for sensing several different measurands, since the birefringence of the fiber (which dictates the state of polarization) may be affected by temperature, side pressure, axial strain, magnetic and electric fields. The practical difficulties associated with POTDR are, of course, that all of these external influences affect the output simultaneously and it is difficult to separate the wanted output from spurious signals. Moreover the signals obtained, even under ideal conditions, are complex and require considerable computerized signal processing to extract the desired information. In order to retrieve the birefringence information along the fiber, it is necessary to measure six separate backscatter waveforms using different launch and analysis polarization optics. These practical difficulties have, to date, precluded its use in real applications. However, research has continued to this day, and the measurements of side pressure and temperature have been demonstrated [6].

POTDR in a polarization-maintaining fiber under special launching conditions does however provide a high-frequency beat signal which is characteristic of the local linear birefringence of the fiber. The polarization-preserving nature of the fiber simplifies the signal output considerably since, given the correct launching conditions, the fiber can be understood in terms of linear birefringence only. However, the signal is now at a very high frequency (typically above 10 GHz) and is thus difficult to detect and amplify with low noise.

4.3.4 Numerical Aperture Effects

According to equation (4.2), the scatter capture-fraction is a direct and strong function of the numerical aperture of the fiber. If a fiber could be designed in such a way that its NA was a well-behaved function of the measurand, then this mechanism can be used, in conjunction with OTDR to form a distributed sensor. The way in which such a sensor operates differs substantially according to the type of fiber in question.

In the simplest case of a step-index multimode fiber, the numerical aperture at the point of interest z_0 may be thought of as the parameter controlling the size of the cone of scattered light re-captured by the waveguide in the return direction. The larger the NA, the larger the cone of acceptance is and the greater the local fraction of the scattered light captured. However, we must remember that this captured light must travel back to the launching end. A step index fiber preserves the angles of the rays launched into it (subject to any mode-mixing due, for example to bends in the fiber) and the size of cone accepted by the fiber should be readable back at the launching end.

However, if, during the return trip, the NA decreases owing to a different state of the measurand locally at, say, z_1, then the highest order modes (those corresponding to rays travelling at the sharpest angle with respect to the fiber axis) will be stripped out. The backscatter intensity from z_0 will thus be representative of the lowest NA encountered along the whole fiber to z_0. This is therefore not particularly useful unless the measurement requirement is only to detect a condition corresponding to a minimum NA.

In contrast in multimode graded index fibers, a change in the fiber numerical aperture can be detected using backscatter techniques [10] because at a reduction in NA, power is converted from low-order modes to high modes and the converse occurs at an increase in NA. By fitting a suitable mode filter at the launching end (essentially a section of fiber of low NA and relatively narrow core diameter), it has been shown [10] that it is possible to retrieve the NA information (subject, again, to no mode mixing taking place along the fiber).

In single-mode fiber, the capture fraction also depends on the NA, but in this case, provided that any changes of NA are adiabatic (i.e. sufficiently gradual that the mode can adapt to the new waveguide without loss), there is complete independence between the NA at all points along the fiber. In particular, there is no requirement to avoid mode conversion (an issue of serious practical importance) nor do points upstream from the point of interest influence the reading at the point of interest. Thus a single-mode fiber implementation based on NA changes would be quite feasible provided that the NA can be made adequately sensitive to the measurand.

In order to modulate the NA, it is necessary for the difference in the refractive indices of core and cladding to vary and clearly it is desired that this should occur in a well-understood manner in response to a certain measurand. For example, if the thermo-optic coefficients between core and cladding are radically different, a means of sensing temperature is achieved. Equally, differences in stress-optic coefficients would produce a distributed strain sensor. In practice, no all-glass fibers have been reported based on these principles, presumably owing to the absence of adequately strong effects in high-silica glasses. However, the cryogenic leak detector described above is an example of a sensor where induced local attenuation is caused by NA changes. In this case, the NA change is caused by the difference between the temperature coefficients of the refractive index of silica forming the core, and that of the plastic material forming the cladding. Equally, in the liquid-core sensor to be described in the following section, there is a very strong NA effect, owing to the difference in the temperature coefficient of the liquid forming the core and that of the surrounding silica. Both this sensor and the cryogenic leak detector are based on large core step-index multimode fibers and thus suffer from the effect described above, namely that only the minimum NA along a particular stretch of fiber may reliably be detected. Thus, in the case of the liquid core sensor, a different effect was used and the NA effect eliminated.

4.3.5 Modulation of the Scattering Loss

Given the limitations of implementing distributed sensors based on mcasurand-modulated loss effects, the absence of materials suitable for using numerical aperture changes and the difficulty in making POTDR measurand-specific, the next option to be examined is the possibility of modulating the scatter loss. The attraction of the modulation of the scattering loss is that the signal processing is in principle simpler than in the case of POTDR, but the number of points which can be addressed is far higher than in the case of total loss modulation. The underlying reason for this is that the power lost to the scattering mechanism contributes directly to the signal for the sensor input.

The first distributed fiber-optic sensor [4] to be demonstrated, shown in fig. 4.2 used a special fiber having a liquid core and utilized the temperature variation of the scattering coefficient of this liquid core to form a temperature sensor.

The fibers were manufactured by drawing hollow silica tubes into capillaries of approximately 300 μm outer diameter and 200 μm bore. A thin coating of polyimide (a dark varnish capable of withstanding temperatures well above 350°C) was applied during the drawing process. In addition to forming a protective layer, the polyimide acted as a cladding-mode stripper. Prior to the drawing operation, a thin layer of ultra-high purity silica was deposited on the inside wall of the tube (which would eventually form the cladding of the fiber) to ensure that higher modes (whose evanescent field extends further into the cladding than low-order modes) have similar attenuation to that of the lower order modes. The deposited layer thus minimized differential attenuation between the modes. The guiding structure of the fiber was formed by filling the tube with ultra-pure hexa-chloro-buta 1,3 diene, a liquid having a higher refractive index than silica and no absorption bands in the wavelength region of interest. Lengths of several hundred meters could be filled in less than one hour using a high pressure syringe. It was necessary to take stringent precautions to ensure that no bubbles or particles were allowed into the fiber core.

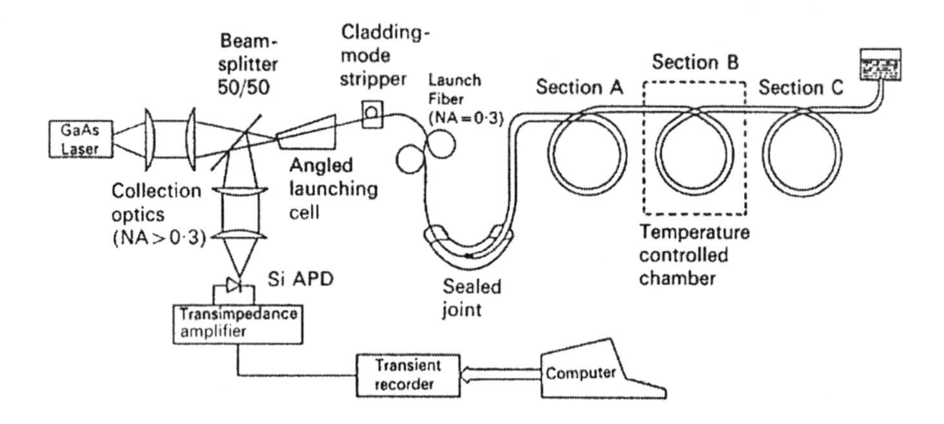

Figure 4.2. Experimental arrangement for the distributed temperature sensor

The resulting fibers exhibited low-loss (13 dB/km at 900 nm) and large numerical aperture (of order 0.5). The loss is entirely dominated by scattering in the liquid core, the Rayleigh scattering coefficient in a liquid being much higher than that of silica at the same wavelength. The high numerical aperture of the fiber, combined with its large Rayleigh scattering coefficient, result in a backscatter factor more than one order of magnitude greater than would be found in a typical telecommunications-grade multimode fiber at the same wavelength. The refractive index profile of these fibers is of course a perfect step-index profile, their dimensions corresponding to the acceptance of several thousand modes.

Backscatter waveforms were measured using a standard OTDR arrangement, based on a 904 nm GaAs laser, an arrangement shown in fig. 4.2. The special arrangements required to deal with the termination and launching into liquid-core fibers may be seen schematically in the figure.

Backscatter waveforms, which were acquired with one section held at various temperatures in the range 13 to 71°C whilst surrounding regions remained at room temperature, are shown in fig. 4.3 and reveal how the local backscatter factor increases with increasing temperature. The sensitivity of the backscatter signal to temperature is caused directly by the increase in scattering exhibited by the core material as the temperature is raised, owing to a greater molecular agitation at the higher temperature, the relative sensitivity of the scattered signal being of order 0.5% $°C^{-1}$.

Referring to equation (4.1), an increase in the scattering loss alters the multiplicative term as well of course as the total loss which appears in the exponential part of the expression. Compensations for changes in fiber loss were implemented from the assumption that the loss is dominated by the scattering and can thus be derived from the integral of the measured temperature up to the point to be corrected for fiber loss.

As noted in the previous section, the fibers also exhibited modulation of the numerical aperture (since refractive index of the core liquid changes with temperature). It is necessary to eliminate this NA effect, since it acts in the opposite direction from the effect of scattering loss changes and could result in unpredictable results. The effect was eliminated in [4] by means of a mode filter which (as shown in fig. 4.2) was implemented by inserting, into the launching end of the liquid core fiber, a fiber of lower NA than that of the liquid core fiber. This has the same effect as an angular stop, but is self-aligned and also forms a launching cell. This filter rejects the highest order modes whose characteristic angle falls between the acceptance angle of the two fibers. Provided that the launch fiber has an NA lower than that found anywhere along the fiber (i.e. lower than the NA at the hottest point along the fiber), a varying NA in the liquid core fiber cannot modulate the backscatter signal by truncation of the acceptance angle. A second-order residual effect of the change in core refractive index does subsist [4]; however this now adds to the scattering loss sensitivity and is therefore beneficial. Typical signals received are shown in fig.4.3.

Analysis, confirmed by a few experimental results, showed that the liquid-core sensor using a semiconductor laser is capable of resolving around 0.1°C over a distance of 100 m with a spatial resolution of 2.5 m after averaging 1000 pulses. Similarly a resolution of 1°C, with 1 m spatial resolution over 100 m of fiber is achievable with 1000 pulses, a measurement time well below 1 s.

The performance achieved at that time in liquid-core fibers is still the best that has been reported to-date (in term of the number of pulses averaged for the temperature resolution achieved at a given spatial resolution) and is attributable to the large signals available in this type of fiber and the quite reasonable temperature sensitivity of the medium. However, the approach is out of favor since liquid-filled fibers are inconvenient to work with, have limited temperature range and unproved life-times. The sensitivity of the scattered signal to temperature in glasses is orders of magnitude lower and different means are thus required for solid fibers.

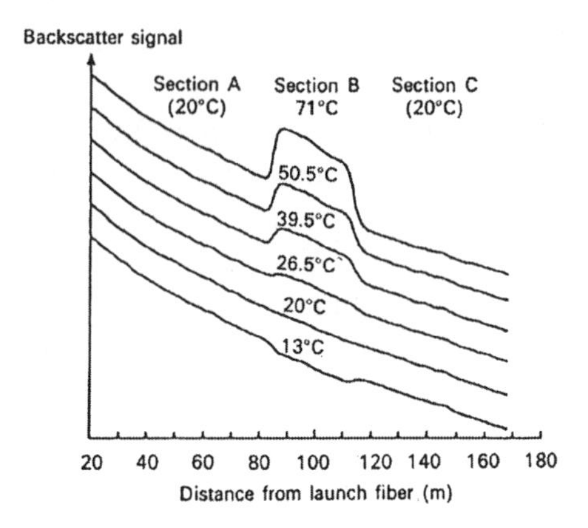

Figure 4.3. Averaged back-scatter waveforms measured in the liquid-core fiber with the temperature distributions indicated

Similar measurements, when performed in solid core fibers, result in measured sensitivities to temperature two orders of magnitude lower than in liquids.

4.3.6 Inelastic Scattering

The scattering coefficient in all-solid optical fibers is caused principally by Rayleigh scattering and is attributable to density and composition fluctuations 'frozen-in' to the material during the drawing process [20]. This type of scattering is largely independent of ambient temperature provided that the thermo-optic coefficients of the fiber constituents are similar.

However, the scattered light spectrum also contains a small contribution to the scattered power from Raman and Brillouin spectral lines which originate in thermally driven molecular and bulk vibrations, respectively. The difference in wavelength between these lines and the incident wavelength reflects a gain or loss of energy by the scattered photon in its collision with the thermally generated phonons. The intensity of these spectral lines is temperature sensitive and the finite sensitivity of the total scattered signal in solid fibers is largely attributable to the contribution of Brillouin and Raman scattering. By selecting only one of these parts of the scattered light spectrum, the sensitivity of the measured signal to temperature can be greatly enhanced. These contributions to the scattered light spectrum are *linear*, in that their intensity is proportional to the incident light intensity, but *inelastic* in that the photon energy is not maintained in the scattering process.

In practice, the Brillouin lines are shifted by only a few tens of GHz from the incident radiation frequency, the exact frequency shift being determined *inter alia* by the wavelength of the source, the material forming the fiber core and the fiber temperature. This makes heavy demands on the linewidth and frequency stability of the source and optical filter and it is only recently that the Brillouin scattering approach has been demonstrated and it has not yet been used in commercial sensors. However, we shall see later in this Chapter that very exciting results have been obtained using stimulated Brillouin scattering and more recently still, using the spontaneous Brillouin emission. In contrast, the Raman spectrum is sufficiently displaced from the incident wavelength that it can readily be separated by means of standard multilayer dielectric optical filters. Unlike the narrow spectra demonstrated by free atoms and molecules, the Raman spectrum of high-silica glasses consists of very broad bands with a 200 cm^{-1} wide band centered around 440 cm^{-1}. One of the major Raman bands of the SiO_2 molecule occurs at this frequency shift and the dominant Raman band for GeO_2, a frequently used additive, overlaps the silica band.

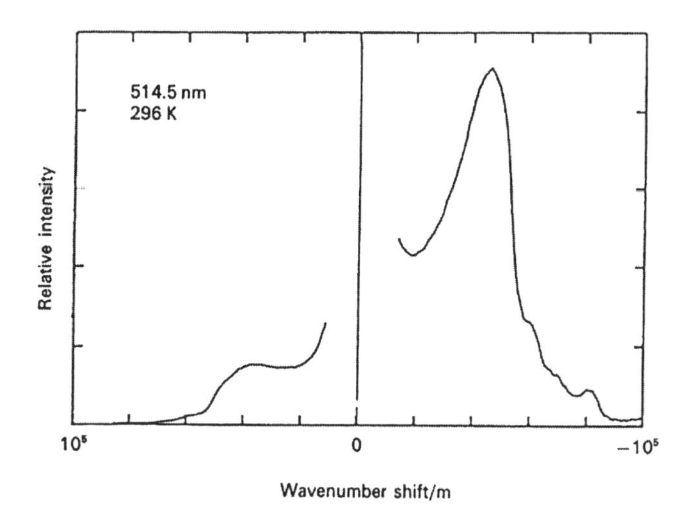

Figure 4.4. Raman spectrum of 50/125μm, 0.2NA optical fiber

Owing to the breadth of the Raman bands in glasses, some of the details which can be used in conventional Raman spectroscopy are thus lost to us in glasses. However the information is sufficient to obtain the temperature distribution along the fiber [21] and to eliminate spurious effects caused, for example, by fiber attenuation.

The Raman spectrum of silica is illustrated in fig. 4.4.

In the typical commercially available distributed temperature sensor (DTS) systems, the shorter-wavelength Raman band (the anti-Stokes band) is used to obtain the temperature information since it has a far higher temperature sensitivity (0.8% $°C^{-1}$ at room temperature) than the longer wavelength, Stokes, band. The temperature dependence of the anti-Stokes band measured in a commercially available, germanium-doped, silica fiber is shown in fig 4.5. The curve has been normalized to unity at 300 K.

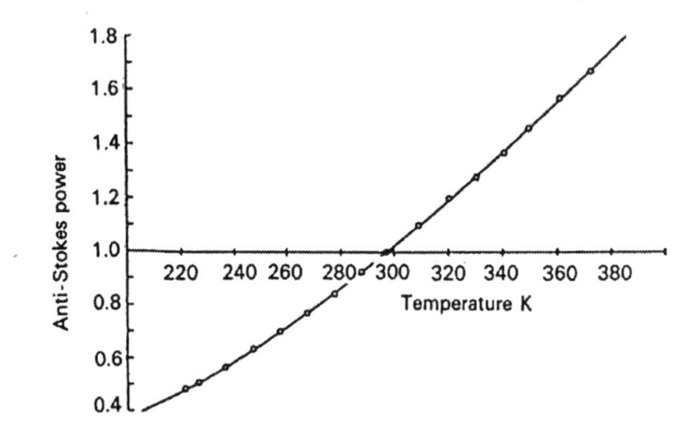

Figure 4.5. Back-scatter signal measured as a function of temperature (points). Results have been normalized to unity at 24°C. Solid line is fitted to experimental data

The Raman OTDR approach provides a practical solution to a number of measurement problems and forms the basis for the York Distributed Temperature Sensor. One of its main advantages is that it can be used with standard multimode telecommunication fiber, which is readily available and relatively inexpensive. By comparison with the liquid-core fiber approach, however, the signal is several orders of magnitude weaker, which implies a far worse performance, all other aspects of the system being equal. Fortunately, the associated technologies have moved rapidly since the work of Hartog [4] in the early 1980s and the performance offered by Raman OTDR is adequate for a number of practical applications. Moreover, the reduction in signal level is partially offset by an increased sensitivity of the signal to temperature.

Several systems have been marketed by Japanese and UK companies, the differences in approach being related to methods used for eliminating the effects of fiber attenuation on the measurement: in a practical system the fiber will in general be subject to kinks and stress which can affect the local attenuation of the fiber. One means of combating these parasitic effects has been to reference the sensitive signal (anti-Stokes Raman band) to a less sensitive one (the Stokes band) but this in itself does not provide adequate accuracy, particularly in long fibers, for the most demanding applications. Residual effects, such as the difference in attenuation between the two wavelengths used, or different sensitivities to modal effects e.g. at bends or at connectors can cause serious loss of accuracy. Sophisticated signal processing methods are used to provide rejection of these effects, such as the measurement of the fiber from both ends [21]. Another method is to employ iterative calculations and correction [22] of the distorting effects of temperature and cable layout-dependency of differential attenuation values.

The first system to be produced commercially [23] used a semiconductor laser source operating at 904 nm and Si APD detectors. Typically, a 2 km fiber could be measured in 12 s to a repeatability (one standard deviation) of 0.4°C and in the years since the release of that system, its performance has been gradually enhanced. Radical improvements in design have allowed much longer lengths of fiber (10 km) to be measured with high resolution (1 m) in second generation systems, based on the use of longer wavelength, more powerful sources which give a stronger initial signal and operate at wavelengths where the fiber attenuation is lower. Such systems will allow of order 10 000 points to be measured on a single fiber.

Second generation systems, now appearing on the market [24], are limited by fundamental considerations, such as the maximum power which may be launched into a fiber before the onset of non-linear effects, the fiber attenuation and its bandwidth. In contrast, earlier systems were limited in their performance by the power which could be launched and the performance of the built-in electronics.

Given the levels of performance achieved and the development effort which has been devoted to it, the Raman technique appears uniquely placed amongst all of the alternative approaches to see commercial exploitation in the short term.

A recent development in the field of Raman-based distributed sensors is their extension to single-mode mode fibers and operated in the low-loss window around 1550 nm. The low intrinsic attenuation at this wavelength has allowed the range to be extended to greater than 30 km.

Single-mode fibers form the basis of modern long distance optical communications systems. The ability to measure the temperature distribution along such fibers is a significant advance in that fibers already installed, for communications purposes, e.g. alongside energy cables, can now be used for monitoring the cable. In contrast, systems based on multimode fibers usually require the installation of a new fiber cable, a proposition which, in the case of energy cables, is uneconomic. Single-mode fiber technology thus brings the opportunity to retrofit the measurement to certain existing cables.

The monitoring of single mode fiber brings specific challenges [25]. In particular, the peak power which can be launched before the onset of non-linear effects is lower (by about one order of magnitude) than is the case in multimode fibers. Moreover the measurement in single mode fiber suffers from a lower scatter coefficient and a lower backscatter capture factor, both effects reducing the Raman signal for equal energy launched. Finally, at the wavelength where the fiber attenuation is at a minimum, InGaAs, rather than Si detectors must be used, resulting in a further degradation in signal-to noise ratio. To some extent, a highly efficient optical design and data acquisition system have offset these limitations, but nevertheless, the spatial resolution achievable to date is limited (in the range 5-10m) relative to that commonly found in multimode systems.

Owing to the very accurate referencing of all propagation losses which is possible in this configuration, the need for double-ended measurements has been eliminated. In particular, losses arising from non-linear effects in the fiber can be fully corrected. The power level in the fiber can thus be optimized for maximum backscatter signal at the remote end.

The optical arrangement used in single-mode DTS systems is shown in fig. 5.6. Without such signal processing schemes, the maximum power is dictated by the errors caused, e.g. in the anti-Stokes/Stokes ratiometric approach, by transfer of power to the Stokes wavelength. The resulting increase in allowable launch power is equivalent to at least 5 km of measured cable.

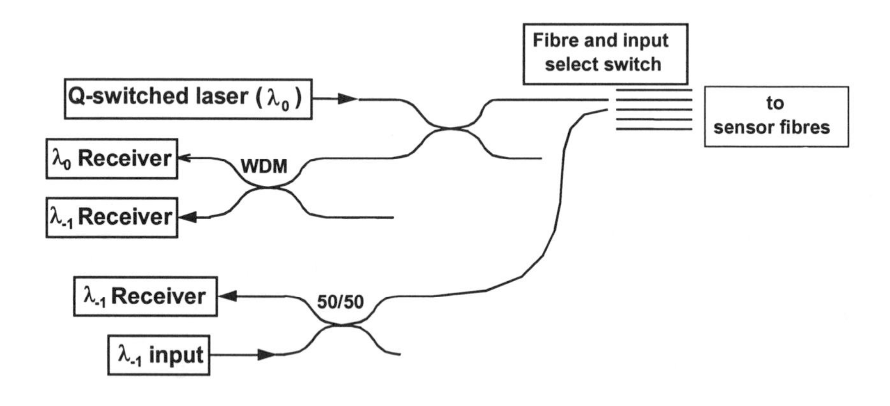

Figure 4.6. Optical arrangement for a single mode DTS system

In order to illustrate the performance of the single-mode systems, a detail of the remote end of a 30km length of fiber is shown in fig. 4.7. In this case, it was arranged for the temperature of successive 100 m sections of fiber, separated by further 100 m sections at room temperature to vary in sections of 100 m from 0°C to 60°C. In this case, the measurement time was 600 s and the spatial resolution was set to 10 m. The spatial resolution is apparent in the sharp edges separating the sections and the temperature resolution is well below 1 °C rms.

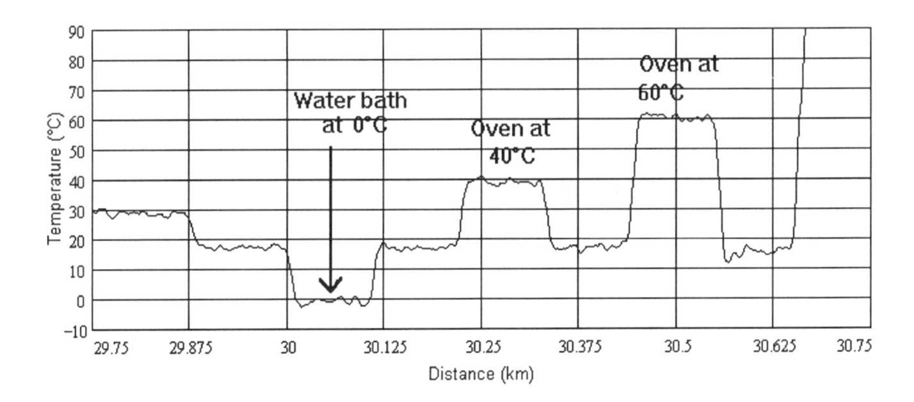

Figure 4.7. Thermal profile induced on a 30km single mode fiber in the laboratory and recorded using a single mode distributed temperature sensor

In recent years, the measurement of the temperature distribution using the spontaneous Brillouin spectrum has been demonstrated. In this case, the ratio of the Brillouin backscatter intensity to that of the Rayleigh backscatter, the so-called Landau-Placzek ratio [20] is used. Demonstrated results were obtained in the long-wavelength transmission band of single-mode optical fibers, at around 1550 nm [26-30]. A very narrow band source (<2 GHz) with a stable center wavelength, yet able to deliver higher power, short-duration pulses is required. In the work published to date, this was achieved using a Q-switched fiber laser, the wavelength-controlling element of which is a fiber Bragg grating [31]. The filter was implemented using an all-fiber Mach-Zehnder interferometer, which is tuned to track the source wavelength. In recent work [30], the Mach-Zehnder interferometer was used in a double pass configuration to improve its rejection. A temperature resolution of 1.4°C, with a spatial resolution of 10m has been reported [30].

The Brillouin (Landau-Placzek ratio) method promises a somewhat better performance than the Raman approach in the same fiber type. Moreover, it is compatible with long-distance communications fiber types.

4.3.7 Fluorescence

One of the drawbacks of the inelastic scattering approaches outlined in the previous sections is that they are based on very weak effects, much weaker than the Rayleigh scattering. For example, the anti-Stokes Raman signal is, at room temperature, typically three orders of magnitude less intense than the Rayleigh component, which results not only in very low signal levels and signal-to-noise ratios, but also in most of the light being wasted (in the sense that it is lost during propagation and does not serve to generate a signal for the sensor). Moreover, where it is desired to measure relatively short fibers at high spatial resolutions, a low scattering loss is of no advantage in terms of reducing significantly the signal degradation along the fiber and is a disadvantage in so far as the signal from the near end of the fiber is weak. It has therefore been proposed to use fluorescence effects to enhance the radiative losses in the fiber in the hope of generating a stronger signal for the sensor to exploit. The fluorescence in certain materials displays a number of temperature-dependent effects, e.g. rate of decay, which may be exploited in a sensor. It is conceivable to incorporate fluorescent elements (e.g. rare-earth ions) into the fiber material to give rise to these effects. Unfortunately, the characteristic decay times of materials compatible with high-silica glasses are orders of magnitude too long: they would restrict the spatial resolution to many km. The approach has thus not been exploited in practice to date.

4.3.8 Non-linear Optic Effects

The very weak nature of backscatter signals, particularly when used for high-resolution distributed sensors, has prompted research into means of providing far stronger signals. In single mode fibers, the tight confinement, long interaction lengths and low propagation losses make for particularly low non-linear thresholds, in spite of the relatively linear behavior of high-silica glasses. Several schemes have been proposed, based on counter-propagating beams in the fiber, which interact through a non-linear process. At least one of the beams must be pulsed to provide the distance resolution. The first experiments in this area [32] used stimulated Raman scattering to give rise to the interaction between the counter-propagating beams, the response to the measurand being provided by the sensitivity of the stimulated Raman scattering process to the relative polarization alignment of the interacting waves. The signal processing algorithms required are more formidable still than in straightforward POTDR, involving not just one, but two waves, traveling in opposite directions in the fiber and at different wavelengths. This approach was not taken beyond initial experiments, the results of which were never processed to the point of showing a measured distribution of the parameter of interest; research is nonetheless continuing in this area [33].

The optical Kerr effect, where the phase of the light travelling in one direction is modulated by optically induced changes in the refractive index has also been tried [34]. In this case the optical length of the fiber was measured interferometrically and a change in this optical path length was generated by an intense pulse. The experimental arrangement used in is illustrated in fig 4.8. Unfortunately, the initial results were disappointing, owing to the very weak sensitivity of the Kerr effect to external parameters; however work is continuing on this principle and very recently results involving Kerr interaction between counter-propagating beam in a polarization-maintaining fiber has been reported [35].

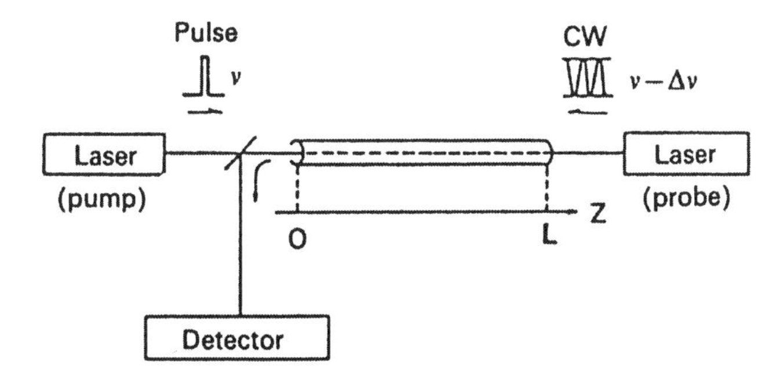

Figure 4.8. Basic configuration for Brillouin optical-fiber time-domain analysis

Finally stimulated Brillouin scattering has been proposed [36, 37], initially as a means of enhancing the performance of OTDR field tests, but more recently, the frequency shift of the Brillouin gain spectrum has been demonstrated to be sensitive to temperature and strain. This work lead to a series of elegant experiments on the determination of axial strain, for example in installed cables [38, 39]. In this case, the fiber is interrogated from one end by a continuous wave source and a pulsed source from the other, the difference in wavelength between these sources being carefully controlled to be equal to the Brillouin frequency shift (of order 10-20 GHz, depending on the operating wavelength), fig. 4.9. The lower frequency wave (which can be cw or pulsed) is then amplified, the gain being dependent on *inter alia* the match between the frequency difference of the two sources and the local frequency shift. From the time variation of the amplified signal, the local gain can be determined, as illustrated in fig 4.10. Independent calibration of the gain spectrum vs. strain or temperature thus allows these measurands to be mapped along the fiber from measured values of the gain along the fiber.

One of the keys to this measurement has been the development of the diode-pumped, single frequency, tunable Nd:YAG ring laser[40], capable of delivering of order 1 mW into single mode fibers with remarkably narrow (5 kHz) spectra. Moreover the temperature tunability of these devices has allowed the gain curves to be explored by varying the frequency difference between the sources placed at each fiber end. Results published to date give spatial resolutions of 50 m with 0.01% strain resolution over >1 km fiber lengths.

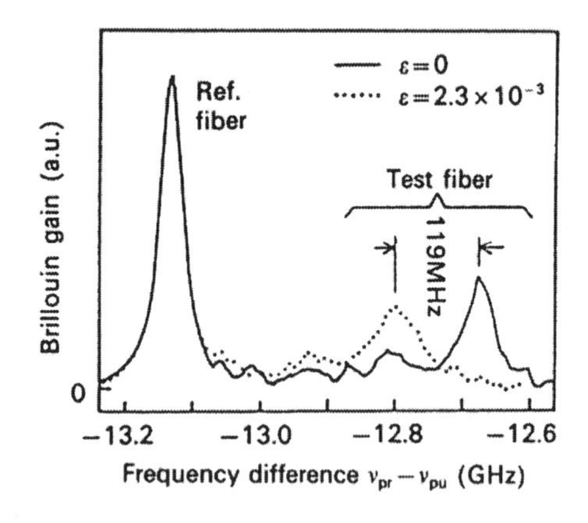

Figure 4.9. Sample Brillouin gain spectra: strain effect

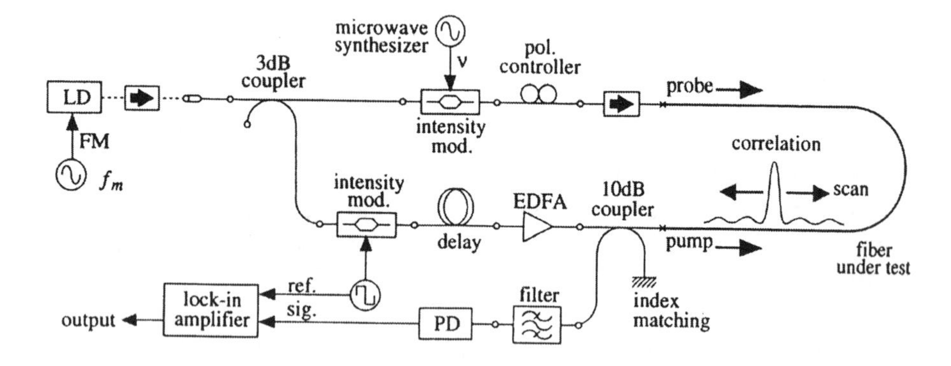

Figure 4.10. System for measuring Brillouin gain spectrum distribution along an optical fiber [36]

4.3.9 Discrete Signal Sources (Quasi-Distributed Sensors)

Where truly distributed sensing has proved elusive, researchers have used techniques similar to those discussed above, but derived the optical input for the detector, not from an intrinsic scattering mechanism in the fiber material, but from an artefact.

The simplest example of this type of sensors is that of a series of reflectors in the fiber. Controlled reflections may be created by cutting the fiber and re-jointing it either with a connector or with an adhesive having a slight refractive index mismatch with the fiber core material. This gives rise to a reflection, which although it could be several orders of magnitude smaller than the 8% reflection intensity expected at a perfect reflection from a dry connector, is still several orders of magnitude stronger than the backscatter signal.

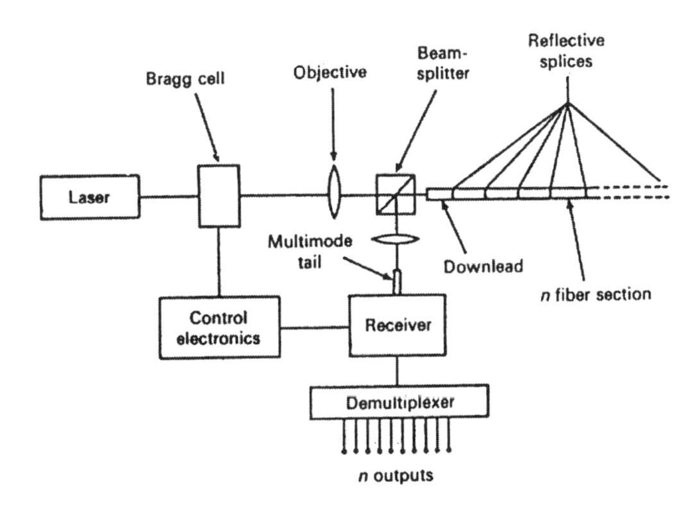

Figure 4.11. Optical fiber hydrophone array [37]

The string of pulses resulting from such a fiber assembly has typically been analyzed for the relative position of these reflectors, which is a measure of the fiber elongation. This measurement, performed on a static fiber results in a quasi-distributed strain sensor. A dynamic measurement provides an acoustic transducer, or hydrophone array. In the case of dynamic measurements, an elegant coherent detection method has been explored [41], as illustrated in fig 4.11. Two pulses, from the same high coherence laser source are launched into the fiber, separated by the time between reflections (in this case arranged to be at equal intervals). The frequencies of these pulses are shifted from each other by means of an acousto-optic deflector (or Bragg cell). The result, see the timing diagram, fig. 4.12, is that at any one time, the detector receives the superposition of a reflection from reflector R_n at frequency f_1 and from R_{n-1} at f_2. These signals mix coherently at the detector to produce a frequency difference signal, which preserves the phase relation between the waves. Electrical demodulation from (f_1-f_2) restores the instantaneous phase between the reflections and hence the elongation of the section of fiber between the two reflectors. Such an arrangement has been used to form arrays of optical fiber hydrophones which have undergone undersea trials.

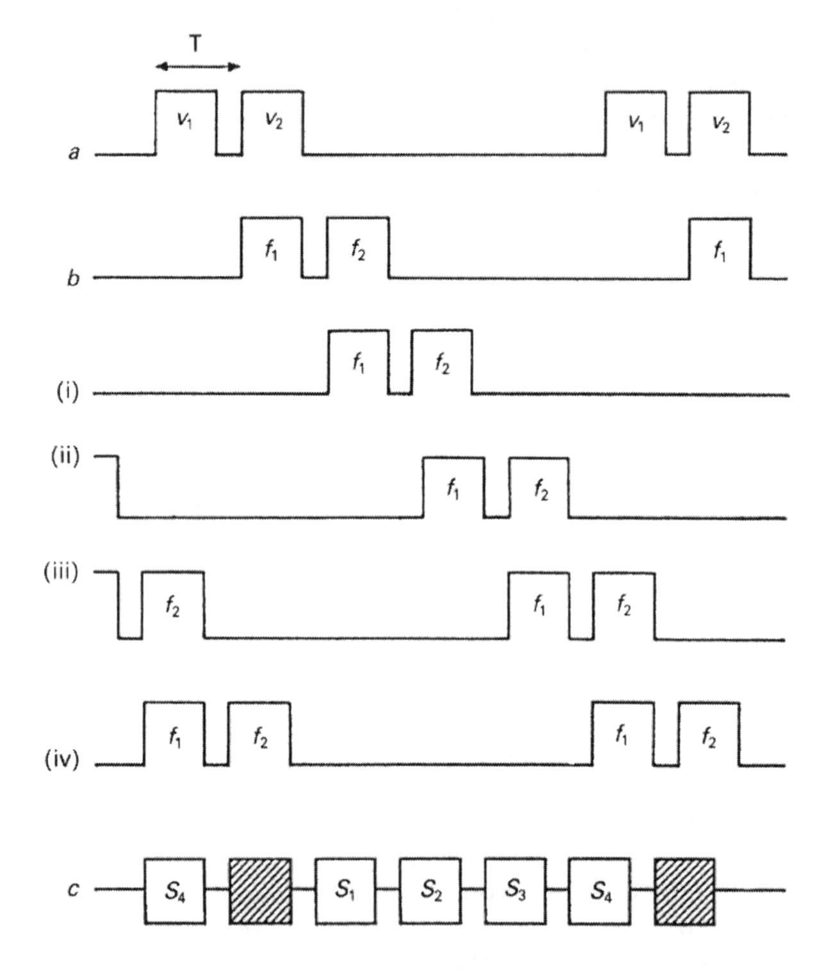

Figure 4.12. Timing diagram for a four-element array. (a) Input to Bragg cell. (b) Reflections from end of: (i) download; (ii) 1st sensor; (iii) 2nd sensor; (iv) 3rd sensor; (v) 4th sensor. (c) Output from photodiode

4.3.10 Forward Scattering Methods

In these approaches, the probe light and the signal travel in the same direction down the fiber. In general, the measurand causes a power transfer to a different mode of the fiber which has a different group velocity from that of the original mode. Examples of implementations include polarization-maintaining fiber (York HiBi fiber) where the probe is launched into one of the eigenstates of the fiber and conversion to the orthogonal state is monitored. In such fibers, the relative propagation velocity difference between the two modes can be as high as 0.1%. An earlier implementation, which used a fiber supporting the two lowest order LP modes, relied on the conversion of power between the modes.

In these schemes, the transit time difference between the probe and the signal provides the location information, whereas the intensity of the signal provides information on the magnitude of the measurand. The time scale is now compressed, with respect to OTDR, by the ratio of the transit time to that of the inter-mode dispersion (typically 1000:1). Therefore, very good time resolution is required. In practice, FMCW and related techniques have been employed [42]. To date, however, little quantitative information on the origin of the external source of mode conversion has been gained - reliable calibration has not been attempted.

4.4 PERFORMANCE OF DISTRIBUTED SENSORS AND ENGINEERING ASPECTS

4.4.1 Performance Criteria in Distributed Sensors

The performance of distributed sensors is judged not only on their accuracy, measurement range and measurement time but also on the length of fiber they can cover and their spatial resolution. The maximum length of fiber is determined by the signal-to-noise ratio required to give the required temperature resolution after signal processing. Thus, as the total loss in the measurement loop increases (with increasing fiber length), so the quality of the measured signal is degraded and the uncertainty on the output increases.

Spatial resolution on the other hand describes the ability of the instrument to distinguish adjacent points in the fiber. The author chose to define the spatial resolution as the distance over which an abrupt temperature transition along the fiber appears to be spread, measured between 10% and 90% points, a definition which is gaining more general acceptance.

The parameters describing the performance of the sensor are inter-related. Clearly in a system where the measurement accuracy is limited by the signal-to-noise ratio, the measurement time (or integration time) will vary as the square of the accuracy required.

Moreover, the criteria specific to distributed sensors also impact the measurement time: as the required fiber length increases, so too does the system loss and, therefore, the signal-to-noise ratio referred to the measurand is degraded. Similarly, as the spatial resolution is made finer, the pulse width of the source must be reduced (which, in peak-power-limited laser reduces the signal proportionately) and the receiver bandwidth must be increased, which degrades the system noise. Thus as the spatial resolution is improved, the measurement uncertainty is increased, for a constant measurement time.

There is therefore a trade-off between the various performance criteria and, for a fixed accuracy in the measurement, the integration time increases rapidly with improving spatial resolution and increasing total loss of the sensing fiber. The exact relationship between these parameters is complex and depends on the details of the instrument design. Typically, however, an improvement in spatial resolution by a factor x must be traded-off for an increase in measurement time by a factor between x^3 and x^5. The trade-off can only by overcome by improving the performance of some of the components of the systems, e.g. by improving the receiver sensitivity, by increasing the output power of the laser and by careful minimization of all optical losses in the system.

Some indication of the relationship between these performance parameters may be gained from Table 4.1, where the temperature resolution attainable from a Raman-type distributed sensor has been calculated as a function of fiber length, for values of the spatial resolution parameter in the range 1 to 10m. In this case, it is assumed that the data is collected after averaging the scattering from 10000 pulses; moreover, the curves model second generation systems, operating at 1064 nm with energy limited (not peak power limited) pulses.

Table 4.1. York sensors model: DTS-800: manufacturer's performance data

Type	Range	Spatial resolution	Temperature resolution / Accuracy (°)	Measurement time
short	5km	1m	1.0/1.0	600s
medium	10km	2m	1.5/2.0	600s
long	20km	4m	2.5/2.0	600s
ultra-long	30km	8m	2.0/2.0	600s

4.4.2 Constraints in the Engineering of Distributed Sensors

In addition to the basic performance criteria outlined above, the acceptance of this technology depends on aspects of importance for the ease and cost of use being addressed by suppliers, such issues having been addressed successfully by the manufacturers of traditional sensors for many years.

Calibration: The system must be capable of being calibrated by the supplier and retaining that calibration after installation. The calibration must remain stable over several months at least and must be easily verifiable by the user.

Interchangeability: If, for any reason, a sensing fiber is replaced in whole or in part, the calibration of the sensing system must be retained or capable of being re-established with minor intervention by the user.

Ruggedness: Optical fibers are still perceived to be fragile objects, far too delicate to be used in the midst of heavy machinery and other unprotected industrial environments. One of the keys to the success of distributed fiber-optic sensor will be the knowledge of how to install the fiber in such a way as to protect it from the environment it is used in and yet to retain desired sensitivity to certain aspects of this environment. For example, a fiber used for temperature monitoring in a process should be cabled or sheathed so as to avoid damage by the material being processed. On the other hand, heavy armouring would slow down considerably the thermal response of the sensing element. The fiber must be capable of withstanding the temperature range it will be exposed to, which in many cases implies coatings not used in the field of optical fiber communications. Fortunately fibers are now available with a range of coatings. For the primary coating (i.e. that deposited directly onto the glass surface) the following materials are used:

- UV-cured acrylates: these are standard telecommunications coating materials, capable of operation up to about 90°C (depending on the duration of the exposure to the high temperature and the exact type of material used).
- silicone rubbers: these materials are capable of continuous operation up to about 150°C. These materials, do not, however, exhibit such a good low temperature performance as do acrylates; below approximately -40°C (again depending on the exact product used) the silicones tend to become hard and impose severe microbending losses on the fiber they are intended to protect.
- polyimide: a varnish-like polymeric material which is capable of operating up to about 350°C.

- metal and ceramic coatings: a range of materials has been developed but these are of only limited availability; aluminum is rated to about 450°C and gold to 750°C. In practice, it is found that the attenuation of certain fiber increases well below the upper limit for gold. Other metals which have been employed include copper and nickel. The main problem found with metal coatings is the difficulty in making a product which does not exhibit increased microbending losses, especially after repeated heat cycling. Non-metallic coatings, including amorphous carbon and silicon oxynitride have been developed for their hermeticity. The presence of OH ions from water tends to accelerate the mechanical failure of fiber under tensile load for extended periods, especially at elevated temperatures.

A range of materials is also available for the next layer of protection, including certain high-temperature polymers (e.g. PTFE or PEEK) and fine-bore stainless-steel tubing.

A common construction for high-temperature applications in rugged environment – such as the monitoring of the external wall of process plant, where mechanical damage to the fiber may occur from unrelated work – is a fiber coated firstly with a carbon coating to provided hermiticity and secondly with a thin layer of polyimide. The coated fiber is inserted into a steel tube, typically 3.2 mm outer diameter with 0.5mm wall thickness. The grade of the stainless steel is selected according to the environment, a corrosive environment requiring the use of type 316L or higher still levels of corrosion resistance.

One company specializing in the installation of fiber into oil wells for distributed temperature sensing, Sensor Highway Ltd [43], uses a process in which a ¼" (6 mm) tube is installed in the oil well during the well completion and the fiber is installed later by fluid drag. Fibers as long as 10km have been installed in this way at rates of a few km/hr.

Connectorization and spliceability: the installation of a distributed sensor will require fiber to be installed in sections jointed via splices or connectors. This requirement imposes restrictions on the types of fiber and components which can be used. Moreover, it is generally desired to use standard components (i.e. in practice those developed for the telecommunications market) for reasons of cost and security of supply. Moreover, those sensors capable of operating using standard fiber (whether for communications or some other sensing application using large quantities of fiber, such as the fiber gyroscope) will have a distinct commercial advantage (of fiber price and quality) derived from the large volumes in which the fiber is produced.

The sensor must be capable of being disconnected and re-connected, or of being repaired after accidental damage with only minor actions on the part of the user; clearly if the fiber length has varied during the repair, some recalibration of the locations of the sensed points will be required.

User interface: the large volume of data produced by distributed sensors requires, in the system design, special attention as to how the information will be used. This may involve post-measurement filtering, signal processing, interconnection with process control equipment, notification of alarms and so on. The human brain is unable to cope with the volume of data produced by currently available sensors (e.g. 4000 temperature readings every few seconds) except in special circumstances such as where the distribution of the measurand is uniform under normal conditions. The measurement equipment will therefore be required increasingly to classify the output of the sensor and even take actions as the result of the measurements it has made.

4.4.3 Alternative Methods of Interrogation and Signal Acquisition

The discussion in section 4.3 is based largely on time domain reflectometry, i.e. the return resulting from launching a pulse into a fiber is analyzed. This requires high peak powers with low duty cycles, whereas semiconductor lasers are better suited to delivering sustained outputs. The relatively weak signals encountered in distributed sensors have resulted in a search for more effective measurement approaches and some of these have been applied to reflectometry problems in sensing or communications and are detailed below.

4.4.3.1 Optical Frequency Domain Reflectometry (OFDR)

In this approach [43], a cw laser is amplitude-modulated with a sine wave the frequency of which is varied. If the modulation frequency is stepped through an equispaced set of values [44], the response of the fiber can be Fourier-transformed to give similar information to that which would have been obtained if a pulse (time domain) method had been used.

Another possibility [43,45] is to apply a linear ramp to the frequency of the modulation as illustrated in fig. 4.13. In this case, at any one instant, the modulation signal being applied to the laser is mixed, electronically, with the signal from the receiver and generated by the scatter return. In the spectrum of this demodulated photocurrent, scattered light from a particular point along is directly related to a particular frequency. For example, if the modulation frequency is swept at a rate of 100 $MHzs^{-1}$, then light scattered at a distance of 1 km from the launching point will be delayed by 0.01 ms and will give rise to a beat signal of 1 kHz. It can be shown that the spatial resolution in such systems is directly proportional to the frequency excursion of the modulation signal, subject to the linearity of the sweep being adequate to avoid blurring through cross-products.

The main advantage put forward for these optical frequency domain schemes is that the average power launched into the fiber is far higher than in the OTDR approach. In particular, as the spatial resolution is made finer, the pulse width in OTDR must be reduced, which in general reduces the energy launched into the fiber; thus the OFDR approaches would be expected to show significant advantage in higher resolution applications. This comparison is of course highly dependent on the characteristics of the sources available in each case; in recent years, with the advent of high-power, cw laser diodes aimed at laser printer applications, for example, this balance has shifted to the benefit of the frequency domain approach. For OTDR, a somewhat broader bandwidth must be used in the electronics and the peak powers available are somewhat higher than the levels which can be obtained for cw operation. Therefore, the performance edge is in general with OTDR at low resolution (say >10m resolution), whereas for resolutions finer than about 1m, the OFDR approach may show advantages. To date, however, relatively little work has been presented using OFDR and the results shown have concentrated on demonstrating the high resolution, of the technique with little application to an actual sensing task. Moreover, none of the commercially available implementations have yet adopted this approach.

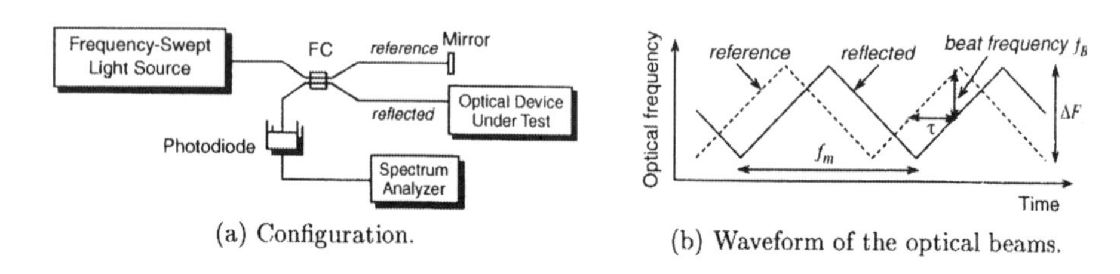

Figure 4.13. Configuration and principle of FMCW reflectometry [48]

4.4.3.2 Pseudo-Random Coding

Pseudo-random coding (sometimes termed pulse compression coding) [36] is a technique (taken from the experience of radar engineers) where a pulse train, rather than a single pulse is launched into the fiber. The backscatter returns from each pulse overlap in time at the detector and result in a more intense, but time-scrambled signal. However, knowledge of the exact sequence of pulses allows the true backscatter signal to be extracted from the detected output using a correlation algorithm. The net result is, in principle, a signal of similar resolution to that which would have been produced by a single pulse, but having a signal-to-noise ratio improved by a factor $1/2\sqrt{N}$, where N is the number of pulses (i.e., of 'ones') in the code sequence. The code sequence should ideally be chosen in such a way that its auto-correlation function contains a single peak (equal to the sum of the pulses in the sequence) and zero elsewhere. Any non-zero values in the autocorrelation function outside this central peak would result in cross-talk between sensor elements.

One of the main difficulties in transferring the technique from the field of radar is that, in contrast to the radar case, where the codes developed consisted of a sequence of positive and negative-going pulses, it is not possible to generate pulses of 'negative light' and as a result, there are no known codes which have the auto-correlation properties described above. This has been dealt with in two ways.

In the first approach is based used initially for OTDR [47], but later for distributed sensing [48, 49] a so-called m-sequence is applied to the laser. An *m*-sequence may be generated very simply using a shift register and a few exclusive OR gates in a feedback arrangement. The codes repeat after m bits, where m is related to the length n of the shift register by $m = 2^{n-1}$, so that long sequences can be generated relatively simply with just a few logic gates. The correlation properties of optical m-sequences give a central peak equal in intensity to the number of 'ones' in the code. The autocorrelation function of an optical *m*-sequence is close to ideal, with a central peak equal to sequence length and the remainder of the correlation response equal to -1.

Experimental results have been shown using pseudo-random coding in distributed fiber sensors[49] but have not to date matched the performance of single pulse systems, probably as a result of the considerably greater development effort which has been devoted to the single pulse approach. One of the drawbacks of the m-sequence approach is the periodicity of the sequence: signal-to-noise analysis indicates that, where long fibers are used, the shot noise generated by near-end backscatter dominates the system performance: it is therefore desirable to send pulse bursts which occupy a fiber length corresponding to only a few dB of fiber loss and yet contain enough 'ones' to provide a significant advantage in signal level.

Unfortunately, there are no known aperiodic, unipolar codes which have the desired correlation properties. Instead, it has been necessary to find sets of complementary codes, for which a cross-correlation algorithm exists yielding the desired zero-sidelobe result. There is naturally a signal processing overhead in this approach, which nevertheless has been successfully applied to the design of optical time domain reflectometers for optical communications [50].

4.4.3.3 Coherent Detection

In systems where inelastic scattering is not used, a number of coherent detection schemes can be employed. These include coherent OTDR, as pioneered at BTRL [51] where the scattered signal is made to beat with a local-oscillator. This arrangement allows the photocurrent to be increased in direct relation to the power of the local oscillator, this ensures that the signal from the detector always dominates the noise of receiver, leaving the shot noise associated with the backscatter as the dominant noise mechanism and providing a nearly quantum-limited receiver performance.

The implementation of coherent OTDR may be carried out by splitting the laser output into two paths, one of which is sent directly to the detector where it provides a local oscillator and the rest is modulated to select a pulse and is launched into the fiber. If the probe laser is frequency-shifted, for example acousto-optically by means of a Bragg cell, a heterodyne system is obtained, in which the information returned from the fiber appears at an intermediate frequency (IF) equal to the frequency-shift applied to the outgoing pulse.

Several points are worth noting in this type of system: firstly, the laser is providing both the signal and the local oscillator, which simplifies the design by comparison with a coherent communications system, but still requires a high degree of coherence from the source, a degree of coherence normally beyond that achieved by standard laser diodes, although line-narrowed devices are capable of meeting the stringent requirements of this measurement technique. Secondly, given that the coherent detection process is essentially an interference process at the detector, the arrangement is intrinsically polarization-sensitive, which is clearly desirable in POTDR-type applications, but otherwise must be circumvented by polarization scrambling. Finally, reflectometry with highly coherent sources results in fading effects [52] caused by the interference of backscattered light from various parts of the fiber which reach the detector with random relative-phase relationships which vary with the source frequency, the fiber temperature and the vibration environment.

4.4.3.4 Frequency-Modulated, Continuous Wave (FMCW)

Frequency-modulated, continuous-wave (FMCW) reflectometry [53, 54] is a technique related to coherent-detection backscatter and to OFDR, in which, instead of modulating the frequency of an amplitude modulation signal imposed on the source output power, the optical frequency itself is modulated, usually with a linear ramp. FMCW has been demonstrated to good effect in a number of distributed sensor configurations, usually over relatively short distances.

The frequency modulation of the carrier wave is usually achieved by varying slightly the bias current applied to a carefully temperature-stabilized, reflection-suppressed, single-mode laser diode. The change in bias current affects a number of parameters in the laser cavity, notably its local temperature (regardless of the stability of its heatsink) and its effective refractive index via the carrier-injection process. Thus, at the price of a moderate change in injection current, and therefore in output current, a wide excursion in the optical frequency can be achieved (typically several GHz) and, since the spatial resolution is directly proportional to the frequency excursion of the source in this scheme, very high spatial resolutions (of order a few cm) may be achieved and have been demonstrated.

The backscattered light is mixed on a detector with a sample of the light currently generated by the laser. Since the detector responds to optical intensity, the two signals mix coherently at the detector to produce sum and difference frequencies, the latter being at radio or microwave frequencies. This coherent detection process, together with the availability of large local oscillator power (in this case the optical signal obtained directly from the laser) ensures that the photocurrent can always be made to dominate the noise of any following preamplifier; very large signal-to-noise ratios can thus be obtained.

The main drawback of the method is that, being coherent, a number of techniques, e.g. spontaneous Raman scattering are precluded, because the coherence is lost in the scattering process. In order to obtain good mixing (or in optical terms, interference) between the two waves at the detector, the fiber must be of the single mode variety and the state of polarization of the two waves must be arranged to be co-linear. There are also practical difficulties in ensuring that the frequency sweep is adequately linear, any departure from linearity resulting in a degradation of the spatial resolution and cross-talk between sensor elements.

Recent development of sources has improved the prospect for the technique, however. In particular, the diode-pumped monolithic Nd:YAG [55] laser has demonstrated linewidths of a few kHz with coherence lengths of order 50 km of fiber. This type of source has been put to use so far only in long range reflectometry applications although the potential for sensors is obvious.

4.4.3.5 Photon Counting

Another enhancement [46], has been to use a photon counting receiver to provide extremely high sensitivity. In this approach, a photomultiplier tube is used as a detector and it produces a single pulse for each photon detected, the pulse amplitude being sufficient to be clearly distinguishable from the noise of the following amplifier. A digital system then follows to average the counts corresponding to different 'time slots', each related to a particular section of fiber. A further enhancement used in [56], is that the resolution was improved by 'time-correlation' where the time of arrival of the photon is detected with extremely high resolution (limited only by the width of the pulse launched into the fiber). A high spatial resolution (<0.5 m) was demonstrated, with the penalty of a very long measurement time. The method is particularly well suited to measurements where a high resolution is required, a short fiber length acceptable and the measurement time not critical.

4.5 APPLICATIONS

Distributed sensor systems have a vast range of applications; almost certainly some of these will not become apparent until the technique is well known in more obvious applications. For the sake of illustration, some of the applications areas are mentioned below, mainly related to temperature measurement.

4.5.1 Power Supply Industry

For some time it has been proposed to insert optical fiber sensors into high-power transformers in order to detect hot spots and determine their temperature. It is the temperature at the hottest point which determines the lifetime of the insulation material and thus of the equipment itself. However, hot-spots may be caused by accidental faults in the winding and their position cannot be determined in advance. A large number of point-sensors is thus required to provide adequate coverage. The fiber-optic distributed temperature sensor is ideally suited to this application since it can provide continuous coverage of the whole of the transformer windings. The present performance is adequate in most respects although improvements in spatial resolution are desirable. The benefits of continuous monitoring depend on the value of the equipment, but more importantly on the cost of purchasing replacement power, in the case where the transformer has failed.

Braendle and co-workers at the Baden Research Center of ABB have carried out extensive work on transformer monitoring using distributed temperature sensing. Figure 4.14 shows the thermal profile of a transformer alongside a photograph of the equipment. Approximately 1 km of fiber was inserted within the copper segments forming the winding of the transformer during its manufacture. Non-uniformities in the temperature of the windings are clearly visible, together with the end effects where excess heat is found. Using distributed temperature sensing, ABB has been able to improve significantly the design of its transformers, thus saving in weight and size for a given capacity. ABB also evaluated of one of its locomotive transformer designs on board a train travelling in the Swiss Alps, the temperature profiling reflecting the load of the transformer as the train ascended and descended.

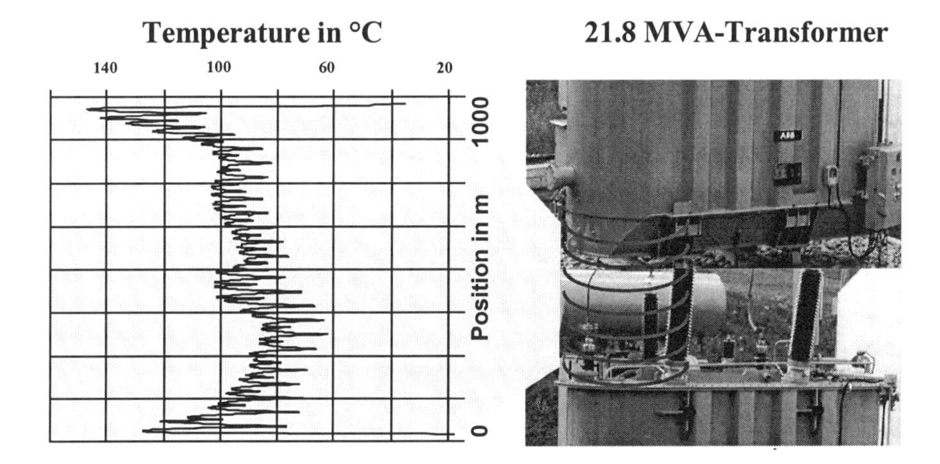

Figure 4.14. Thermal profile in a high-voltage transformer

In addition to a design tool, the technology is offered to transformer customers for the monitoring of their transformers in service, which allows the user to be confident of operating the unit within its specified rating. Another application in the same industry is the monitoring of large generators, where the risk is primarily from blocked cooling pipes. The very high spatial resolution required can be obtained by coiling the fiber into a number of separate sensing coils to monitor each cooling pipe. In this case the cost of a failure has been put at over $3M [57].

The capacity of high-voltage transmission lines depends on their temperature and it has been proposed to install a sensing fiber inside one of the conductors for monitoring purposes. In this case, the range (maximum fiber length) must be increased as much as possible. It is believed that the range of the DTS system can be increased considerably, with additional development, and that such a scheme would be economically attractive, provided that the ability to carry additional peak loads is considered to be a real benefit. It has been estimated [57] that the value of an additional capacity of 100 MW over 10 miles at 230 kV is around $3.5 M per year.

The problem with estimating the rating of an energy cable is that its thermal dissipation depends on the thermal conductivity of the back-fill surrounding the cable and the soil temperature. Neither of these is well known, given differences in soil conditions and the possible presence of other services, such as district heating pipes or other cables. The actual measurement of the cable all along its length lifts some of these uncertainties and allows the ampacity of the cable to be estimated far more accurately.

The measurement of energy cable thermal distributions is now becoming accepted good practice in the industry, at least for cables of 110 kV and higher. As a result, most new cables are now laid with fibers within the cable (or attached to its outside surface) specifically for cable monitoring. One reason for this accelerating awareness is the failure in February 1998 of the power supply to the Auckland Central Business District as a result of successive cable breakdowns; business was disrupted for several weeks. Although several reasons for the problem appeared in the enquiry, one cause was the fact that the cables were designed for a set ground thermal conductivity, whereas the actual conductivity was worse in places by a factor of at least four. The operators thus believed that they were operating well within the capacity of the cables, whereas in fact this may have been exceeded, once the first cables failed for unrelated reasons.

In the more advanced installations, the data from the DTS systems is fed to a real-time thermal rating system, which uses the data and other inputs, such as air temperature to provide continuously updated information on the cable capacity. Such data are usually in the form of the maximum current which can be transmitted for given durations and is primarily made available in case a failure elsewhere in the network increases the load to be passed through the remaining cables. The operator can then decide, in an emergency, whether to exceed the nominal rating of the cable for a short period, knowing that once a surge in demand has passed, the cable will be allowed to cool down.

The ability confidently to measure from one end only of the fiber, together with the achievement of 30km range (with 10m resolution for an averaging time of 600s), has allowed very long cables (up to 60km), especially sub-sea cables, to be monitored over their entire length. This total range is achieved by interrogating the fiber from separate instruments sited at each of the shore stations. The first such sub-sea single-mode system to be operated on an energy cable in commercial operation was brought into service in 1996 between the Island of Penang and the Malaysian mainland.

Results on a single-mode land-based cable of similar length in the Netherlands between the towns of Vorburg and Delft (data obtained by the author in collaboration with KEMA and the utility EZH) are shown in fig. 4.15. In this case, the cable was lightly loaded and measured on a cold winter's day. A number of features appear in the thermal profile and these have been related to changes in the environment of the cable, such as changes in depth (e.g. where the cable passed below a motorway junction, marked A13 in the figure) or where it crosses other services, such as a canal or a road, or other cables. Although, in the case of power lines, transformers and generators, good models exist of the thermal effects inside the plant, there has been reluctance to use the full predicted capacity without independent temperature measurement, owing to the enormous repercussions which would result from an equipment failure.

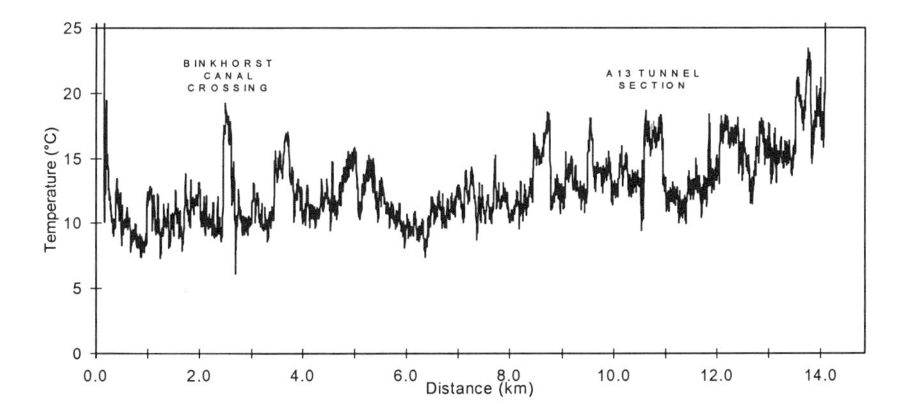

Figure 4.15. Distributed temperature measurement on a live energy cable using single mode fiber integrated in the cable

Finally, in thermal power stations, the DTS system can be used to monitor high pressure steam pipes for leaks which are extremely hazardous as well as disruptive and experiments to test its ability to detect high-pressure steam leaks have been conducted satisfactorily.

The use of this type of equipment, particularly in the power supply industry, will require the development of considerable knowledge, in addition to the design of the sensor itself. For example, new methods have been developed to install the fiber in each of the applications outlined above and a number of problems have been solved. These include how the fiber is inserted within the windings of the transformer, in the conductors of the generator or in the power cable with minimal effect on the design, performance, reliability of this well-established equipment. Electrical power authorities are increasingly specifying that new plant must have fiber built-in for monitoring purposes and installations for commercial use have been made in a number of European countries and in Japan.

For applications in the power supply industry, the fiber must have the correct dielectric properties, be non-conductive, non-tracking, non-contaminating and avoid the creation of locally enhanced electric fields.

Suitable coatings are required to achieved these objectives and at the same time protect the fiber and withstand the temperature and the chemical attacks they will be subjected to (e.g. transformer oil). All dielectric fiber constructions which are likely to meet all of these requirements have been produced but no long-term results have been published to date.

Finally, the DTS presents the user with orders of magnitude more information than has previously been available. If this new data is to be used efficiently, the methods for monitoring and control of large electrical plant must inevitably be modified. However, experience in other applications areas, e.g. in process control, have demonstrated the possibility of interfacing DTS sensors to traditional process control computers and making use of the information provided. Likewise, it is hoped and expected that, once the control systems in the electrical power are able to use DTS data, the overall system will be more efficient and reliable.

4.5.2 Petrochemicals

4.5.2.1 Pipeline monitoring

The DTS has aroused considerable interest in a number of pipeline temperature monitoring applications. In some, the users are concerned with detecting leaks of the fluid. (In the case of gases, cooling will result from the expansion of the gas leaving the pipeline.) In other cases, materials are heated just sufficiently to reduce their viscosity (excess heating can degrade the material being transported and increases the cost of heating). Failure of the heating system or of the insulation can cause the pipeline to seize up, a situation which is very costly to rectify, in addition to the cost of any production lost.

4.5.2.2 Storage Vessels

Used in the industry are normally required to be monitored for leaks. This is especially the case in the storage of liquefied natural gas (LNG), where vast quantities of the material are stored at a few sites, and would result in a hazard should leaks occur. A single optical fiber is capable of monitoring the whole of the area under a storage tank and of detecting the abrupt drop in temperature caused by a possible leak. Several DTS systems are now installed to monitor LNG tanks. In addition, trials of distributed sensors have been carried out in methane ship terminals, again for leak detection.

4.5.2.3 Upstream Oil and Gas Industry

In the past four years, a number of applications of distributed temperature sensing have been identified in the *down-hole environment* in the field of oil and gas exploration and production, but also for a number of other geophysical applications, such as hydrology or geothermal energy extraction. Within the field of oil and gas alone, the following applications of distributed temperature measurement have been identified.:

4.5.2.3.1 Well Management

Well management includes:

- Flow measurement,
- Electric Submersible Pump (ESP) monitoring,
- Gas lift optimization,
- Scale management,
- Leak detection)

4.5.2.3.2 Well Bore Analysis.

Well bore analysis includes:

- Identifying flowing zones,
- Trending of perforation performance,
- Cement integrity

4.5.2.3.3 Reservoir Monitoring

Reservoir monitoring involves:

- Water/gas breakthrough,
- Gas coning,
- Flow rate determination

The data and illustrations below were obtained by courtesy of Sensor Highway Ltd, an oil field services company specializing the application of optical sensing techniques, and notably distributed temperature sensing to the oil and gas exploration and production industries [43]. Sensor Highway have established themselves as a leading installer of distributed optical fiber temperature sensors in the upstream oil and gas industry. The distributed temperature sensor has been installed on over 120 wells in the past three years.

The majority of these installations have been in the heavy oil markets monitoring the effectiveness of steam injection programmes. This has proved very effective and the cost savings seen have meant that these systems are considered essential on new installations. Indeed, if government regulations permit, it is likely that by using the distributed temperature profiling system on the producing wells it will be possible to eliminate the need to drill observation wells, which are regulatory authorities required to ensure efficient management of the natural resources. This experience is now being extended to the 'light' oil market with an established product and installations have already been made Wytch Farm (Dorset, UK) and BHP systems (Liverpool Bay).

4.5.2.3.4 Steam Flow Monitoring

One of the most actively developed application in the field of oil and gas exploration, is the monitoring of the effect of steam injection in enhanced oil recovery. Many of the wells where steam injection is used are only marginally cost effective and a better management of the energy use in this type of oil field can substantially improve the profitability, allow a greater proportion of the reserves to be extracted and reduce the CO_2 emissions resulting from the extraction activity. The traditional approach to this problem is to drill a very large number of observation wells in addition to the production wells and log their thermal profiles using a wirelogging tool. In certain oilfields, well may be found every few meters. This approach is very labour intensive and thus only infrequent logs are made. In addition, horizontal wells cannot be easily be logged by a drive-by service.

In contrast, the DTS allows a fiber to be installed permanently in the well (including in horizontal wells) and it can be interrogated at suitable intervals by simply connecting a measurement system, obtained the entire thermal profile and then moving on to the next well to be logged. In shallow oil fields, the capacity of DTS800 systems is sufficient that the fibers from a number of wells can be connected in series and logged simultaneously. In other areas, the operators prefer to install the monitoring equipment permanently on the well and log automatically at pre-determined intervals, the data being sent by a communications link to the desk of the reservoir engineer.

One example of a data series obtained on a horizontal well during steam injection is shown in fig. 4.16. The lowest curve shown was obtained prior to steam injection. With increasing time from the start of the injection process, the temperature profiles are seen to rise, which provides the field operator with a wealth of information, including rate of penetration of the steam, the position of the heat front and the temperature at the point where the oil is extracted. Using this type of information, especially if collected over an entire field, the recovery efficiency can be improved dramatically, which will contribute to increasing the economically recoverable reserves in heavy oil or tar deposits.

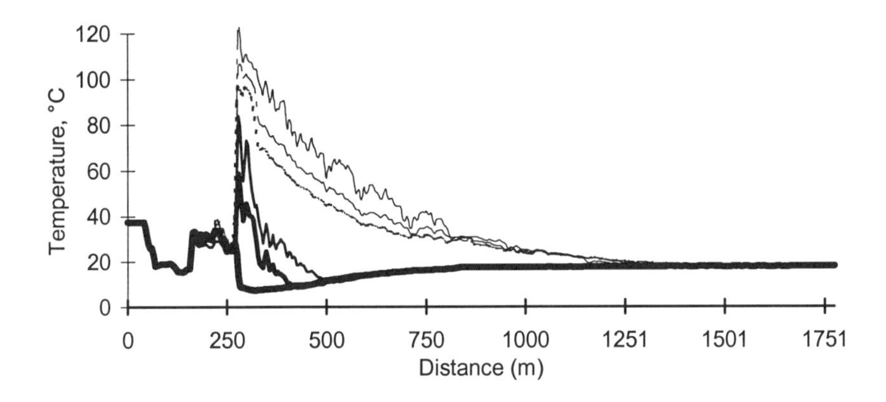

Figure 4.16. Family of thermal profiles recorded during steam injection in an horizontal well

4.5.2.3.5 Electrical Submersible Pump (ESP) Monitoring with Fiber Optic Sensors

Sensor Highway have installed a distributed temperature profiling system on many wells that have included an ESP installed as part of a secondary recovery programme. Experience has shown that as the temperature of the ESP will vary according to its operating conditions. These temperature changes provide information about events in the well that affect the performance of the ESP. The lifetime of an ESP is directly affected by the way it is maintained and operated. By monitoring the temperature changes at the ESP it is possible to identify events that are causing the ESP to operate beyond its specification. Temperature is a clear and direct measure of changes in the ESP operating conditions, for example:

- an ESP will heat up if sufficient flow past the motor is not maintained
- circulating fluids around an ESP will create a rapid temperature increase
- gas breakout at a pump will show up as a change in temperature
- a poorly running pump section will display an increase in temperature
- waxing/scale occurs below a certain temperature and pressure

A recent example of the value of having a distributed temperature system on an ESP is shown in figure 4.17. Sensor Highway installed a distributed temperature sensor across an ESP and then carried the fiber on across the well's perforated section. The plot below shows profiles taken :

- during the pump start-up
- a short time after the ESP was stopped
- after the ESP had cooled
- as the fluid ran back into the perforations

Looking closely at the figure, the temperature at the ESP at start-up, at the point the ESP was switched off and then when it had cooled back to its normal temperature are all apparent. It is estimated that, in the absence of distributed temperature monitoring on this ESP, the unit would have burnt out within 1½ hours of start-up. The cause of the problem was later traced to the premature failure of a burst disk used to purge the ESP of well fluids prior to being retrieved. This resulted in the produced fluid being circulated round the pump rather than being pumped to surface, which caused a loss of the cooling normally expected from a continuous flow past the pump. These plots clearly show the ESP heating rapidly to 290°F (~143°C), at which point the ESP was switched off.

The real time use of a DTS allowed Sensor Highway to recognize the problem before it became catastrophic, thus saving the operator the cost of a new pump.

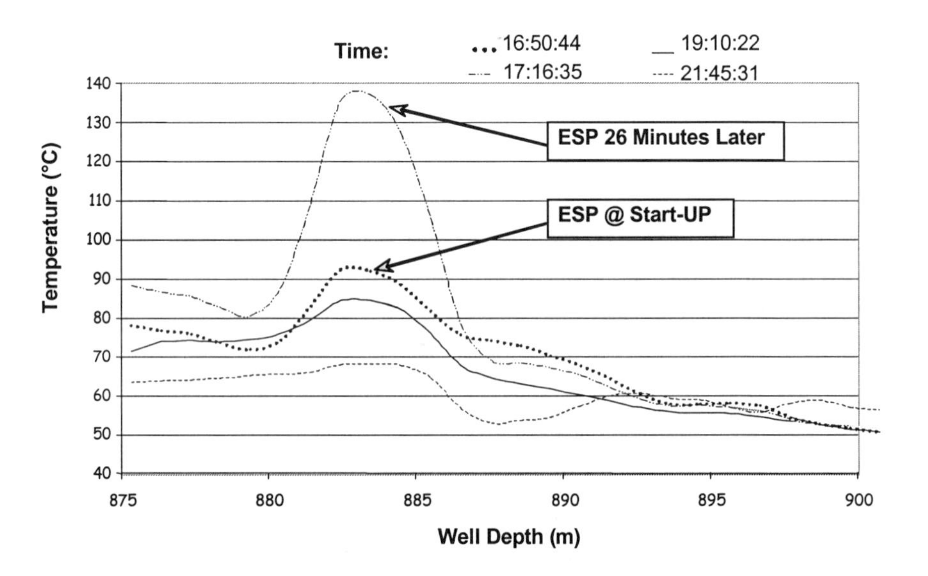

Figure 4.17. Temperatures recorded on a submersible pump

The second interesting event, figure 4.18, that was captured by the DTS system occurred across the perforations. The profiles shown below are the readings taken just after the ESP was shut off. These clearly show the heated fluid travelling back down the well and exiting the tubing through the perforations. The extent to which the fluid returns into the formation is a measure of the permeability of the specific perforation.

The amount of fluid that is exiting back into the perforations is reflected in the profiles and will provide the operator with information of the overall performance of the wells.

These two examples provide an idea of the economic benefits that can be derived from using the distributed temperature monitoring system as an ESP management tool. The cost of replacing one ESP system would probably be five times cost of installing the DTS system.

4.5.2.3.6 Reservoir Monitoring using fiber optics.

This can be considered under several different areas, as discussed below.

– **Monitoring the Near Well Bore Area**

Very recent observations on DTS measurement in oil wells could prove to be the most important application of this technology to the oil and gas industry. If the fiber optic distributed temperature sensor is placed across the length of the perforations (aperture in the well which allow the oil to flow into the bore and the collected by the well), the temperature of the perforations can be plotted as the well produces. This will provide the operator with information that can improve the recovery rates from his well and also from the reservoir.

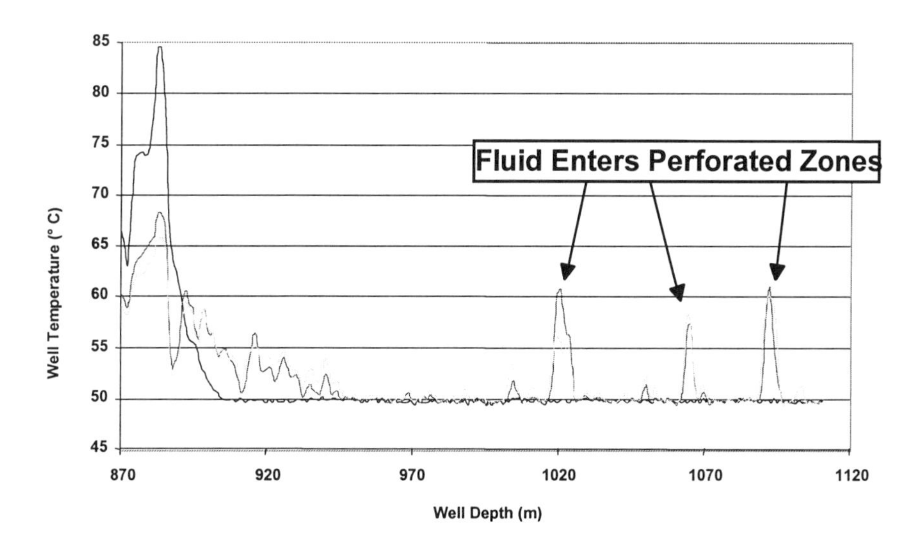

Figure 4.18. Temperature profile resulting from fluid flowing through perforations, after the pump has been switched off

– **Flow Rate Measurement**

In the same manner that flow is measured with differential pressure measurements it is possible to calculate flow from the change in temperature along the length of the well. Figure 4.19 shows an example of such data, where the change in temperature – over a 5 hour period – has been calculated and plotted as a function of position. Note the very high temperature resolution obtained in this case where the data is significant to better than 0.1°C and also the very long distance (4.5km for this data subset) over which the information has been gathered (in fact a total of 11km of fiber was installed in this well in a double-ended configuration). The very slight trends of the temperature allow those zones where the flow of oil originates to be clearly identified. This data will therefore allow the operator to review the effects on individual zones of changes in production.

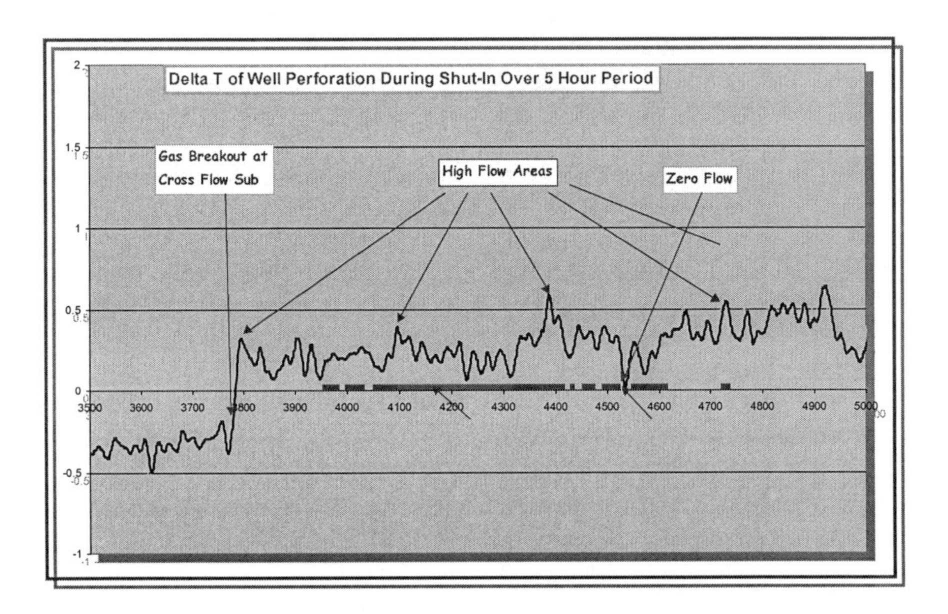

Figure 4.19. Differential temperature profile in a long horizontal oil well

Flow determination from DTS data allows most of the hardware required to make the flow measurements downhole to be removed, which reduces the cost of downhole instrumentation; moreover the fiber optic temperature measurement is far less intrusive than conventional instrumentation, especially where access beyond a flowmeter is required. Measuring the flow using temperature rather than pressure eliminates the need for a venturi section to be included in the well design.

A further method of analyzing the DTS data in oil wells is a comparison of the profile when oil is flowing with the natural thermal gradient, which exists where fluids are static. This differential curve may be analyzed to provide an estimate of the mass flow rate of the fluid as a function of position in the well.

– **Identifying Water Breakthrough**

Extracting water from an oil well is highly undesirable. In general, when water breakthrough occurs there is a change in the fluid temperature. This is especially true in the case of injected water breakthrough, but is also true of reservoir water as well. Reservoir water will generally breakthrough from a lower aquifer via 'water fingers', because of this the water entering the well bore will generally have a slightly higher temperature than the produced fluids above it. Detecting this small change in temperature indicates the onset of water breakthrough.

– **Flow Behind the Casing**

A temperature change detected at a point in the perforated section of a well, but at some distance from the perforations, is an indication that there is fluid flow behind the casing. This is especially useful where specific areas of the section have been isolated to prevent the production of water. If the distributed temperature profile shows that there is indeed flow at this point, the direction in which the fluid is moving can be determined and, from that information the location of the well integrity problem can be located.

– **Gas Cap Tracking**

In a similar manner to the detection of flow behind the casing, the progress of the gas cap can be tracked by monitoring the temperature of the casing string. As the gas cap descends, the corresponding temperature changes. This information allows the operator to try and prevent the gas cap penetrating into a well and if this is not possible he can preparc for the inevitable production of gas.

The examples given above illustrate how the DTS data, when interpreted by an experienced reservoir engineer can provided totally new data: the equivalent of a DTS does not exist with conventional instrumentation. Thus, whilst wirelogging tools can provide a temperature profile, they disturb the well conditions and thus distort the information they are collecting. The ability to log thermal profiles continuously during well production provides a new perspective on the well management.

4.5.2.3.7 Reliability of Distributed Optical Sensors compared with Electrical Sensors

Fiber optic sensors have two main benefits to oil field operators considering installing downhole permanent monitoring systems. Firstly they are more reliable than an electronic system owing to the simplicity of the downhole components. Secondly, because the downhole components are essentially passive instruments (i.e. no downhole processing) they can operate at higher temperatures and pressures. Other tangible benefits are also very important, e.g. the lower cost for multiple systems owing to the low cost of downhole components and the ability of the opto-electronic instrumentation to be multiplexed between multiple fibers.

Permanent monitoring systems have been considered very important tools with which to understand the reservoir. Although, the reliability of the electronic systems have proved to be poor, over 250 stand alone electronic permanent monitoring systems are installed annually. Current estimates are that between 25% and 50% of these systems are likely to fail within the first year.

The reason for this lies in the products design. Electronic sensors require electronic components to be installed downhole in order to send the signal from the gauge to the surface. Unfortunately hot, humid, environments are hostile to the electronics. It is generally the electronics of the signal processor or the sensor itself that fail prematurely. With fiber optics, there are no downhole components requiring electrical power or local electronics associated with them. All processing of information is carried out at the surface once the light signal has returned to the central processing unit.

4.5.2.3.8 Installation of fiber optic sensors

Fibers are installed in oil wells in one of two standard ways. The first is the standard hardwiring system, which as in the case of electronic systems, involves lowering a sensor cable into the well. However, Sensor Highway have also developed another method in which a control line (or sensor highway) is pre-installed during the completion phase of the well and the fiber optic sensor itself is installed at a later date by means of fluid drag (the fiber is carried along the tube by fluid pumped into the control line). This means that the downhole connections are all purely hydraulic – a technology familiar to the oil field services companies.

The importance of installing the fiber in this way lies in its simplicity and adaptability. Because the system is designed to have the fiber inserted *after* the sensor highway has been put in place, the technical issues that are associated with optical connections are removed. Although fiber optic connections can be very reliable they are obviously more complex than hydraulic connection and therefore they will have a higher potential to failure.

The sensor Highway system allows for replacement sensors to be installed at any time during the life of the field. Even if a fiber is damaged during the wells installation it is possible to retrieve the fiber and pump in a replacement fiber without any effect on production. The ability to install new sensors into the well with no workover means that system reliability is even greater than purely the difference in the mean time to failure of the two types of initial sensor.

Additionally, new fibers can be installed at any time without the stopping the production flow. It follows that the sensing system can be modified and improved with no incremental cost other than the cost of the improvement. New sensor designs can be installed at any time during the life of the field, allowing the operator to take advantage of technical advances in fiber optic technology as they happen.

The oil and gas industries would also benefit from distributed strain measurements for monitoring structures (e.g. platforms, pipelines, ships), distributed hydrocarbon gas measurement (leak detection) and distributed pressure measurement for analyzing distribution networks. However, the author is unaware of experimental data reported on other distributed measurands to date in the oil and gas industries.

4.5.3 Process Industry

There are a number of industrial applications to which DTS is ideally suited, such as the monitoring of long thermal curing or drying processes [58], where the temperature gradient along the oven is important.

In addition, the technique can be used for monitoring machinery to detect over-heating (e.g. in bearings) before damage to the equipment has occurred. In some cases, the scale of the equipment is such that multiplexing traditional sensors is not an option, such as in the monitoring of bearings on conveyor belts, where thousands of points may be required to checked.

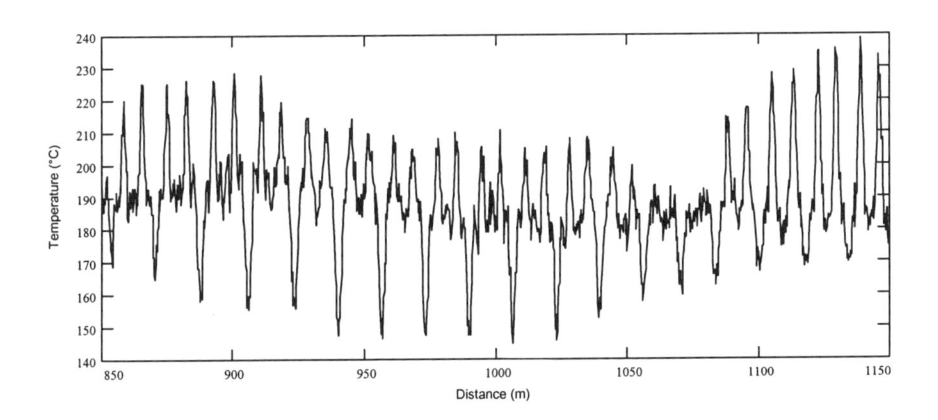

Figure 4.20. Measured surface temperature distribution of a secondary reformer

The heat, pressure and corrosive environments in certain large scale processes require monitoring of the health of the walls and linings of the vessels. For example, if the state of the linings of a blast furnace can be deduced from a thermal profile of its wall, significant cost savings can be gained from reduced and better-planned down-time. Other chemical processes are, out of necessity, run close to the maximum thermal ratings of the containment vessels and any instability in the process could result in temperature excursion sufficient to weaken the strength of the material forming the enclosure. Several York Sensors instruments are presently monitoring the walls of such vessels. One such example is shown in fig. 4.20, and is a subset of a thermal profile of a fiber wrapped around a refractory-lined vessel, known as a secondary reformer. In this case the refractory lining was poorly installed and it contains cracks which allowed the hot gases from the process to move behind the lining onto the inner surface of the steel vessel. This accounts for the very strong periodic appearance of the thermal profile, since the fiber crosses the hot line for every turn. Over most of the fiber length the perimeter is of order 16m, but given the conical shape of the vessel, the period is reduced at the left-hand side of the profile.

4.5.4 Geophysical Measurements

The ability of the DTS to monitor down boreholes has led to its application for geological research, where rapidly obtained thermal profiles of wells (2km depth being the deepest achieved so far, using a York Sensors DTS-80) allow the geologist to understand the structure below ground. This has been applied to the determination of the suitability of waste disposal sites; fig. 4.21 shows such a profile logged in a salt-rock formation as part of a survey for waste disposal.

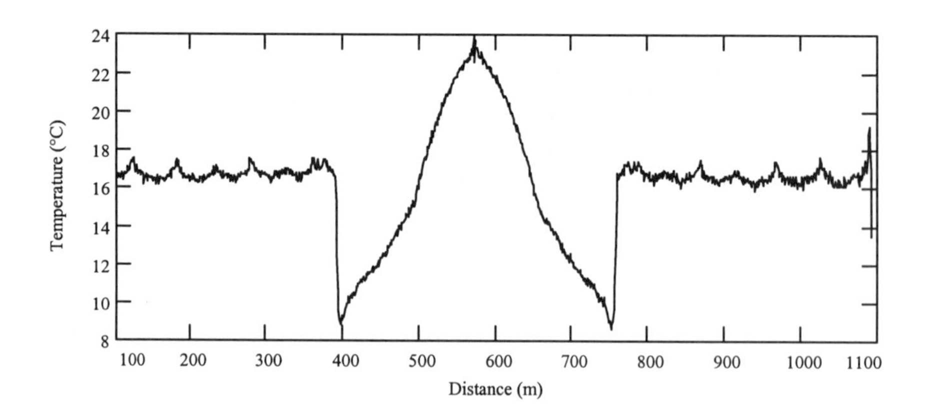

Figure 4.21. Thermal profile in a salt-rock formation

Similar techniques are used to assist in the prediction of earthquakes and volcanic activity: small changes in the thermal distribution along a borehole can reveal, long before surface measurements, small precursor movements of the earth's crust. The physical mechanism causing the temperature changes is thought to be the variation of the flow in underground water channels. The profile of fig. 4.22 was measured in a deep (>2km) borehole for just such a purpose.

In both cases, very high temperature resolution (±0.3°C) is required and has been demonstrated.

An experimental distributed temperature sensor has been deployed at the bottom of the sea, at a water depth of 3000m, to help study the hydrothermal activity at mid-ocean ridges, where minerals pour out of the center of the earth into the sea [59], following trials in lakes and near-shore seawater [60].

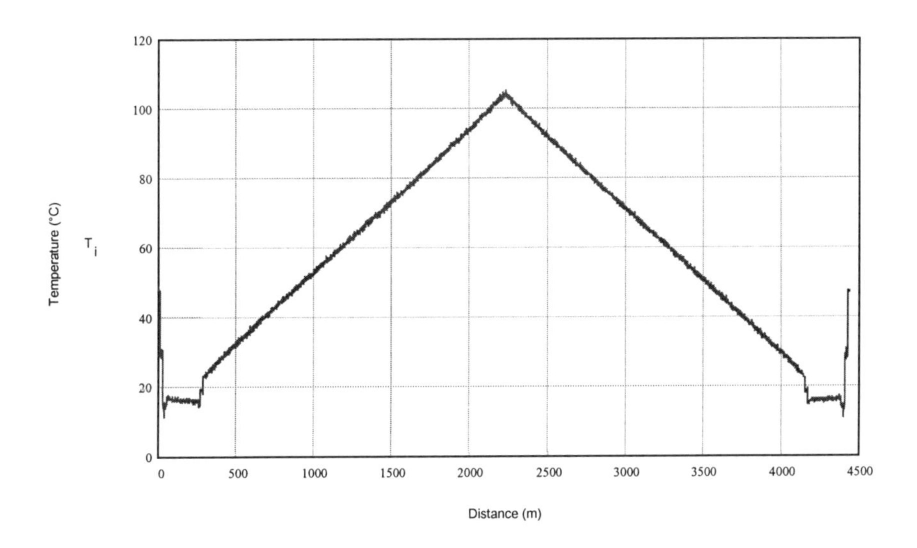

Figure 4.22. Double-ended thermal profile measured in a deep borehole

4.5.5 Tunnels

The DTS has important applications inside buildings and tunnels as a fire-alarm system. For example, fires in railway tunnels frequently start in cable trays and can smoulder for a long time undetected before a fire breaks out. When a fire does occur, the consequences are of course truly terrifying. A fiber installed along the tunnel could detect the outbreak of a fire very quickly, wherever it occurs. A fiber placed in the cable tray could detect overheating before a fire actually occurs and thus help to avoid the incident.

Likewise, fires in road tunnels, possibly started by accidents or fuel spillage, require immediate alerting of the emergency services and actions to be taken to divert incoming traffic. Trials have taken place to develop knowledge of both applications [61] and DTS systems have now been installed permanently in railway tunnels – for example in Milan Underground. Two DTS 800 units from York Sensors have been used to monitor temperature of about 18 kilometres of optical fibers; one is placed in Lancetti station and the other in Repubblica station. Each DTS800 is connected to three loops of optical fiber.

Other systems are used to monitor fires in escalators leading to underground trains. The accumulation of dirt and grease in the escalator machinery has been a contributory cause of fires, such as that at King's Cross in London in 1987.

4.5.6 Construction

In buildings, the DTS can be used not only as a fire alarm, but also as a means of measuring the temperature in each area to provide inputs to heating and air conditioning systems. Of course, the fire alarm and environment monitoring functions can be combined in a single system. One major advantage of the DTS in these applications is that the amount of wiring is reduced considerably. With dedicated software, it should also be possible to reduce the incidence of false alarms since an unusual condition can be rapidly qualified as a real emergency or as equipment malfunction. A break in the fiber, for example, can be identified by the processing software as a fiber break (and its location given) and not as a fire alarm. The DTS is currently being integrated in at least one suppliers building control system.

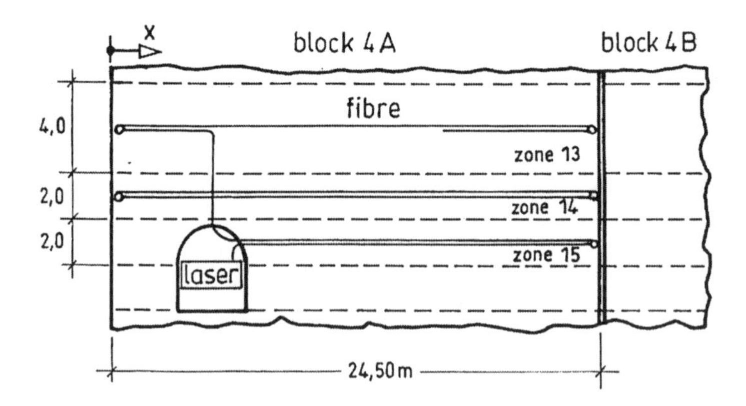

Figure 4.23. Birecik Dam - order of the fiber inside block 4A

Extensive trials have taken place to evaluate the DTS as a means of monitoring the exothermic curing of concrete in dams: excessive heat generated during this process can result in weakened structure and it was shown that by burying an (expendable) fiber within the structure, its thermal profile can be determined in three dimensions. This allows the rate of addition of new material to be adjusted to ensure that no thermal run-away can occur. Although the original trials were carried out in Japan at the Miyagase dam in the 1990-91, the German company GTC Kappelmeyer has recently re-started this type of measurement on a commercial basis. Their work on the Birecik dam in Turkey is illustrated in figs. 4.23 - 4.25. The layout of some of the fiber installed is shown in fig. 4.23. Fig 4.24 shows the temperature rising in Zone 15 in the first few days after the concrete was poured, showing a temperature of some 17°C resulting from the chemical reaction; six months later the increased temperature is still apparent in the center of the structure, although the section of fiber closest to a inspection gallery – which can channel away the heat has fallen considerably. Fig. 4.25 shows the temperature profile at three different heights in the dam, five months after the pouring of the concrete. Again the residual heat in the center of the structure is plainly visible.

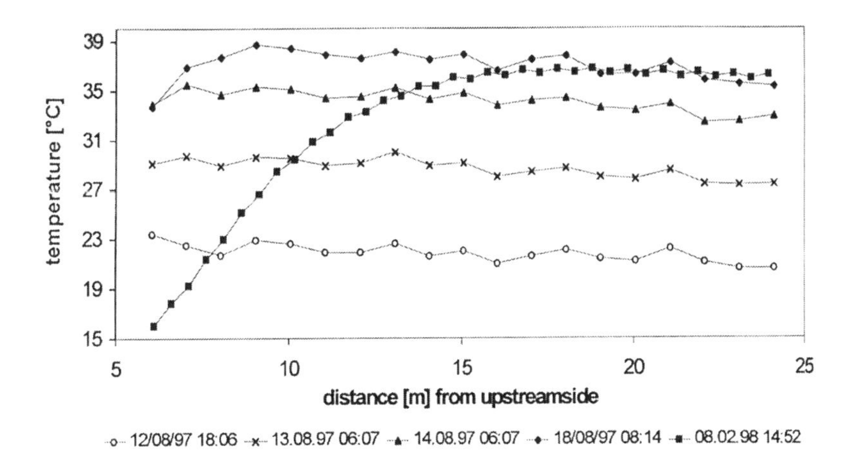

Figure 4.24. Curing temperature development in concreting section 4.15. The left hand side is adjacent to the inspection gallery

GTC Kappelmeyer have also used the DTS technique to detect leaks in dams and water channels: a fiber is installed below the lining of the structure and their method relies on a temperature difference between the ground and the water flowing from a leak. Even though the temperature differences can be small, a fraction of 1°C is sufficient to indicate the existence and location of an imperfection in the lining.

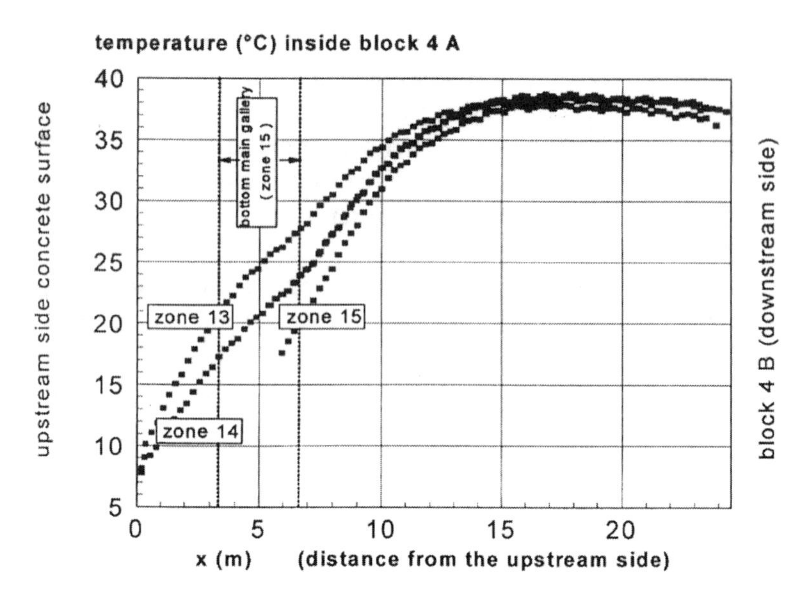

Figure 4.25. Birecik Dam - temperature distribution five months after concreting

A distributed strain sensor is thought to have a large applications area in construction, where many structures (bridges, dams, tall buildings) would benefit from detailed monitoring for strain and crack growth. Trials involving the embedding of fiber-optic strain gauges into concrete bridges have already taken place, but have not yet involved distributed strain sensors.

4.5.7 Transportation

Distributed temperature sensing has received much interest from aviation for fire detection, skin temperature measurements and for the monitoring of the curing in composite panels.

In the latter case, the fiber is embedded in the composite material and the exact temperature reached by the inside of the panel accurately determined. Similarly, there is interest is a strain measurement method to monitor the time dependence and the uniformity of the stress applied by the manufacturing process. Later in the life of the component, a built-in strain measurement can determine whether excessive loads are applied to the component, e.g. to a panel during flight. This is the so-called 'smart-skin' concept which has attracted considerable interest in the past few years. It promises structures where feedback and software techniques may be used to push the components to their design limits with health monitoring providing accurate estimates of the remaining life of components, making the whole process of preventative maintenance more efficient and yielding lighter structures for a given design strength.

Interests in the area of shipping have to date concentrated on fire and flooding detection as well as measurements associated with on-board power generation and machine health monitoring, and tests have been carried out on static vessels.

4.6 EXAMPLES OF PRACTICAL IMPLEMENTATIONS OF DISTRIBUTED SENSORS

In this section, a few distributed sensor designs which have been taken to the stage of commercial products or field trials will be described briefly. In recent years, the number of distributed sensor installations has grown rapidly. The author estimates that the total number of distributed temperature sensors sold to date based on Raman scattering alone exceeded 250 units worldwide in mid 1998. York Sensors alone has achieved an experience of 600 machine-years by the end of the twentith century. The latter figure applies only to systems having a diode-pumped solid state laser; it excludes sensors based on semiconductor lasers as the source.

4.6.1 Herga Pressure Mats

For a number of years, Herga Ltd has supplied a simple sensor based on microbending and used *inter alia* within pressure-sensing mats to protect personnel working close to robots or to detect collisions of remotely controlled vehicles [62]. In this case, the entire length of a section of standard graded-index multimode fiber is made pressure-sensitive by winding a thin plastic spiral around it at a pre-determined pitch. Very slight pressure at any point along the sensitized fiber will cause a large fraction of the power to be lost. Although not a truly distributed sensor, it is capable of detecting an external pressure anywhere along the fiber and could clearly be adapted using some of the techniques described above to be truly distributed. Instead, the manufacturers have chosen to design a sensor, the electronics of which are extremely simple and thus capable being analyzed for failure modes, an essential attribute in such a safety-critical application, where the sensor forms part of the primary safety system of the machinery it protects. Moreover, the approach leads to a design fully able to compete on cost with any other safety sensor.

4.6.2 Distributed Cryogenic Leak Detection System

The transmission of light by an optical fiber is critically dependent on the maintenance of a core:cladding refractive index ratio greater than unity. However, in certain fibers, this ratio, and consequently the fiber guidance, varies as a function of temperature. By appropriate choice of core and cladding materials, it is possible to arrange for the refractive indices of core and cladding to converge with decreasing temperature [17]. At temperatures below -55°C, the refractive index of the cladding material becomes higher than that of the core, thus causing the light to be guided preferentially in the cladding. The inferior optical quality of the cladding material, together with the sensitivity to surface quality, ensures that this light is either scattered and lost from the fiber or severely attenuated. The net effect is a linearly distributed temperature switch, sensitive over its entire length, which ceases to transmit light when in contact with fluids below -55°C anywhere along its length.

Traditionally, sensors for LNG leakage detection have been predominantly thermocouples which, being point sensors, need to be deployed in large numbers to achieve satisfactory cover. To achieve a cover equivalent to that provided by, for example, 100 meters of optical fiber would require in the region of 4000 thermocouples, with the associated cabling and electronics costs. In real installations, therefore, thermocouples are distributed sparingly and the reduced level of cover is accepted. For, say, 12 thermocouples or 100 meters of optical fiber sensor, the overall costs, dominated by the installation costs are similar.

The Cryogenic Leak Detection system comprises an electronic monitoring unit located in a safe area control room, a two-core optical fiber link cable to the sensing area and a sensor fiber protected by a small diameter stainless steel tube. Leak alarms are initiated by increases in optical fiber attenuation greater than 1.5 dB, which is equivalent to a 30% reduction in received light. Extensive testing of laboratory, prototype and commissioned industrial systems have demonstrated response times of typically less than 10 seconds for 25 ml of liquid nitrogen.

The electronic monitoring unit, which is capable of operating with several kilometres of optical fiber, has been designed to incorporate a high level of fail-safety. Clearly the fiber itself is fail-safe: damage to it is equivalent to the effect of a leak and a number of checks are performed in the electronics to hence the dependability of the device.

By 1991, there had been 15 systems installed on British Gas LNG storage sites and predictions of high system reliability and a calculated mean time between failures (MTBF) of 14 years having being borne out by operational experience, approaching 100 instrument-years. In addition, long-term optical fiber condition monitoring, since the project inception in 1981, has shown no evidence of optical fiber degradation in correctly installed systems.

4.6.3 Distributed Temperature Sensor Based on Raman Backscattering

The first fully distributed temperature sensor system to be produced commercially (1987), the DTS-II supplied by York Ltd [23], was designed around the principle of spontaneous Raman scattering (see section 4.3.6) and incorporated a semiconductor laser source operating at 904 nm and Si APD detectors. Typically, a 2 km fiber could be measured in 12 s to a repeatability (one standard deviation) of 0.4°C with a spatial resolution of 7.5 m and in the years since the release of that system, its performance has been gradually enhanced to cover ranges typically greater than 4 km in a similar measurement time.

In 1991, an improved system (DTS-80) was released by the same company and improvements to the design have resulted in a reduction of the spatial resolution to approximately 1m, whilst the length of the sensing fiber has been increased to beyond 10km. The increase in performance was achieved in large measure by the incorporation of a diode-pumped solid-state laser delivering higher-power, but narrower pulses into the fiber at a wavelength at which the loss is half that experienced by the DTS-II. The progress in the optics was matched by higher bandwidth electronics, high-speed analog-to-digital conversion and digital averaging circuitry, capable of operating at 100 megasamples/s and of course by increased computing power (32 bit processor) to analyze the raw data into temperature results inside the instrument itself.

EU regulation, in particular on electromagnetic compatibility resulted in 1995 in an updated system, DTS 800, which is offered in a variety of configurations, including single-mode fiber versions capable of a range of up to 30km.

The performance of these sensors has continued to improve substantially since the initial launch of the DTS80, and much experience has been gained as to the installation and calibration of these sensors, as well as in the requirements of the display, storage, interpretation and usage of the results. The integration into SCADA systems, as well as remote communications, e.g. over telephone lines has been addressed.

The technology is rapidly gaining acceptance in a number of applications, notably energy cable monitoring. Counting only systems based on diode-pumped solid state lasers (i.e. ignoring laser-diode based systems), York Sensors alone has delivered over 150 systems which now have a cumulative experience approaching 600 machine-years.

4.6.4 Distributed Cable Strain Monitor

Engineers at NTT have developed, over recent years, a distributed strain measurement technique, which is described in section 4.3.8 above. The technique is based on the measurement of the gain of the stimulated Brillouin scattering process as a function of the frequency offset between two counter-propagating waves. In spite of the apparent delicacy of the measurement (two sources required to be tuned in frequency to accuracies of a few tens of MHz), the use of diode-pumped solid state lasers, allowed field trials to be carried out successfully. In particular the strain distribution applied to a fiber optic cable during and after its installation was monitored, although with only moderate resolution.

4.6.5 Fiber-Optic Hydrophone Array

In section 4.3.9, the operation of an acoustic sensor based on reflections from suitably positioned reflective splices was described. This device (implemented as a hydrophone array) has been developed by Plessey Ltd at least to the stage of successful sea-trials. Several stages of development were reported and one of the major problems which the engineering team had to solve was that of achieving a source of adequate coherence.

4.7 SAFETY OF DISTRIBUTED SENSORS

4.7.1 Explosion hazards

Much is made of the intrinsic safety of optical fiber sensors and, although in most cases it is not in doubt that the sensors are inherently safe, it is possible to ignite flammable vapors by radiative heating, under certain conditions [63].

In 1986, when the first tests of a distributed fiber-optic temperature sensor were carried out on a British Gas site, there was, and still is, an absence of standards relating to safe optical power levels in flammable atmospheres. However, an extrapolation of the latest results, from work funded by OSCA in the UK [64], suggested that a 50 μm diameter particle absorbing 6.6 mW cw from a 50 μm core optical fiber might cause ignition in a fuel-rich diethyl ether/air mixture. This extrapolation has in fact never been corroborated and in later work using electrically heated particles, it has been impossible to create ignitions at powers below 55 mW with objects of diameter less than 0.25 mm; an exception exists, however for catalytic particles, where the critical minimum power level is reduced to approximately 37 mW [65].

In subsequent research, Hills et al [66] have achieved ignitions in hydrogen/air mixtures using optical sources with power levels around 120 mW cw, by irradiating a pile of coal dust particles of less than 38 μm diameter, through a 10 μm core singlemode fiber. This currently represents the lowest optical power with which an ignition has actually been achieved.

In a review of the safety aspects of distributed temperature sensing, a detailed analysis of the power capabilities of DTS-II showed conclusively, that even under worst case conditions [64], it was not capable of causing an ignition. The subsequent experimental work suggests an even greater margin of safety thereby confirming the inherent safety of DTS-II and in fact even of DTS-800 with its higher power output, the latter having been reviewed by a third-party body and recommended for use, at least in Zone 2 applications.

No similar analysis has been published for other distributed fiber-optic sensors. Their ability to be used safely in hazardous environments will depend primarily on the average power emitted (rather low in pure OTDR methods, but increases with pulse compression-coding techniques and certain non-linear optical methods) and on the size of the fiber core. Safety officers, in addition will want to evaluate the maximum output which could be generated under fault conditions. Clearly before a system is accepted into an area where a risk of optically induced explosions exists, careful consideration of the optical output at a possible fiber should be thoroughly analyzed.

Work funded by the EU [67] has resulted in a series of recommendations on the maximum power to be launched in fibers to avoid causing ignition hazards in the event of a fiber break. For multimode fibers, under most practical conditions, the recommended limit is an average power level of 35mW. It should be noted that under certain conditions, in particular if single-mode fibers are involved, the limit may be lower.

4.7.2 Laser Safety Considerations

There is a further hazard in the use of distributed fiber optic sensors which must be considered, namely the damage which might be caused to humans, especially to their eyes, if exposed to high power optical radiation. Since fiber optic sensors will frequently be used in uncontrolled areas, where people untrained in the safe use of lasers may be expected to work, the laser safety of fiber optic sensor systems is of real importance.

Until a few years ago, the only published standards were those established for free space laser beams (e.g. BS 4803 (UK), ANSI Z136.1 or

IEC 825). These standards were often felt by those working in the field of fiber-optic sensors to be unduly restrictive to fiber optic applications of lasers since they did not adequately take into account the divergent nature of the light emerging from the fiber end, or they were perhaps excessively severe when dealing with repetitively pulsed sources. Recently, however, much progress has been made towards these two problems, thanks in no small part to the work of Petersen and Sliney [68].

The result has been an update of the standards to reflect more accurately the hazards of fiber-optic systems and which grade the hazard levels more finely than previously. It is now recognized that in normal operation a fiber-optic system encloses the power transmitted and that a hazard therefore only arises at the end of fiber, e.g. at a break or at an open connector. For example, an addition to the ANSI standard, Z136.2, defines a series of service groups for fiber-optic systems and takes into account the divergence of the beams emerging from unterminated fiber ends. The service group classification may then be used to determine in which areas the system may be deployed, what labeling and safety devices will be required. The British standard has also been revised (to BS7192), as has the IEC standard (modification 1).

No general comments can be made here as to whether a distributed system will be outside of the inherently safe category (Class 1 or Service Group 1); it is sufficient to comment that components are presently commercially available with ample power to exceed the limits of inherently safe operation. It is therefore incumbent on the designers and users of such systems to check the power output of their sensors to be able to classify and restrict access as appropriate. In practice, the precautions required of the standards may be relatively simple to carry out, depending on the optical power involved, e.g. label connectors, enclose these in tamper-proof boxes, train operators in the potential hazard presented by the equipment.

4.8 FUTURE PROSPECTS

The performance of existing distributed fiber optic temperature sensors is already sufficient for a number of applications, but falls short of the requirements for many others. In particular, the spatial resolution will need to be improved for many industrial applications and work is progressing in this area. This will involve the development or adaptation of more suitable sources, refinements of the electronics and possibly of the fiber itself.

Other applications such as pipeline monitoring, will demand extreme range and it is projected that systems spanning about 20 km of fiber should be achievable without unduly sacrificing performance in other respects;

longer range has already been achieved for lower spatial resolution and/or longer measurement time.

Attention is also turning to the measurement of other physical parameters such as strain, magnetic and electric fields, detection of gases and chemicals for pollution, monitoring acoustic waves and so on. This will almost certainly involve the development of special fibers with tailored sensitivity. It is not yet clear whether some of the very advanced methods described above (e.g. non-linear optical processing [69]) will find their way into commercial sensors. However, there is a clear requirement to sense a multiplicity of measurands and the research is being driven actively.

Although, for most measurands, the technical and commercial challenges are still in the basic measurement technology, the ability to apply these new methods to solve practical industrial problems will increasing dictate whether or not the distributed sensor concept will progress beyond the laboratory stage for measurands other than temperature, where the body of applications knowledge is growing rapidly.

4.9 REFERENCES

1. Barnoski, M. K. and Jensen, S. M. (1976) *Appl. Opt.*, **15**, 2112-5.
2. Hartog, A. H. and Gold, M. P. (1984) *J Lightwave Technol.*, **LT-2**, 76-82.
3. Neumann, E-G (1980) *AEU*, **34**, 157-60.
4. Hartog, A. H. (1983) *J. Lightwave Technol.* **LT-1**, 498-509.
5. Farries, M. C. et al (1986) *Electron. Lett.* **22**, 418-9.
6. Ferdinand, P. (1990) *PhD Thesis*, Nice.
7. Ross, J. N. (1981) *Electron. Lett*, **17**, 596-7.
8. Mickelson, A. R. et al (1982) *Appl. Opt.*, **21**, 1898-909.
9. Conduit, A. J. et al (1981) *Electron. Lett.*, **17**, 308-10.
10. Gold, M. P. et al (1982) *Electron. Lett.*, **18**, 489-90.
11. Gold, M. P. et al (1984) *Electron. Lett.*, 20, 338-40.
12. Theocharous, E. (1986) *IEE Colloquium on distributed optical fibre sensing*, London, May 1986 (Digest N° 1986/74)
13. Claus, R. O. et al, (1985) *Proc. SPIE*, **566**, 243-8.
14. Bergqvist, E. A. (1991) An optical fiber cable for detecting a change in temperature. *European Patent Application* 0 490 849A1 26 Nov. 1991.
15. Michie, W. C. et al (1994) Optical Fiber Grout Flow Monitor for Post Tensioned Reinforced Tendon Ducts. *Proc 2nd European Conf. On Smart Structures and Materials*, Glasgow, 1994, 186-9.
16. Gottlieb, M. and Brandt, G. B. (1981) *Appl. Opt.* **20**, 3867-73.
17. Pinchbeck, D. and Kitchen, C. A. (1985) *Proc Electron in Oil and Gas*, London.
18. Hartog, A. H. et al, (1980) *Proc 6th European Conf. Optical Communication* (post-deadline session) York, UK.
19. Rogers, A. J. (1980) *Electron. Lett.*, **16**, 489.
20. Fabelinskii, I. L. (1968) *Molecular light scattering*. Plenum Press.
21. Hartog, A. H. et al (1985) *Electron. Lett*, **21**, 1061-3.

22. Tanabe, Y. et al (1989) *Proc. OFS*, Paris, Sept. 1989, 537.
23. York Sensors Ltd, Chandler's Ford, UK, DTS System-II
24. http://www.york-sensors.co.uk
25. Hartog, A. H. (1995) Distributed fiber-optic temperature sensors: technology and applications in the power industry. *Power Engineering*, June 1995.
26. Lees, G. P. *et al* (1998) Recent Advances in Distributed Optical Fiber Temperature Sensing using the Landau-Placzek ratio. *SPIE Distributed and Multiplexed Fiber Optic Sensors VII*, accepted for Publication, November.
27. Wait, P. C. and Newson, T. P. (1996) *Optics Commun.*, **122**, 141-6.
28. Wait, P. C. and Newson, T. P. (1996) *Optics Commun.* **131**, 285-9.
29. Souza, K. De *et al* (1996) *Electron. Lett*, **32**, 2174-5.
30. Lees, G. P. *et al* (1998) *Photonics Technol. Lett.*, **10**, 126-8.
31. Lees, G. P. *et al*(1996) *Electron. Lett.*, **32**, 1299-300.
32. Farries, M. C. and Rogers, A. J. (1984) *Proc. 2nd Int. Conf. Opt. Fiber Sensors*, Stuttgart, Sept. 1984, 121-32.
33. Valis, T. *et al*, (1988) *Proc. SPIE*, **954**, 83.
34. Dakin, J. P. (1987) *Proc. SPIE*, **798**, 149-55.
35. Ahmed, S. U. *et al*, (1992) *Opt Lett.*, **17**, 643-5.
36. Shibata, T. Opt. Lett. vol. 12, p 269-271, 1987.
37. Culverhouse, D. *et al*, (1989) *Electron. Lett.*, **25**, 913-4.
38. Horiguchi, T. *et al*, (1989) *IEEE Photon. Technol. Lett*, **1**, 107-8.
39. Tateda, M. *et al*, (1990) *J. Lightwave Technol.*, **8**, 1269-72.
40. Trutna, W. R. *et al*, (1987) *Opt Lett*, **12**, 248-50.
41. Dakin, J. P., Wade, C. A., Henning, M. L. (1984) *Electron. Lett.* **20**, 53-4.
42. Franks, R. B. *et al*, (1985) *Proc SPIE*, **586**, 84-9.
43. Nakayama, J. *et al*, (1987) *App. Opt*, **26**, 440-3.
44. MacDonald, R. I. (1981) *Appl. Opt.* 20, 1840-4.
45. Venkatesh, S. and Dolfi, D. W. (1990) *Appl. Opt.*, **29**, 1323-6.
46. Healey, P. (1981) *Proc. 7th Eur. Conf on Opt. Commun.*, 5.2-1-5.2.4.
47. Bernard, J. J. *et al*, (1984) *Symposium on Opt. Fiber Measurements, National Bureau of Standards*, Boulder Co, 1984, NBS Publication p683, 95-98.
48. Bernard, J. J and Depresles, E. (1987) *Proc SPIE*, **838**, 206-9.
49. Everard, J. K. A. (1989) *Electron. Lett.*, **25**, 140-2.
50. Nazarathy, M. *et al*, (1989) *J Lightwave Technol*, **LT-7**, 24-38.
51. Healey, P. *et al*, (1982) *Electron. Lett.* **18**, 862-3.
52. Healey, P. (1984) *Electron Lett*, **20**, 30-2.
53. Eickoff, W. and Ulrich, R. (1981) *Appl. Phys. Lett.*, **39**, 693-5.
54. Uttam, D. *et al* (1985) *J. Lightwave Technol.*, **LT-3**, 971-6.
55. Sorin, W. V. and Donald, D. K. (1990) *Symposium on Optical Fiber Measurements*, Boulder, Co 1990, NBS Publication 792, 27-30.
56. Stierlin, R. *et al*, (1987) *Appl. Opt.*, **26**, 1368-70.
57. Stein, J. (1989) *EPRI survey*, Electrical Power Research Institute, Palo-Alto. Ca, June 1989.
58. Marcus, M. *et al* (1989) *SPIE Boston*, Sept. 6-8, 1989.
59. Nishimura, K. *et al* (1995) Development of seafloor thermometry system using optical fiber distributed temperature sensor for study of mid-ocean ridges. *Proc. Ocean '95 MTS/IEEE*, October 1995.

60. Nishimura, K. and Matsubayashi, O. (1996) Application of an Optical Fiber Distributed Temperature Sensor to Marine and Lake Surveys. *J of the Japan Soc. For Marine Surveys & Technol.*, **8**, 17-31, March 1996.
61. Norman, S. R. *et al*, (1992) *IEE Colloquium on Fiber optics Sensor Technology*, London, May 29, 1992.
62. Oscroft, G. J (1987) *Optical Fiber Sensors*, **2**, 269-79.
63. Moore, S. R. and Weinberg, F.J. (1983) *Proc R. Soc.*, London A.**385**, 373.
64. Tortoiseshell, G. (1985) *Fiber Optics '85*, London 1985, SPIE Proc. **522**.
65. Tortoishell, G. (1990) The safety of Optical Systems in Hazardous Areas. *SPIE Proc.* **1266**.
66. Hills, P. C. *et al* (1991) *Proc. 7th Int. Conf Optical Fiber Sensors*, Sydney, Australia.
67. European Commission: Report EUR 16011 EN, *Optical techniques in industrial measurement: safety in hazardous environments*
68. Petersen, R. C. and Sliney, D. H. (1986) *Appl. Opt.*, **25**, 1038-47.
69. Rogers, A. J. (2000) Nonlinear optics and optical fibers. *Optical Fiber Sensor Technology 5*, Eds Grattan, K. T. V. and Meggitt, B. T., Kluwer Academic Publishers, 187-238.

5

Referencing Schemes for Intensity Modulated Optical Fiber Sensor Systems

G. Murtaza and J. M. Senior

5.1 INTRODUCTION

For nearly two decades, intensity modulation has remained as one of the most extensively investigated forms of optical signal modulation for sensing applications [1-10]. The simple reason for the extensive and diversified usage of this modulation scheme is a multitude of potential benefits that include the inherent simplicity, reliability, flexibility and relatively low costs. Although intensity modulated optical fiber sensors have been fabricated in many different designs and with varying degrees of complexity, the essential building blocks of a simple optical fiber sensor system are depicted in figure 5.1. Light from an optical source, such as an LED, is coupled into an optical fiber for transmission to the optical sensor where it can be modulated in accordance with the state of a measurand. When using reflection-mode sensing the modulated optical signal is retroreflected into the same optical fiber for transmission to the photodetector [8]. However, in transmission-mode sensing a second optical fiber is normally used for the transmission of the modulated signal to the photodetector.

One example of a simple intensity modulated sensor is the optical microswitch which measures the ON and OFF states of a device by monitoring the continuity of an optical beam [2,11]. An extension of the optical microswitch principle that has evolved into a digital sensing technique is its use in conjunction with, for example, a Gray-coded disc [12] which forms the basis of an optical digital encoder [13]. Whereas the advantage of such a digital sensor is that it does not require any referencing, the cost of components can be significant for this configuration. In general, an n-bit system utilizes n optical microswitches and n discrete optical signals to interrogate them [13].

K.T.V. Grattan and B.T. Meggitt (eds.), Optical Fiber Sensor Technology, 303–336.

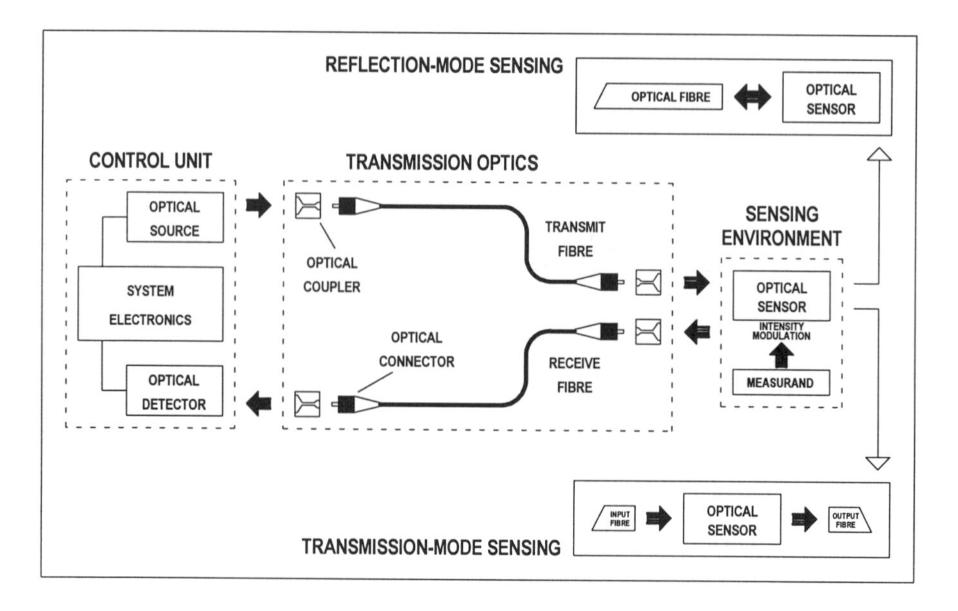

Figure 5.1. Essential components of a basic optical fiber sensor system and the arrangement for reflection-mode as well as transmission-mode sensing

In analog measurement applications of intensity modulated optical fiber sensors, there is the requirement that the output from a system should be truly related to the measurand alone. In practice, this condition cannot be easily satisfied due to the variable losses within the optical components (e.g. optical fiber leads, optical couplers and connectors shown in figure 5.1). Furthermore, additional measurement unreliability can arise from the instability of the optoelectronic components such as optical sources. Typical spectral emission variations of a near-infrared LED are illustrated in figure 5.2 where the numbers adjacent to the curves indicate the corresponding junction temperature in degrees Celsius. It is impossible to eliminate these variations in any optical fiber system design but compensation may be applied by monitoring the undesirable optical signal losses. The usual approach is to generate at least one additional (reference) signal which may then, in conjunction with the measurand signal, be used to make a relative measurement that is free from these so called common-mode variations. Both signals need to be separately identified, and as illustrated in figure 5.3, a distinction is normally achieved by employing either spatial separation, wavelength separation, temporal separation or frequency separation or a combination of these methods.

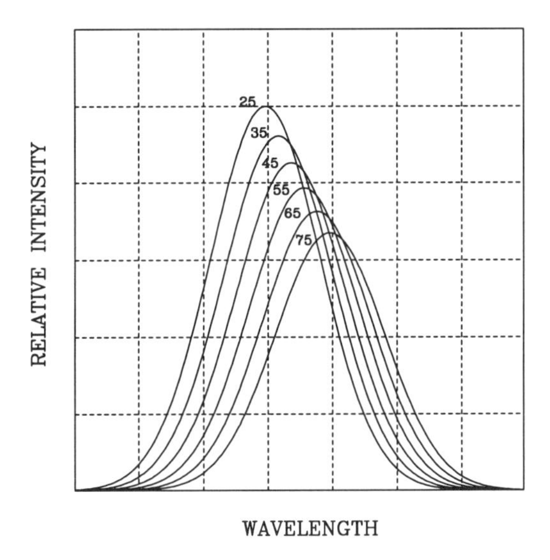

Figure 5.2. Typical spectral emission variations resulting from the junction temperature changes (indicated in degrees Celsius) of a near-infrared LED source

Spatial separation is required when the optical signals have the same spectral constitution as in the case where they are generated from a single optical source (which is usually an LED but it can also be a laser or a lamp) used in conjunction with a Y-coupler or from two similar optical sources. Such signals can only be identified by separation within different transmit and receive components so that they can remain physically isolated as shown in figure 5.3(a). In the case when each signal is contained within a separate spectral band both signals can be transmitted on the same optical fiber link, as depicted in figure 5.3(b). The presence of spectral spacing between the signal channels enables wavelength separation by the use of simple wavelength demultiplexing elements. However multiplexing components may also be necessary when the optical signals are launched from separate light sources. Temporal separation can be employed when the signals are generated from separate optical sources as illustrated in figure 5.3(c). It is achieved by time division multiplexing (TDM) of the two optical signals. The optical signals are then transmitted within separate time slots and therefore they can be separately identified from their time of arrival at the point of detection. As illustrated in figure 5.3(d), temporal separation can also be produced between pulses of light obtained by dividing the optical power in a single parent pulse into two pulses followed by the introduction of a time delay between them. An alternative signal separation method called the frequency division multiplexing (FDM), shown in figure 5.3(e), is sometimes employed instead of the TDM method. The FDM operation involves electrical modulation of each optical source at a different carrier frequency. The optical signals are then separated by the use of electrical filtering of the photocurrent into the respective frequency components. An advantage with both TDM and FDM signals is that they can be allowed to propagate into the same optical fiber link as well as enabling detection with a common photodetector.

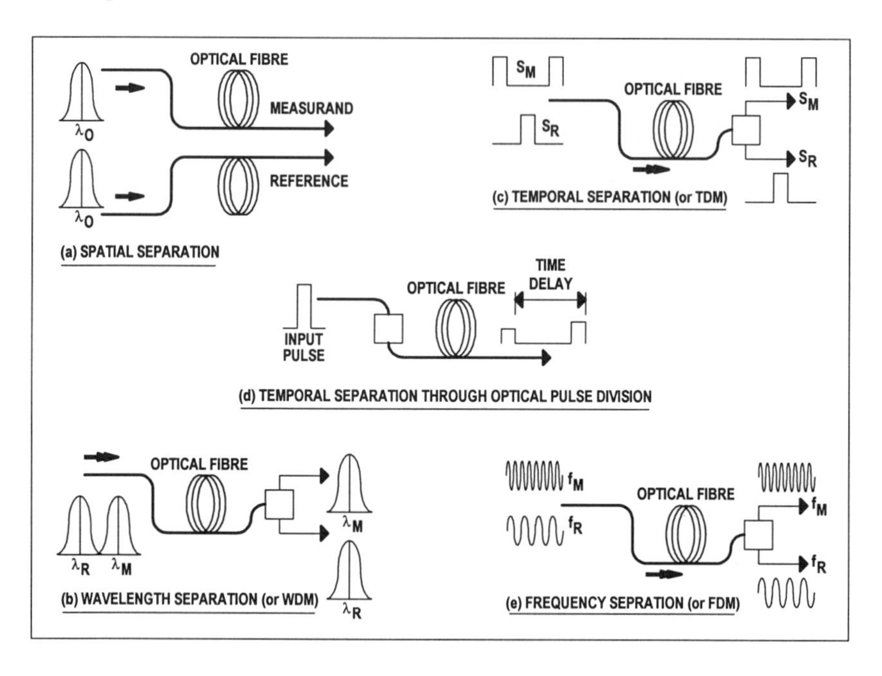

Figure 5.3. Major techniques used for the measurand and the reference signal separation

This chapter presents a review of the major referencing techniques used to provide immunity against the undesirable variable optical losses in analogue intensity modulated optical fiber sensor systems. The referencing schemes are classified according to (a) the method employed for separating the reference and measurand signals and (b) the optical configuration or mechanism used to obtain a referenced output from the optical fiber sensor system. The most popular strategies are investigated using a simple analytical approach to assist in the evaluation of each scheme for its referencing effectiveness. A further assessment of these techniques is also made for their operational efficiency in the areas where predominant common-mode variations are experienced namely those associated with optical source variations (see figure 5.2) and those caused by optical propagation factors. Finally, possible improvements are also suggested in each case as well as other considerations such as the suitability to reflection-mode or transmission-mode sensors and the potential for application to multiplexed sensor arrays.

5.2 IMPORTANT DESIGN CONSIDERATIONS

Two major sources of common-mode error can be identified as resulting from the optical transmitter and the transmission optics: the former is caused by the thermal instability of the optical source which produces intensity fluctuations of the optical signal with changes in the ambient temperature as well as the electrical heat dissipation within the device (see figure 5.2) whereas the latter is due to changes affecting the optical path which also leads to intensity variations of the optical signal. Both of these variations are in addition to the intensity changes introduced by the modulating device (or the measurand). The resulting error signal can therefore be a combination of errors caused by the optical source fluctuations, variable losses in the connecting fibers and couplers, sensitivity changes at the detectors and the sensitivity of the system to environmental perturbations.

A good intensity modulated optical fiber sensor referencing technique would seek to provide a mechanism whereby the intensity variation at the sensing head may be distinguished and measured independently of all the common-mode variations. This may be accomplished by ensuring that the measurand and the reference signals follow the same optical path through all parts of the system except at the point of measurement (i.e. the optical sensor head). However, because the two signals may be generated from independent sources and they can be detected by separate photodetectors then further errors may arise from the operational differences between these components. It may not be possible therefore to eliminate all sources of uncertainty in a system. Nevertheless precautions can be taken to enhance the design of a system by reducing these errors to a level acceptable for the desired system accuracy. Some general guidelines in relation to providing reliable and accurate referencing for intensity modulated optical fiber sensors are as follows:

- Both signals should follow the same optical path at every point in the system except at the point of sensing. The signals will then be subjected to the same common-mode variations. Hence the use of a common optical fiber between the transmitter and the receiver units is indicated.
- The sensing head should be configured such that the measurand and reference signals are separated just before the modulation point and then be allowed to immediately recombine. This will ensure that only the measurand signal is modulated by the parameter to be measured and that

both signals are subjected equally to all the remainder environmental effects and other constraints acting on the system.

- As observed in figure 5.2, optical source output can drift considerably with temperature fluctuations and therefore wherever possible a single source based configuration should be used to remove the possibility of differential effects likely to arise from source-to-source variations. Alternatively, a scheme must be devised to either cancel these differential effects or to control the behavior of each optical source.
- The optical detector may change its responsivity at the channel wavelengths due to environmental changes (e.g. temperature). Although this variation may be quite small, the resultant differential effect can be significant when two signals are detected by separate photodetectors. Whenever possible, a single photodetector based configuration should therefore be used.
- More significant than the photodetector responsivity fluctuation is the susceptibility of the two signals (the measurand and the reference) to the likely perturbations within the transmission optics (see figure 5.1). Irrespective of the referencing strategy employed, both signals should be analyzed carefully to ensure that their response to the common-mode variations would be identical. This requirement can present a different set of problems for each separate referencing scheme.
- Whenever possible, LEDs should be used instead of lamps or laser sources. Conventional lamps (e.g. tungsten types) suffer from relatively short lifetimes, poor electrical-to-optical conversion efficiency as well as being unsuitable for efficient coupling of light into optical fibers. On the other hand, the use of a laser is likely to create safety concerns and unnecessary additional optical power variations due to undesirable effects such as Fresnel losses at interfaces between optical components, fluctuating modal power distributions [14] and uncertainties arising from coherence effects. A laser source should therefore only be used in cases when LEDs provide an inadequate optical power level or the system design can benefit from the coherent output of the laser.

There may be other considerations specific to each referencing mechanism depending on the particular components utilized in the system as well as the specific system configuration.

5.3 REFERENCING MECHANISMS

A classification of the intensity modulated optical fiber sensor referencing schemes can be made by dividing these according to the method employed for separation between the reference and measurand signals.

Therefore, all existing referencing techniques can be grouped into three distinct classes, these being spatial, temporal and wavelength referencing techniques as illustrated in figure 5.3. The other two methods of TDM and FDM (see figure 5.3) are almost invariably used in conjunction with one of the aforementioned signal separation techniques to simplify the electrical processing of the two signals.

Further classification of the most widely published referencing techniques for intensity modulated optical fiber sensors that have been proposed in the recent years can be achieved by grouping these according to the method employed to obtain a separate reference signal. These include (1) optical bridge balancing [14-21], (2) optical signal tapping [22-26], (3) bypass fiber monitoring [27-33], (4) temporal signal recovery [34-36] and (5) dual wavelength referencing [37-48]. Although different in the approach adopted, all five methods essentially provide at least one additional signal which may, in conjunction with the measurand signal, be used to obtain a differential or ratiometric output that is directly related only to the state of the measurand.

5.4 SPATIAL REFERENCING

All the techniques described in this section incorporate optical signal tapping followed by the introduction of spatial separation between the subsequent signals. In section 5.4.1 the optical bridge techniques rely upon optical tapping to provide four signals from two optical sources although all four can also be produced by using a single LED source. The technique described in section 5.4.2 utilizes optical tapping that is performed within the sensing head whereas with the bypass fiber monitoring technique described in section 5.4.3 the optical tapping is performed prior to the launching of the two signals on separate optical fibers.

5.4.1 Optical bridge balancing

The balanced bridge referencing scheme is based on a four-port sensing head (SH) configuration with two input ports and two output ports. Each input port is supplied from an identical LED source (or identical optical signals derived from a common source) and each output port is connected to an identical photodetector. Therefore, as with the electrical currents fed into a wheatstone bridge, the optical power from each input is divided between the arms of the optical bridge thereby introducing spatial separation between the output signals before their passage through to the output ports, as shown in figure 5.4. Subsequently each photodetector receives two similar optical signals which are separated by electrical TDM or FDM operation of the LED(s). Multimode optical fiber links are used to connect each of the sensing head ports to the LED source(s) and photodetectors. Although there are only four separate optical signals, each signal path becomes common to two different signals that can then be divided to remove the associated common-mode variations. In this geometry therefore a distinct reference channel does not exist. An algebraic division amongst the four signals to produce a final balanced output achieves the elimination of the common-mode variations. Thus the scheme combines spatial separation with TDM or FDM to distinguish between the signals whilst improving its effectiveness by optical power distribution from the input ports to the output ports of the bridged sensing head. The sensing head configuration varies between different implementations depending on the approach employed to perform the required optical cross-coupling of signals. A precision machined prism structure, an arrangement of five GRIN-rod lenses, several off-the-shelf optical fiber couplers, a combination of directional couplers as well as a series of fused Y-couplers have been used separately in the design of such a sensing head [14-21].

An optical bridge technique has been proposed, as shown in figure 5.5 [16]. In this scheme, two optical sources (S_1 and S_2) supply light to the optical bridge via two independent optical fibers connected to its input ports. A symbol C_{ij} is introduced as the cross-coupling function to designate the portions of light coupled from the ith optical source to the jth photodetector. Light from source S_1 is coupled into the first incoming fiber (F_1^i) and passed through the measurand arm (M) as well as through the element C_{12} of the bridge. Similarly, light from S_2 is injected into the second incoming fiber (F_2^i) which is coupled to the arms (C_{22} and C_{21}) of the optical bridge. Two additional fibers are used to transmit the light from the output ports of the bridge to the detectors (D_1 and D_2): F_1^0 interfaces the first output port of the bridge to the detector D_1 whilst F_2^0 interfaces the second output port to the detector D_2. Either frequency or time domain electrical modulation of the sources is necessary to recover all four signals (two from D_1 due to S_1 and S_2 as well as the two from D_2 by the same).

This optical bridge configuration is now analyzed by designating each element label S, F, M, C_{ij} and D as the corresponding *transfer function*. The expressions for four intensity signals (I_{ij} where i represents the source and j the detector) can be obtained by following the optical path of each signal through the system. For example, the expression for the signal transmitted by the LED S_1 to the photodetector D_2 is given by:

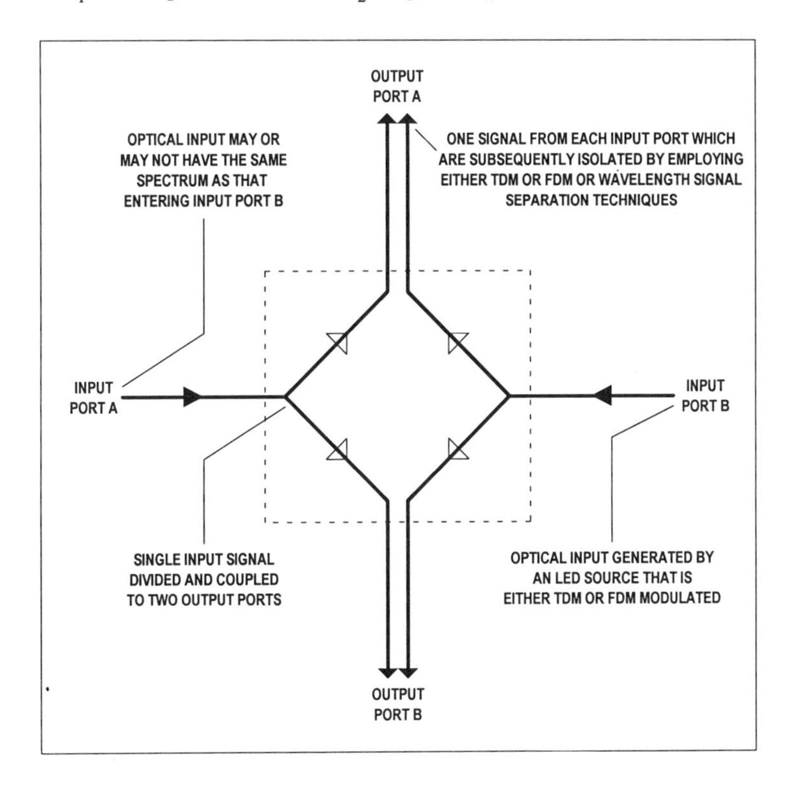

Figure 5.4. Characteristic four-port configuration of all balanced bridge schemes and optical signal path directions in the network

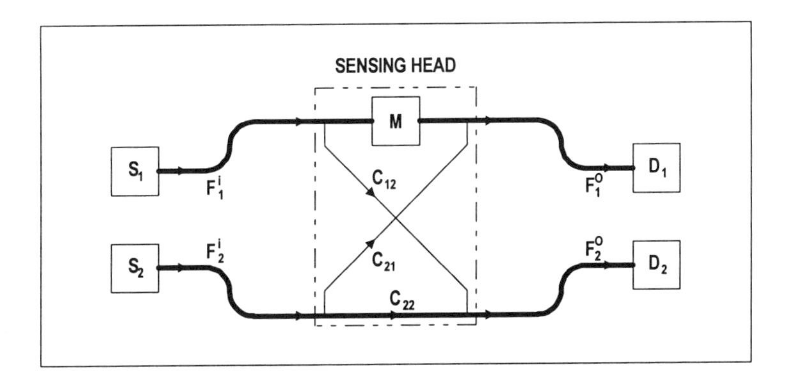

Figure 5.5. Layout of the optical bridge balancing network where four separate signals are obtained by transmission through the sensing head

$$I_{12} = D_2\ F_2^o\ C_{12}\ F_1^i\ S_1 \tag{5.1}$$

The processed output in terms of M can then be represented as:

$$V = (C_{22}/C_{12}C_{21})\ M \tag{5.2}$$

It should be noted that $V \propto M$ if $(C_{22}/C_{12}C_{21})$ remains constant under the constraint of different environmental conditions acting on the system. Furthermore, it should be noted that the functions used in the analyses are primarily dependent on wavelength, e.g. $C = C(\lambda)$. Therefore, equation (5.2) can be rewritten as;

$$V = [C_{22}(\lambda_2)\ /\ C_{12}(\lambda_1)\ C_{21}(\lambda_1)]\ M \tag{5.3}$$

Equation (5.3) implies that in the case when two different LEDs are used, the constant of proportionality (which is composed of three independent components) may exhibit significant variation due to the likely spectral output fluctuations of either source. Consider the case when $\lambda_1 = \lambda_2$ (i.e. two LEDs with the same spectral output characteristics): the constant of proportionality may be fixed and equation (5.3) will reduce to equation (5.2). However, because two separate LEDs are employed, their identical operation (i.e. the same optical output at all times) cannot be ensured without the use of stringent environmental control. The differential spectral effects resulting from such a situation can therefore lead to inaccuracy in the proportionality constant and subsequently V will no longer be proportional to M alone. Another major drawback of this technique is the requirement for several fibers and fiber couplers which makes it a relatively costly method of referencing.

An alternative technique is outlined in figure 5.6 [17]. Again two optical sources are used to provide four independent optical signals by utilizing a bridge configuration for spatial separation between the signals. Light from each LED source is coupled into a separate optical fiber for transmission to the sensing head. The sensing head has a variable-ratio beamsplitter incorporated as the rotary sensing element. Both the reflected and transmitted portions of the incident beam on the beamsplitter element are collected and transmitted to separate photodetectors. Each photodetector is arranged to receive a reflected and a transmitted signal. Although the LEDs have different emission spectra, the need for wavelength demultiplexing is reduced by operating these in the time domain. Intensity modulation is introduced by the rotary movement of the beamsplitter that causes a change in the reflected and transmitted portions of the incident beams. The arrangement is that of a bridge network with a total of four intensity signals at the two detectors [17,18] which are processed to obtain the final output V as given by:

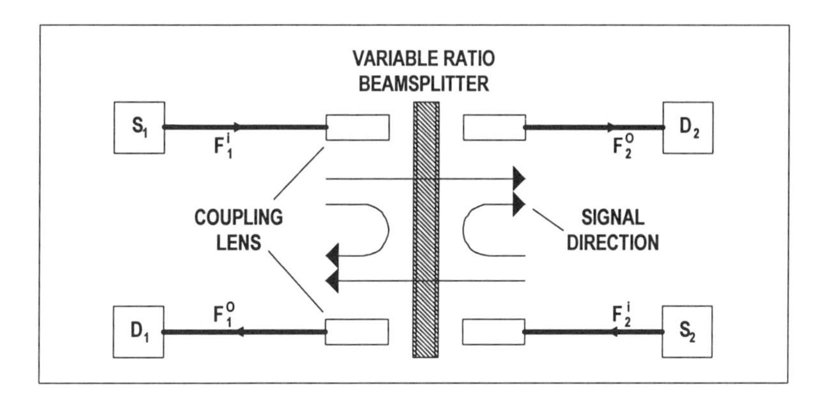

Figure 5.6. Balanced bridge configuration where the four signals are obtained by collecting all the reflected and transmitted portions of light

$$V = \sqrt{T(\lambda_1)\,T(\lambda_2)\,/\,R(\lambda_1)\,R(\lambda_2)} \qquad (5.4)$$

where T and R are the transmission and reflection functions of the variable-ratio beamsplitter which acts as a rotary displacement sensing mechanism. However, it is apparent from equation (5.4) that the output is susceptible to errors resulting from the spectral drifts of the source output.

Both configurations shown in figure 5.5 and 5.6 are therefore equally prone to errors associated with different source drifts as well as the subsequent differential transmission effects. However, there is an important difference between the two arrangements: whereas the former configuration only allows a single optical signal to interact with the measurand (see figure 5.5), the latter (see figure 5.6) provides a measurement through the interaction of all four optical signals with the rotary sensing element. An advantage of such a differential measurement with two or more optical signals is that it offers some additional potential compensation against the cross-sensitivity of the sensing head to variables other than the intended measurand.

Some of the above considerations have been examined experimentally during the long term stability testing of a pressure sensor system referenced by using a balanced bridge configuration where two 850 nm LEDs were operated in the TDM mode [21]. It was found that whilst the system worked perfectly under the protected environment of a laboratory, the referencing provided by the balanced optical bridge was very unsatisfactory in field trials. Additional stabilization measures had to be employed against the long term drifts in order to achieve a stability level of 0.2%. In particular, the investigations revealed temperature-dependent drifts associated with the optical bridge structure, the electronic transceiver components, different spectral drifts of the LED sources and variations of coupling coefficients within the optical components. Whilst some suppression of the drift errors was achieved by a careful modification of the sensing head design, a compensation against the additional temperature effects was provided by normalizing the system output with the digitally stored data (digitized look-up table) from earlier calibration process. Nevertheless, long term stability (tested over a period of two years) of 0.2% was only realized by using adaptive filtering techniques to suppress residual slow output variations.

It is therefore clear that in spite of very careful and stringent referencing design, the balanced bridge configuration remains susceptible to differential spectral effects because of its reliance on two separate optical sources even when using two LEDs with the same spectral output characteristics. As significant errors can arise from the use of two separate optical sources, it is important to consider the possibility of devising a single-LED based referencing arrangement. A simple strategy for providing two spatially separate signals to the optical bridge could be implemented by using a single LED in conjunction with a Y-splitter. However, it is then essential to provide a mechanism whereby the two signals arriving at each photodetector from separate arms of the optical bridge may be distinguished. This can be easily achieved by employing PZT devices to induce external modulation of each optical signal (produced from the Y-splitter) at a different electrical frequency. This process makes it possible to employ the conventional FDM technique to separate the optical signals at each photodetector. An alternative approach may be adopted to realize a single-LED based optical bridge configuration by using an optical switch [49,50] instead of the Y-splitter. A 1 x 2 optical switch can be used to provide dynamic spatial switching of an input optical signal to a required output link [50]. Consequentially, each photodetector connected to the optical bridge receives only one signal at a time as enabled by the TDM function of the optical switch.

A practical implementation of a single-LED based balanced bridge configuration has been demonstrated for a temperature sensing system [14] where the TDM operation of two optical signals was provided by connecting the LED through an optical switch. Although, this system utilized a quartz birefringent crystal as the temperature sensing element, the use of a polarizing cube beamsplitter in the sensing head ensures that the polarization modulation was converted to intensity modulation before the optical signals are returned to their respective photodetectors. The design of the sensing head allowed a differential modulation of the orthogonal polarization components such that the final processed output was obtained as:

$$V = \sqrt{(I_{11}\, I_{22})/(I_{12} I_{21})} = (M_1/M_2) \tag{5.5}$$

A fully compensated output was thus obtained which is independent of all optical source drifts as well as all transmission loss variations. The drifts of the uncompensated and compensated output voltage over a three hour period for a full scale temperature range of 65 °C were reported as being 5 °C and 0.2 °C, respectively. It is therefore apparent that the single-LED configuration can achieve a high degree of referencing effectiveness without the need for very complex and very stringent additional stabilization mechanisms that are required in the dual-LED variants.

5.4.2 Optical signal tapping

The technique outlined in this section can be described as a single optical source variant of the balanced bridge method illustrated in figure 5.6. The major difference between the techniques (figures 5.6 and 5.7) is the way in which the intensity modulation is produced. The underlying principle associated with this optical signal tapping scheme depends on a beamsplitter that divides an input optical beam into two portions that can be compared in order to compensate for the common-mode variations. As the two signals are obtained by dividing the optical power from the same input signal an output free from the source fluctuations can be produced [22-26]. The optical tapping (or optical monitoring of the light source output) can be performed within the sensing head [22-24] or at any point between the optical source and the sensing head [25,26]. Furthermore, such referencing is usually applied to Fotonic® type optical fiber sensors with one or more return signal fiber links [23,24]. However, because separate optical fibers are used for the transmission of the reference and measurand signals to the photodetectors, the two signals remain susceptible to different optical propagation errors.

An example of this technique is depicted in figure 5.7 [22]. Light from an optical source S is passed through a Y-coupler and transmitted to the sensing head on an optical fiber F_t. At the sensing head, a 4:1 partially reflecting mirror acts as the intensity modulating element which reflects 80% of the signal power as the measurand signal and allows 20% to be transmitted for the reference signal. The reflected beam propagates back along the transmit fiber F_t whilst the transmitted portion of the signal is propagated on a separate return fiber F_r.

As the reflected beam is designated as the measurand signal and the transmitted beam as the reference signal, the modulation imposed by the partial mirror onto the measurand and the reference signals may be written as M_M (subscript refers to the reflected portion being designated the measurand signal) and M_R (subscript refers to the transmitted portion being designated the reference signal), respectively. Therefore, the signals at the two detectors are:

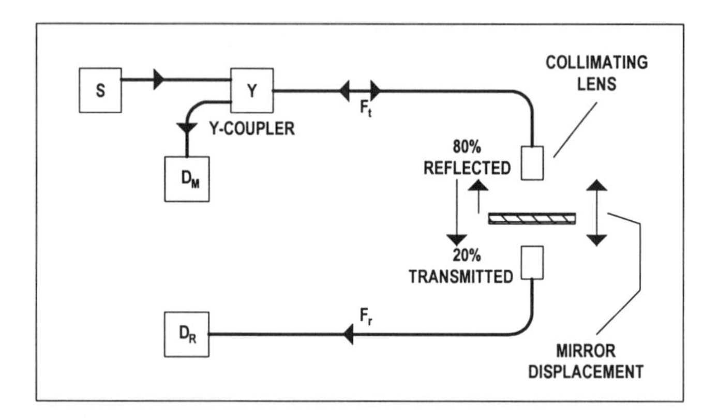

Figure 5.7. Optical signal tapping applied to obtain a reference and a measured signal by employing a partial mirror in the sensing head

$$I_M = S\,Y^2\,F_t^2\,M_M\,D_M \tag{5.6}$$

$$I_R = S\,Y\,F_t\,M_R\,F_r\,D_R \tag{5.7}$$

For a ratiometric output (assuming $D_M \approx D_R$ because a common LED source is used and any unwanted spectral modification of either signal through the system is neglected), then the output is given by:

$$V = I_M / I_R = Y\,F_t\,M_M \,/\, F_r\,M_R \tag{5.8}$$

Equation (5.8) indicates that although the effects of source fluctuation are eliminated, the final output remains susceptible to the variable fiber losses, the Y-coupler loss variation and changes of the transmitted power by the intensity modulating element in the sensing head.

5.4.3 Bypass fiber monitoring

Bypass fiber monitoring is most suitable for referencing in microbend OFS because of the intrinsic modulation that makes it very difficult to generate a reference signal within the same optical fiber [27]. The scheme requires the transmission of a reference signal on a separate optical fiber to the one used for the measurand signal. The main advantage of this scheme is that it can be used in almost all types of optical fiber sensors whether they are intrinsic [27] or extrinsic [28-31], and transmission-mode [28-30] or reflection-mode [31]. The two optical fibers may be sleeved together along their entire length to ensure similarity in their response to environmental effects. At the sensing head (SH), however, the reference fiber is bypassed so that it remains isolated from the effect of the measurand. Furthermore, it has been possible with some configurations to incorporate a 'dummy' SH in the reference fiber link using this scheme [30,31]. Both sensing heads are then subjected to similar environmental conditions but the dummy SH is kept isolated from the measurand. Such an implementation enables an output related to a single physical variable. In practice, however, the optical fibers may still exhibit a different susceptibility to some perturbations (e.g. microbending effects within the sleeve). Therefore an important difference between the optical signal tapping discussed in section 5.4.2 and the bypass fiber monitoring strategies is that in the latter the optical signals within the two fibers are largely subjected to the same environmental factors.

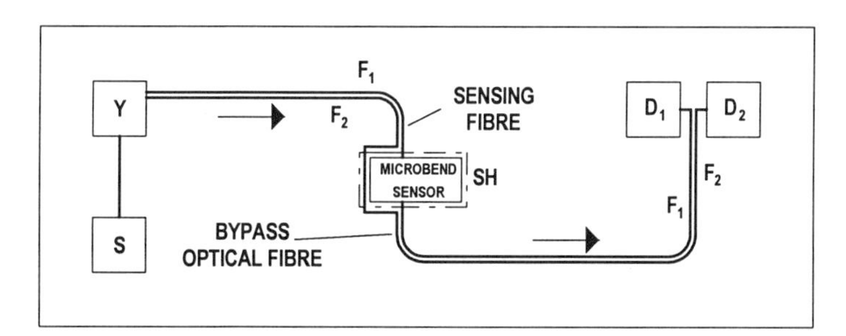

Figure 5.8. The bypass fiber monitoring scheme used in conjunction with a microbend sensor

A microbend OFS has been demonstrated for the measurement of pressure where common-mode variations were compensated by using the bypass fiber scheme shown in figure 5.8 [27]. Light output from an LED source S is passed through a 3dB Y-coupler and transmitted to the sensing head on two separate optical fibers F_1 and F_2. Whereas the light propagating in the first fiber F_1 is intensity modulated by the pressure-to-microbend transducer in the sensing head, the second optical fiber F_2 is left unperturbed at this point. Both optical fibers then carry their respective signals to separate photodiodes D_1 and D_2 respectively (see figure 5.8). The expressions for the two signals can be written by following their optical path through the system so that:

$$I_M = S\,Y\,F_1\,M\,D_1 \tag{5.9}$$

and

$$I_R = S\,Y\,F_2\,D_2 \tag{5.10}$$

Therefore, the ratiometric output will be:

$$V = I_M / I_R = F_1\,M\,D_1 \,/\, F_2\,D_2 \tag{5.11}$$

It can be observed from equation (5.11) that the optical source fluctuations and the effect of the 3dB Y-coupler are cancelled. As expected from the use of separate measurand and reference optical fiber links, the system remains sensitive to the loss variation within the optical fibers and to any responsivity variation of the photodetectors. However, if care is taken in the common sleeving of the two optical fibers and provided that there is not a significant spectral modification of the signals during propagation then the errors will be minimized. Nevertheless, there is a cost associated with doubling the length of the optical fiber cable.

As an example of the referencing capability of this referencing approach, an optical thermometer using the standard telecommunication grade single mode fiber is considered. In this system, a short bend with a radius of 0.5 mm was produced by thermal moulding in a 5/125 μm singlemode fiber which then acted as the temperature probe [32]. The optical loss of the probe depends on the size and the radius of the bend and it will vary due to the temperature-induced changes of the refractive indices [32]. The small bend was coated with a polymer resin before being encapsulated in a glass capillary to form the probe. Both ends of the single mode temperature probe were then spliced to multimode fibers such that the resulting two-fiber probe was supplied with light at one end and detected at the other. An 820 nm LED was used in conjunction with a Y-coupler in order to simultaneously couple light into the probe and a reference fiber. A spatial referencing configuration similar to the one shown in figure 5.8 was thus created. The optical signals from the probe and the reference fiber were detected by two separate Si-PIN photodiodes followed by a process of analogue to digital conversion. A microprocessor was then employed to produce a ratiometric output in accordance with the digitally stored calibration data in order to display the actual probe temperature. The system was tested over a temperature range from 5 °C to 170 °C and it exhibited a measurement error below 0.5 °C. The long term stability of this device has been reported to be within an uncertainty level of ±2 °C.

5.5 TEMPORAL REFERENCING

The temporal signal recovery method has been implemented in two different forms [34]; a Fabry-Perot type optical loop and a recirculating optical loop. In both cases a single short duration optical pulse is launched into the loop and then caused to spread temporally into a train of pulses by its continued propagation. As consecutive pulses then suffer an equal level of attenuation for each additional trip around the loop, any of the two adjacent pulses received at the photodetector may be compared to obtain the value of the measurand. However, in practice, the value obtained represents the value of the measurand plus the insertion loss of the optical loop as will become apparent in the following analysis of the scheme.

5.5.1 Temporal signal recovery

A Fabry-Perot type of loop is formed by placing an optical fiber between two partially reflecting mirrors. An optical pulse is injected through the first mirror and upon reaching the second mirror a certain portion is transmitted to a photodetector whilst the remaining optical power is allowed to recirculate between the two mirrors. This referencing technique can be implemented by using a single source, an optical fiber, two partially reflecting mirrors and a single photodetector [35]. The method can be incorporated into transmission or reflection-mode optical fiber sensor systems. The configuration of the transmission-mode optical fiber sensor is shown in figure 5.9.

A pulse of light is injected into the system through mirror A, which transmits a portion of the total incident power P_i. The pulse encounters additional losses caused by propagation in the optical fiber F, the intensity modulation by the measurand M and at the second mirror B through which a portion of its power is transmitted to the detector whilst some is reflected back into the system. The reflected portion of the pulse propagates between the two mirrors producing a series of pulses until all the energy in the circulating pulse is depleted. Subsequently, the optical detector receives a series of pulses of gradually decreasing amplitudes. A single pulse of light is thus temporally dispersed into several pulses and the amplitudes of these can be processed to recover the value of M.

To analyze mathematically the principle of operation, let R_A and T_A be the reflection and transmission functions respectively of the mirror A with R_B and T_B the corresponding functions for mirror B. The optical loss suffered by each pulse can be computed by following its path through the series of elements. The expressions for the power of the first two pulses (P_1 and P_2) received by the detector are thus given by:

$$P_1 = T_A\ T_B\ M\ F\ P_i \tag{5.12}$$

and

$$P_2 = T_A\ T_B\ M^3\ F^3\ R_A\ R_B\ P_i \tag{5.13}$$

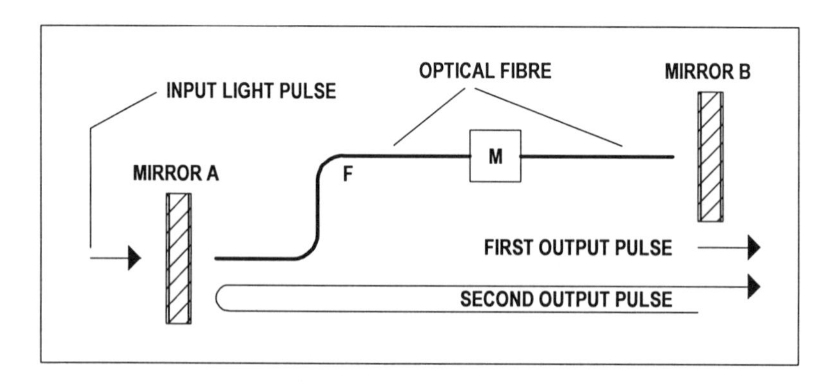

Figure 5.9. Fabry-Perot type optical loop used for introducing temporal separation between the first and the secondary optical pulses at the output

For the elimination of the common-mode variations, the two signal powers are divided to obtain:

$$P_2 / P_1 = R_A R_B M^2 F^2 \tag{5.14}$$

Alternatively, for any two consecutive pulses *n* and *n*+1 then:

$$P_{n+1} / P_n = R_A\ R_B\ F^2\ M^2 \tag{5.15}$$

Hence the measurand may be expressed explicitly as:

$$M = \sqrt{P_{n+1} / P_n R_A R_B F^2} \tag{5.16}$$

It is therefore clear from equation (5.16) that although the technique eliminates the effect of source fluctuations, it is still susceptible to intensity variations along the optical path. The dependence on the optical fiber propagation loss originates from the fact that whereas the first pulse makes a single pass through the fiber, the second pulse undergoes a further round trip before reaching the photodetector. Using a shorter length of optical fiber can reduce the associated error. However, as the optical fiber also acts as the mechanism for introducing the time delay between the pulses, subsequent pulses arriving at the photodetector become very difficult to resolve. It can be estimated that the temporal separation between two pulses one meter apart in an optical fiber is only 5 ns. Although normal PIN photodiodes can have rise times faster than 1 ns, it should be noted that optical receiver sensitivities diminish with the requirement for such high bandwidths. On the other hand, secondary pulses of light received at the photodetector can have very small optical power levels. This illustration explains the inherent difficulty associated with such temporal separation techniques.

5.5.2 Self-referenced multiplexing

The scheme outlined in section 5.5.1 has also been proposed as a 'self-referencing multiplexing' technique where the injected pulse of light is allowed to recirculate in an optical fiber loop formed by using splices and connectors [36]. Analysis of this implementation is as follows. If T_C is the transmission factor of the coupling element L_C, T_S the transmission factor of the splice L_S and T_M the transmission of the intensity modulating element M_N, then processing of the following pulses as in equations (5.12) to (5.15) yields:

$$(P_2 / P_1)_n = 0.81\ T_S^3\ T_C^2\ T_M\ M_n \tag{5.17}$$

where the subscript *n* refers to the *n*th optical fiber sensor (OFS) in the multiplexing arrangement, as illustrated in figure 5.10. Unfortunately, the ratio of the two pulses is dependent on the transmission properties of the splices as well as connectors. Therefore, the largest contribution to signal error will arise from a variation of the transmission of the sensing element as well as a somewhat reduced contribution from the insertion loss variation of each of the splices and the coupling elements. An overall improvement can be obtained by maximizing the transmission of these elements so that the error contribution associated with the higher order terms becomes negligible. Of course, this implies making the insertion loss of each element as small as possible.

5.6 DUAL WAVELENGTH REFERENCING

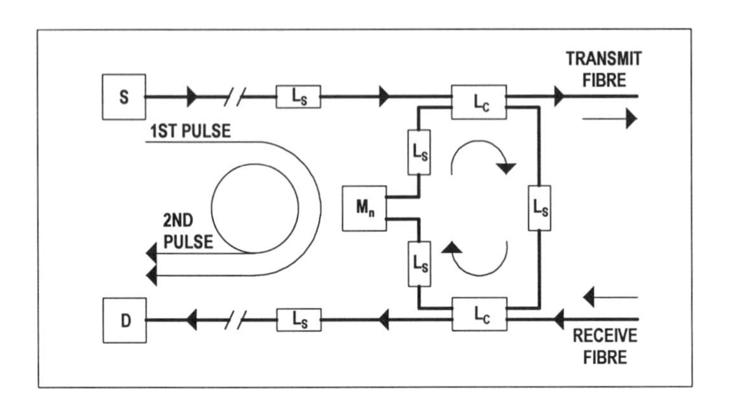

Figure 5.10. Self-referencing multiplexing technique using a recirculating optical loop to provide temporally separated signals for a compensated output

Dual wavelength referencing relies upon the measurement being made by using two optical signals placed within different wavelength bands which are then compared to eliminate any common-mode variations. Hence, a referenced measurement can be obtained by a differential intensity modulation of the two separate wavelength band signals or, alternatively, by subjecting one of the optical signals to intensity modulation by the measurand whilst the other signal (within a different wavelength band) acts as the reference [46]. The signal wavelength bands should be closely spaced to minimize differential spectral variations between the optical signals. However, it is important to consider the associated increase in the optical cross-talk as the separation between the wavelength bands is decreased [46]. Generally, a compromise exists depending upon the desired overall measurement accuracy of the system. There are several different configurations within this scheme that arise from the variety of options available for the generation and detection of the optical signals within separate wavelength bands. In analyzing the dual wavelength referencing scheme, it is assumed that any wavelength multiplexing or demultiplexing components used are 100 % efficient. This assumption implies that optical power is not leaked between the wavelength channels and hence no account is taken of optical crosstalk.

5.6.1 Basic configuration

The most basic approach is where the light output from two independent sources (S_1 and S_2) is multiplexed and coupled into an optical fiber F_i that carries it to the sensing head. Both wavelength signals (λ_1 and λ_2) may then be differentially modulated at the sensing head or alternatively one wavelength signal is arranged to remain unaffected by the modulation which then becomes the reference. The same optical fiber (in the reflection-mode OFS) or a different optical fiber F_o (in the transmission-mode OFS) will then carry the signals to a demultiplexer (DEMUX) where the different wavelength band signals are separated and received by two different detectors (D_1 and D_2). An example of such a transmission-mode OFS is shown in figure 5.11 [37]. Assuming that the signal wavelength bands are chosen close together to ensure their similar behavior throughout the system, then the expressions can be written as:

$$I_1 = D_1 F_o M_1 F_i S_1 \tag{5.18}$$

and

$$I_2 = D_2 F_o M_2 F_i S_2 \tag{5.19}$$

The ratiometric output is then given by:

$$V = I_2 / I_1 = (D_2 S_2 / D_1 S_1) [M_2 / M_1] \tag{5.20}$$

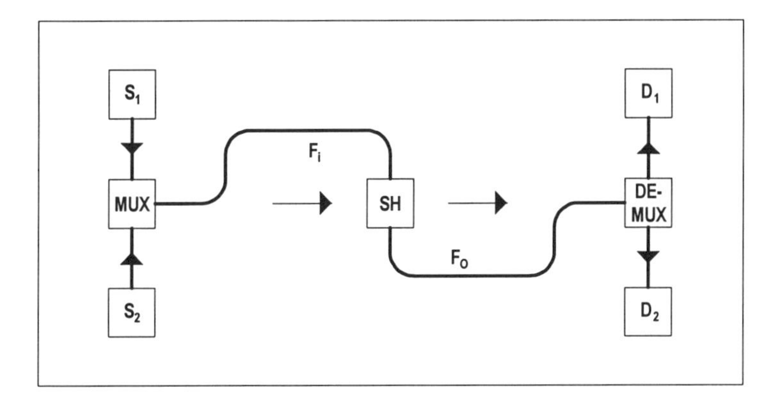

Figure 5.11. Basic dual wavelength referenced intensity modulated sensor system employing two photodiodes and two separate LED sources

The major drawback with the configuration of figure 5.11 is that it is prone to errors arising from the instability of the two optical sources even though it is referenced against optical path intensity fluctuations (subject to a careful selection of the two channel wavelength bands). The effect of LED fluctuations is two-fold in that not only the optical levels in each channel can vary but that the response of each photodiode can also change due to the spectral shifts of either channel. Moreover, it is also assumed that the wavelength multiplexing and demultiplexing operations are passive at all times. However, if the spectral content of each channel is altered due to the LED fluctuations, then each of the two aforementioned operations may further modify the optical power levels in the channels.

5.6.2 Single photodetector configuration

A simple modification of this technique makes the optical demultiplexing element redundant. The modified version, shown in figure 5.12, employs two independent sources operated and monitored using electrical TDM and FDM signal separation techniques [38-41]. Furthermore the scheme benefits from the use of a single detector but it also has similar drawbacks to the previous method. Hence equations (5.18) to (5.20) also apply in this case.

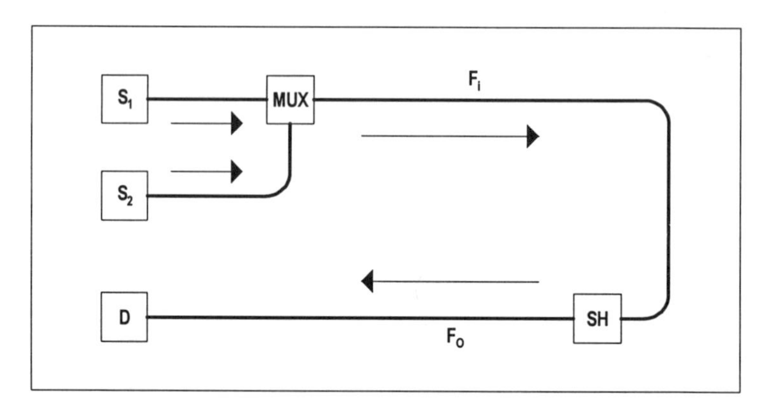

Figure 5.12. Dual wavelength referenced optical fiber sensor system where TDM operation of the LEDs enables detection of both signals with a single photodetector

5.6.3 Dual wavelength bridge

A technique where full referencing, except for spectral variations of the sources, is demonstrated can be observed in figure 5.13. In this scheme, two sources (S_1 and S_2) are switched in the time domain and their outputs are fed to a four-port coupler where wavelength signals centered at λ_1 and λ_2 are combined [42,43]. A fraction of the combined optical signal power is passed to a reference detector D_R which is used to monitor the output power from either source. The remaining light is transmitted to the sensing head via the input fiber F_i where wavelength band signals centered at λ_1 and λ_2 are differentially modulated. Then the modulated signals at wavelengths λ_1 and λ_2 are transmitted to the measurand detector D_M along a second optical fiber F_o. The four output signals obtained were analyzed using a microprocessor to obtain a referenced output [42]. The final output voltage V may be written as:

$$V = \left(\frac{I_R^1 I_M^2}{I_R^2 I_M^1}\right) = \frac{D_R\, C_1\, S_1 \,.\, D_M\, F_o\, M_2\, F_i\, C_2\, S_2}{D_R\, C_2\, S_2 \,.\, D_M\, F_o\, M_1\, F_i\, C_1\, S_1} = \left[M_2 / M_1 \right] \tag{5.21}$$

As in the other cases considered in this section, it has been assumed that the system responds similarly to both wavelength band signals. The center wavelengths used were 885 nm and 930 nm (spectral widths were not reported) which, being reasonably close together, tended to minimize differential spectral effects in the transmission optics. However, the immunity of the referencing technique to the source variations is subject to the properties of the optical coupling element, C, (i.e. the spectral characteristics of each channel may change with the environmental behavior of this element) and further changes may result if the spectral output of either LED source fluctuates. Finally the cost of the components would be another consideration in the implementation of this method.

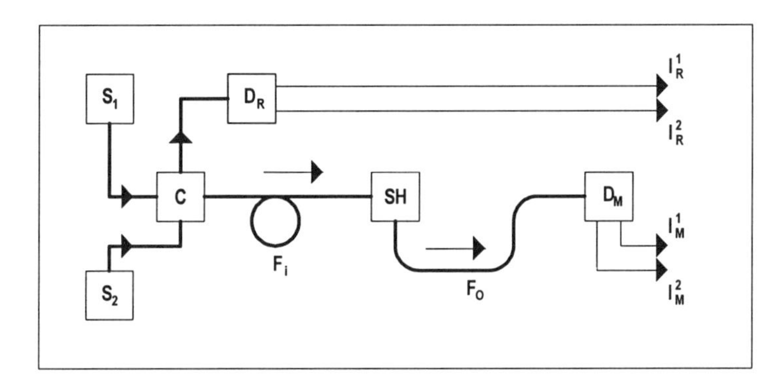

Figure 5.13. Dual wavelength referenced optical fiber sensors configuration incorporating optical power output monitoring from each source to provide a fully compensated output

It can be noted that the configuration of the aforementioned scheme (see figure 5.13) is very similar to the balanced bridge techniques discussed in section 5.3.1. Moreover, the schematic of figure 5.13 can be easily redrawn so that it resembles the balanced bridge configuration of figure 5.3. The use of two LED sources emitting within different spectral bands provides a three-fold improvement to the balanced bridge configurations described in section 5.3.1. Firstly the four-port arrangement can be set-up by using either a single 2 x 2 passive fiber coupler or a single directional coupler (e.g. a beamsplitter cube). Secondly the four-port coupler can be positioned in the controlled environment of the terminal unit. Finally a single optical fiber link can be utilized for the parallel transmission of both wavelength band signals.

5.6.4 Single LED configuration

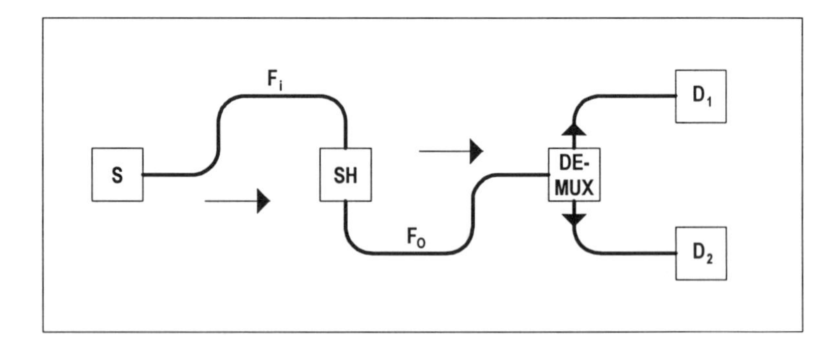

Figure 5.14. Single LED based dual wavelength referenced optical fiber sensor system where both signals are extracted by the use of optical filtering elements

Another possible dual wavelength referencing technique that benefits from the use of a single source at the expense of somewhat reduced referencing effectiveness is depicted in figure 5.14. In this approach, both wavelength band signals are extracted from a single source and therefore source-to-source intensity variations due to temperature drifts are eliminated. The wavelength bands for the reference and the measurand signals may be extracted by employing an optical dichroic filter which divides the spectral output from an LED source into two halves. These signals can be separately analyzed by two detectors and then compared to nullify the common-mode variations. The mathematical analysis of this strategy is again depicted in equations (5.18) to (5.20) because single LED (S) supplies both signals S_1 and S_2. Hence, no safeguard exists against spectral variations in the source output.

However very much improved referencing effectiveness has been achieved in a single-LED configuration that utilizes spectral-slicing of the LED emission to provide two narrowband wavelength signals [44]. In this scheme, the effect of LED source fluctuations is further minimized by positioning the signal wavelength bands on the LED spectrum so that both optical signals behave in a similar manner. The differential variations between the two optical signals are therefore negligible with temperature-dependent changes in the LED emission [48]. The concept is illustrated in figure 5.15 which shows the typical thermal behavior of a GaAlAs LED. Various curves (solid lines) of steadily decreasing amplitude correspond to LED spectral emissions for operating temperatures between 25°C and 75°C. It can be observed that both the initial intensity and its variation with changes in the LED operating temperature are functions of wavelength. This behavior enables the selection of two different thermally-matched wavelength bands which may have different absolute optical power levels but will display the same temperature-induced signal variations [48,51]. For example, a spectral slice A centered at a wavelength of 850 nm displays almost the same optical power variation as an equal width spectral slice B centered at the wavelength of 900 nm, as indicated by the curves labeled slice A and B in figure 5.15.

This wavelength thermal matching strategy provides two wavelength band signals which are largely independent of the LED thermal effects [48,51]. Hence the referencing effectiveness of wavelength referenced intensity-modulated optical fiber sensor systems can be enhanced through the use of this technique. The optical configuration to implement this scheme is the same as the single-LED system depicted in figure 5.14. However, the LED source is common to both signals which are now thermally matched to exhibit similar optical power fluctuations and the final output is therefore referenced for all major common-mode variations.

Another single-LED referencing strategy [47] which is better suited to reflection-mode sensors involves the use of two identical fiber Bragg gratings (FBGs) [52]. In this scheme, light from an LED source is directed to the sensing head through a passive directional coupler so that the optical signals returning from the reflection-mode sensing head are directed to the demultiplexing optics for the required wavelength separation followed by their detection by two separate photodetectors in the same manner as the arrangement already depicted in figure 5.14. The light directed to the sensing head passes through a fiber Bragg grating (FBG) element which reflects a spectral slice of the optical spectrum as the reference signal whilst the transmitted spectrum is returned after it has been intensity modulated by the measurand. At the receiver-end, the light passes through another 3dB directional coupler. The second FBG element (ideally identical to the one in the sensing head) is fitted into one of the output ports of the coupler to allow the transmission of the measurand signal to the first photodetector whilst the spectral slice representing the reference signal is directed to the second photodetector. The final output V is then obtained as a ratio between the two voltages as given by [47]:

$$V = \left(\frac{V_R}{V_M}\right) = \left(\frac{T}{b(1-T)k}\right)M \tag{5.22}$$

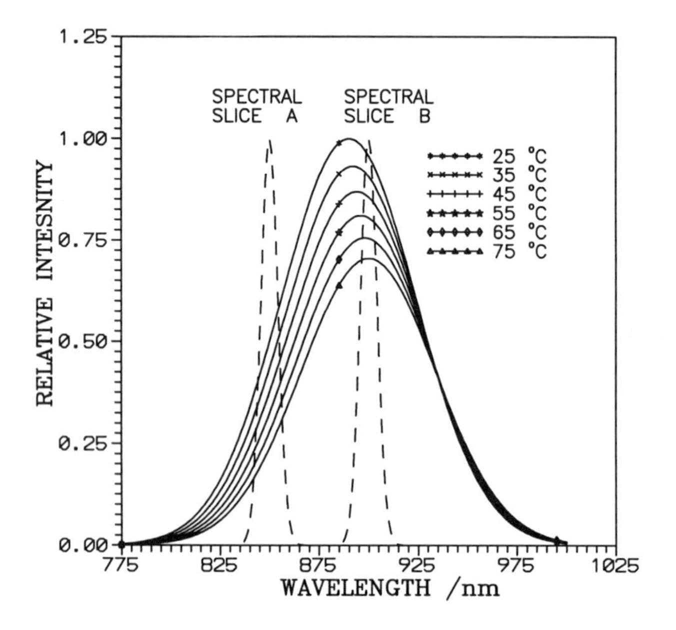

Figure 5.15. Thermal behavior of a typical GaAlAs LED and the pair of wavelength thermal-matched spectral slices selected to eradicate differential optical signal level variations in dual wavelength sensor systems

where T is the transmissivity of the gratings, k is the coupling ratio of the directional coupler being used at the receiver-end, b is the optical power loss factor of the same directional coupler and M is the required measurand. The dependence on the transmissivity, T, of the fiber Bragg grating is said to be unimportant as it is found to be very stable [47] whereas locating the housing of the receiver-end coupler within the protected environment of the terminal control unit provides additional safeguard against the possibility of any temperature-related errors.

Nevertheless, differential temperature changes between the two FBGs can cause significant errors because the spectral slice returned by the FBG element located in the sensing head will drift with temperature and consequently an unreliable reference signal will be produced. This difficulty is recognized and overcome in this system through a servo-control circuitry that continually tunes the receiver FBG to the spectral slice arriving from the sensing head. This is achieved by prestraining the receiver FBG with the help of a PZT device. In spite of this control, some variations are still expected in both reference and measurand signals because of the unpredictably changing position of the reference-signal spectral slice in relation to the LED emission spectrum [53]. The demonstration system used a 1523 nm ELED with a spectral width of 75 nm in conjunction with FBGs that had a center wavelength of 1524 nm, a spectral width of 0.44 nm and a reflectivity of 90%. Servo-controlled tuning was employed to test the referencing capability of the system against LED drive current variation from 100 mA to 80 mA. The resulting deviation of the ratiometric output has been reported as being less than 1.6% whereas the corresponding uncompensated measurand signal variation was around 20% over the full scale [47].

Finally most of the referencing schemes discussed in sections 5.3-5.6 may be enhanced by using optoelectronic feedback to stabilize the LED operating current [54]. This is generally achieved by taking a monitor signal from the LED emission to a photodetector and then feeding back the subsequent amplified photocurrent to the LED current supply to form a negative feedback control loop [34]. However, whereas a constant optical power output level may be achieved by continued adjustment of the LED supply current, there will be an accompanied variation of the electrical heat dissipation within the device. Subsequently, the LED emission may exhibit spectral drifts (i.e. peak wavelength changes as well as the spectral linewidth variation) that can still produce erroneous signals.

5.7 COMPARATIVE ASSESSMENT

In comparison with the other methods the optical bridge techniques provide adequate referencing against all major sources of common-mode variations. However, this is achieved through the use of relatively large number of components and a complex sensing head configuration. Furthermore the requirement for four optical fiber links for the connection of the sensing head to the terminal control does limit the practical usefulness of this strategy. In addition, there still remains the possibility of a measurement error from the sensitivity of the optical bridge elements to environmental factors. Although, differential spectral effects can remain a potential source of errors, it is nevertheless possible to reduce these by incorporating additional stabilization mechanisms. However, a similar level of reduction can be achieved through a more elegant strategy that enables all four optical signals to be derived from a single LED source.

The main advantage of the bypass fiber monitoring strategy is its inherent simplicity but it does remain susceptible to differential signal losses occurring between the measurand and reference channels. However, it may be possible to minimize the measurement error associated with this technique by using a twin-core fiber [55-57] in conjunction with an integrated optics sensing head [54,58]. The referencing effectiveness will be improved when the similarity between the optical paths traversed by the measurand and reference signals is increased through these modifications.

The referencing configurations for temporal signal recovery and optical signal tapping remain strongly susceptible to optical fiber losses. Whereas with the former a shorter optical fiber loop helps to reduce the amount of the associated error, the recovery of the signals then requires rather complex electronic circuitry to provide isolation between the measurand and reference signals due to the reduced temporal separation between the consecutive pulses. Nevertheless the temporal signal recovery scheme is attractive because it requires a minimum number of simple optical components for its implementation. In addition the referencing effectiveness of the optical signal tapping technique can be improved by using the dual-LED configuration illustrated in figure 5.6. This strategy allows compensation for errors associated with both the transmit and receive optical fiber links.

From a practical viewpoint the dual wavelength referencing technique shown in figure 5.13 offers equally effective referencing to that provided by the balanced bridge schemes. This configuration combines dual wavelength referencing with optical feedback for the monitoring of the launched optical signals and thereby exhibits the full benefits of the optical bridge strategy. Furthermore the number of optical fiber links is halved, and the four-port structure used as the optical bridge is shown positioned in the terminal control which leaves full flexibility for a variety of sensing head designs. The major drawback with other conventional dual wavelength referencing methods is that the common-mode variations associated with the optical source(s) are not compensated. However, the single-LED schemes has been demonstrated to have the potential for providing almost complete referencing by combining conventional dual wavelength referencing with matched spectrally-sliced optical wavelength band signals.

Table 5.1. Comparative features of the spatial referencing schemes

Classification	Spatial Referencing			
Referencing Scheme	Dual-LED Optical Bridge Balancing	Single-LED Optical Bridge Balancing	Optical Signal Tapping	Bypass Fiber Monitoring
Basic Principle	Optical cross-coupling between signal channels	Optical cross-coupling between signal channels	Partial mirror divides the input optical beam into two	Monitor signal fiber bypassed at sensing head
Referencing Mechanism	LEDs TDM operated to produce four signals for full compensation	An LED and optical switch to produce four signals for full compensation	Reflected and refracted optical beams compared	Signals from measurand and monitor fibers compared
Component Requirements	2 LEDs, 2 PINs, 4 fibers, 4 port special sensing head, couplers and connectors	Single LED, 2 PINs, 4 fibers, 4 port special sensing head, couplers and connectors	Single LED, 2 PINs, 2 fibers, Y-coupler, partial mirror and connectors	Single LED, 2 PINs, 2 or 4 fibers, Y-coupler and connectors
Referencing Effectiveness	Major source fluctuations and optical propagation losses compensated	All source fluctuations and optical propagation losses compensated	Only the source fluctuations compensated	Major source fluctuations and optical propagation losses compensated
Residual Susceptibility	LED spectral shifts and the sensing head elements	Sensing head elements	Optical fiber loss and the Y-coupler	Differential loss between optical fibers
Complexity Level	High (complex sensing head configuration)	High (complex sensing head configuration)	Medium (simple optics)	Medium (basic strategy)
Sensing Mode Suitability	Extrinsic or intrinsic transmissive	Extrinsic or intrinsic transmissive	Extrinsic transmissive types	Intrinsic and extrinsic transmissive
Multiplexing Potential	Low due to component requirements	Low due to component requirements	Medium when combined with WDM	Medium in extrinsic mode by using WDM
Major Drawbacks	Cost and component requirements	Cost and component requirements	No referencing for optical path losses	Inefficient use of optical fibers
Possible Improvements	Reasonable simplification possible by using WDM with two different LEDs	Possible to incorporate a dummy sensor in the SH to remove residual drifts	Use a twin-core fiber to combine transmit and receive functions	Use a twin-core fiber and an integrated-optics sensing head
References	[15-21]	[14]	[17,22-26]	[27-31]

Table 5.2. Comparative features of the wavelength and temporal referencing schemes

Classification	Wavelength Referencing			Temporal Referencing
Referencing Scheme	Dual Wavelength Referencing	Thermally Matched Wavelength Signals	Servo-Controlled FBG Detection	Temporal Signal Recovery
Basic Principle	Two separate wavelength signals on a single fiber link	Two spectrally sliced wavelength signals on a single fiber link	Spectrally sliced reference signal with servo-controlled detection	Recirculating pulse inside an optical fiber loop
Referencing Mechanism	Optical signals from both wavelength channels compared	Optical signals from both wavelength channels compared	Optical signals from both wavelength channels compared	Consecutive pulses sampled and compared
Component Requirements	1 or 2 LEDs, PINs and fibers, with or without MUX & DEMUX and connectors	Single LED, PINs and fibers, with DEMUX and connectors	Single LED, PINs and fibers, two matched FBGs, two couplers and connectors	Single LED, optical fiber, single PIN, partial mirrors, couplers and splices
Referencing Effectiveness	Can be complete but often only the transmission losses compensated	Complete when the spectral slices are thermally matched	Can be complete with careful design and suggested improvements	Only the source fluctuations compensated
Residual Susceptibility	Optical source variations and LED-to-LED differences	Possible differential drifts of spectral slicing elements	Spectral output variations of the LED source overlooked	Optical fiber loss and beamsplitter elements
Complexity Level	Configuration dependent but can be high	Configuration dependent but can be low	Medium but requires exactly matched FBGs	High (complex receiver electronics)
Sensing Mode Suitability	Extrinsic transmissive and reflective	Extrinsic transmissive and reflective	Reflective with bandpass-reflecting FBG elements	All common types
Multiplexing Potential	Medium when WDM combined with TDM	Medium to high depending on the available sources	Medium to high depending on the available sources	Already proposed (see Ref. 22)
Major Drawbacks	No referencing for source variations	Component requirements	Cost of couplers and FBG elements	Complex electronics
Possible Improvements	Use a single LED source to provide narrowband signals	Integrated optics design and bandpass coated optics	Use two spectral slices and combine with wavelength thermal matching	Reduce sensitivity to fiber loss by using a shorter fiber-loop
References	[37-48]	[44-46,48]	[47]	[34-36]

Finally a more detailed comparative assessment of the various strategies is presented in summary form in Table 5.1 (spatial referencing schemes) and Table 5.2 (wavelength and temporal referencing schemes).

5.8 SUMMARY

The techniques which have been discussed offer differing levels of referencing effectiveness that are usually related to level of design complexity and cost. Overall the methods that offer the most accurate referencing for intensity modulated optical fiber sensors are also relatively complicated and utilize many components which tends to make them unattractive for widescale utilization. However, it is evident that with the steadily maturing optical technologies, it is becoming feasible to achieve cost-effective full referencing capability. Some simpler methods have been suggested but these generally provide only partial referencing against common-mode variations (i.e. they either provide a nulling of the source fluctuations or that of the optical path intensity variations).

Bypass fiber monitoring is considered to be a simple and useful referencing technique which may be further enhanced by the use of a twin-core optical fiber cable to minimize any differential losses between the measurand and reference signals during propagation. Optical signal tapping in a single LED configuration, however, proves inadequate for full referencing against all common-mode variations and it becomes somewhat cumbersome when developed to the dual-LED configuration to enable full referencing. Although some referencing improvement may be possible in the temporal signal recovery techniques, the lack of referencing for several optical transmission elements limits their usefulness to provide full referencing for intensity modulated optical fiber sensors.

Dual wavelength referencing can provide comprehensive referencing effectiveness against all major common-mode variations when the emission from each optical source is also monitored. Then the configuration becomes similar to that of the balanced bridge. However, the dual wavelength equivalent of the bridge provides several benefits such as the reduction in the number of optical fiber links and a simpler four-port fiber coupler arrangement which is moved from the sensing head to the terminal control unit. Other dual wavelength referencing configurations that lack optical source output monitoring are usually ineffective against variations arising from the instability of the optical source(s). However, a wavelength thermal matching strategy for minimizing the effect of the LED instability in a dual wavelength referencing scheme that utilizes a single LED source has been demonstrated. This technique may prove as effective as some of the more complex optical balanced bridge schemes.

5.9 REFERENCES

1. McGlade, S. M. (1981) Optical sensors for displacement measurement. *The Marconi Review,* second qtr., 119-36.
2. Giallorenzi, T. G., Bucaro, J. A., Dandridge, A., Sigel, G. H., Cole, J. H., Rashleigh, S. C. and Priest, R. G. (1982) Optical fiber sensor technology. *IEEE J. Quantum Electronics,* **QE-18** (4), 626-65.
3. Grover, D. J. (1984) Fibre optics inventions assigned to the British Technology Group from universities in the United Kingdom, *Proc SPIE,* **468,** *Fibre Optics '84,* (Sira) London, 28-48.
4. Nakayama Takashi (1984), Optical sensing technologies by multimode fibers. *Proc SPIE,* **478,** *Fiber Optic and Laser Sensors II `84,* 19-26.
5. Pitt, G. D., Extance, P., Neat, R. C., Batchelder, D. N., Jones, R. E., Barnett, J. A. and Pratt, R. H. (1985) Optical Fibre Sensors. *IEE Proc,* **132,** *Pt. J, No 4,* 214-48.
6. Krohn, D.A. (1986) Intensity Modulated Fiber Optic Sensors Overview. *Proc SPIE,* **718,** *Fiber Optic and Laser Sensors IV,* 2-11.
7. Medlock, R. S. (1986) Review of modulating techniques for fibre optic sensors. *J. Opt. Sensors,* **1**(1), 43-68.
8. Senior, J. M., Murtaza, G., Stirling, A. I. and Wainwright, G. H. (1989) Dual wavelength intensity modulated optical fibre sensor system. *Proc SPIE,* **1120**, *Fibre Optics '89*, 332-7.
9. Kersey, A. D. (1996) A review of recent developments in fiber optic sensor technology. *Optical Fiber Technology*, **2**, 291-317.
10. Othonos, A. (1997) Fiber Bragg gratings. *Rev. Sci. Instrum.*, **68**, 4309-4341.
11. Jones, B. E. (1985) Optical fibre sensors and systems for industry. *J. Phys. E: Sci. Instrum.,* **18,** 770-81.
12. Jones, B. E. (1977) Instrumentation, Measurement, and Feedback. *McGraw-Hill Book Company (UK) Ltd..*
13. Gardiner, P. T., Edwards, R. A. (1987) Fibre optics sensors (FOS) for aircraft flight controls. *Proc Applications of Light in Guided Flight, Royal Aeronautical Society*, 42-63.
14. Jin, X., Liao, Y., Lai, S. and Zhao, H. (1995) Single-LED optical fiber sensor system using a novel polarization-modulated compensation technique. *Proc SPIE,* **2594,** *Self-Calibrated Intelligent Optical Sensors and Systems,* 243-8.
15. Culshaw, B., Foley, J. and Giles, I. P. (1985) A balancing technique for optical fibre intensity modulated transducers. *Proc. Optical Fibre Sensors '85,* 117-20.
16. Giles, I. P., McNeill, S. and Culshaw, B. (1985) A stable remote intensity based optical fibre sensor. *J. Phys. E: Sci. Instrum.,* **18**, 502-4.
17. Beheim, G. and Anthan, D. J. (1986) Loss-compensation of intensity-modulating fibre-optic sensors. *Proc SPIE,* **718**, *Fibre Optic and Laser Sensors IV,* 259-65.
18. Beheim, G., Anthan, D. J., Rys, J. R., Fritsch, K. and Ruppe, W. R. (1988) Modulated-splitting-ratio fiber-optic temperature sensor. *Proc SPIE,* **985,** *Fibre Optic and Laser Sensors VI,* 82-8.

19. Bois, E., Huard, S. J. and Boisde, G. (1988) Loss Compensated Remote Fiber Optic Displacement Sensors for Industrial applications. *Proc. EFOC/LAN 88,* 246-50.
20. Martens, G., Kordts, J. and Weidinger, G. (1989) A Photo-elastic Pressure Sensor with Loss-Compensated Fiber Link. *Springer Proceedings in Physics,* **44,** *Optical Fiber Sensors,* 458-63.
21. Bing, Q., Wei, P., Shunping, R. and Junxiu, L. (1996) Studies on the long-term stability of fiber optic pressure sensor. *Proc SPIE,* **2895,** *Fiber Optic Sensors V,* 445-50.
22. Senior, J. M. and Cusworth, S. D. (1987) Intensity Modulated Optical Fibre Sensors Employing Graded Index Rod Lenses. *IOP Short Meetings Series* **7**, *Fibre Optic Sensors, Glasgow,* 89-93.
23. Shaik, M. A. (1989) Design and analysis of fiber optic position sensor. *Proc SPIE,* **1169,** *Fiber Optic and Laser Sensors VII,* 473-84.
24. Cockshot, C. P. and Pacaud, S. J. (1989) Compensation of an optical fiber reflective sensor. *Sensors and Actuators,* **17**, 167-71.
25. Moiseyev, V. V. and Potapov, V. T. (1988) Analysis of the Stability of Reflection-Type Fiber-Optic Sensors. *Telecommunications and Radio Engineering Part 2,* **43,** *No 9,* 72-5.
26. Corke, M., Gillham, F., Hu, A., Stowe, D. W. and Sawyer, L. (1988) Fiber Optic Pressure Sensors employing reflective Diaphragm Techniques. *Proc SPIE,* **985,** *Fiber Optic and Laser Sensors VI,* 164-71.
27. Berthold, J. W., Ghering, W. L. and Varshineya, D. (1987) Design and Characterization of a High Temperature Fibre-Optic Pressure Transducer. *Journal of Lightwave Technology,* **LT-5**, *No 7,* 870-6.
28. Spillman, W. B., Fuhr, P. L. and Kajenski, P. J. (1988) Self-referencing Fiber Optic Rotary Displacement Sensor. *Proc SPIE,* **985,** *Fiber Optic and Laser Sensors VI,* 305-10.
29. Iwamoto, K. and Kamata, I. (1990) Pressure sensor using optical fibers. *Appl. Optics,* **29**(3), 375-8.
30. Ayub, M., Spooncer, R. C. and Jones, B. E. (1988) Environmentally compensated photoelastic pressure sensors with optical fibre links. *Proc SPIE,* **1011**, *Fiber Optic Sensors III,* 130-5.
31. Ramakrishnan, L., Unger, L. and Kist, R. (1988) Line loss independent fiberoptic displacement sensor with electrical subcarrier phase encoding. *Technical Digest Series,* **2,** *Optical Fiber Sensors,* 133-6.
32. Kalinowski, H. J., Valente, L. C. G. and da Silveira Jr. I. I. (1992) Optical thermometer using a short bend of single-mode fiber. *Proc SPIE,* **1795,** *Fiber Optic and Laser Sensors X,* 261-5.
33. Williams, B. A. and Dewhurst, R. J. (1995) Differential fiber-optic sensing of laser generated ultrasound. *Electronics Letters,* **31**(5), 391-2.
34. Adamovsky, G. (1986) Time domain referencing in intensity modulation fiber optic sensing systems. *Proc SPIE,* **661**, *Optical Testing and Metrology,* 145-51.
35. Lammerink, T. S. J. and Fluitman, J. H. J. (1984) Measuring method for optical fibre sensors. *J. Phys. E: Sci. Instrum.,* **17,** 1127-9.
36. Spillman, W. B. and Lord, J. R. (1987) Self-Referencing Multiplexing Technique for Fibre Optic Intensity Sensors. *Journal of Lightwave Technology,* **LT-5**(7), 865-9.
37. Bacci, M., Brenci, M., Conforti, G., Falciai, R., Mignani, A. G., Scheggi, A. M. (1986) Thermochromic transducer optical fibre thermometer. *Appl. Optics,* **25**(7), 1079-82.
38. Jones, B. (1986) The pig that looks after railway lines. *Sensor Review,* **6**(4), 199-201.

39. Scheggi, A. M., Bacci, M., Brenci, M., Conforti, G., Falciai, R., Mignani, A. G. (1987) Thermometery by optical fibers and a thermochromic transducer. *Optical Engineering,* **26**(6), 534-7.
40. Conforti, G., Brenci, M., Mencaglia, A. and Magnani, A. G. (1989) Fiber-optic thermometric probe utilizing GRIN lenses. *Appl. Optics,* **28**(3), 577-80.
41. Liu, X. P., Spooncer, R. C. and Jones, B. E. (1991) An Optical Fibre Displacement Sensor with Extended Range Using Two-wavelength Referencing. *Sensors and Actuators,* **A 25**(1-3), 197-200.
42. Schoener, G., Bechtel, J. H. and Salour, M. M. (1985) Novel fiber coupler for optical fibre temperature sensor. *Proc. Optical Fibre Sensors '85,* 203-6.
43. Dakin, J. P., Wade, C. A. and Withers, P. B. (1988) An Optical Fibre Sensor for the Measurement of Pressure. *Fiber and Integrated Optics,* **7,** 35-46.
44. Senior, J. M., Murtaza, G., Stirling, A. I. and Wainwright, G. H. (1992) Single LED based dual wavelength referenced optical fibre sensor system using intensity modulation. *Optics & Laser Technology,* **24**(4), 187-92.
45. Wang, G. Z., Wang, A., May, R. G., Barnes, A., Murphy, K. A. and Claus, R. O. (1995) Stabilization for intensity-based sensors using two-wavelength ratio technique. *Proc SPIE,* **2594,** *Self-Calibrated Intelligent Optical Sensors and Systems,* 41-51.
46. Murtaza, G. and Senior, J. M. (1995) Dual wavelength referencing of optical fibre sensors. *Optics Communications,* **120,** 348-57.
47. Cavaleiro, P. M., Ribeiro, A. B. L. and Santos J. L. (1995) Referencing technique for intensity-based sensors using fibre optic Bragg gratings. *Electronics Letters,* **31**(5), 392-4.
48. Senior, J. M. and Murtaza, G. (1995) Optical fibre sensor system. *European Patent No. EP 0 470 168 B1*
49. Thylen, L., Karlsson, G., and Nilsson, O. (1996) Switching technologies for future guided wave optical networks: potentials and limitations of photonics and electronics. *IEEE Communications Magazine,* 106-13.
50. Adams, M. J, Barnsley, P. E., Burton, D. A., Davies, D. A. O., Fiddyment, P. J., Fisher, M. A., Mace, D. A. H., Mudhar, P. S., Robertson, M. J., Singh, J. and Wickes, H. J. (1993) Novel components for optical switching. *BT Technical Journal,* **11**(2), 89-97.
51. Murtaza, G. and Senior, J. M. (1994) Wavelength selection strategies to enhance referencing in LED based optical sensors. *Optics Communications,* **112,** 201-13.
52. Hill, K. O. and Meltz, G. (1997) Fiber Bragg grating technology fundamentals and overview. *Journal of Lightwave Technology,* **15**(8), 1263-76.
53. Murtaza, G. and Senior, J. M. (1997) Influence of LED thermal drifts on optical cross talk in spectrally sliced WDM systems. *Microwave and Optical Technology Letters,* **14**(3), 153-5.
54. Dakin, J. and Culshaw, B. (1989). Optical fibre Sensors: systems and applications. **vol. I,** *Artech House, Inc.,*
55. Romaniuk, R. S. and Dorosz, J. (1989) Multicore micro-optics. *International J. of Optoelectronics,* **4**(3/4), 201-19.
56. Cozens, J. R., Green, M. and Gu, Y. (1988) Special Fibres for Sensing. *Proc SPIE,* **1011,** *Fiber Optic Sensors III,* 62-6.
57. Kociszewski, L., Stepieh, R. and Buzniak, J. (1988) New manufacturing method of sensor oriented optical fibres. *Proc SPIE,* **1011,** *Fiber Optic Sensors III,* 71-80.
58. Grosskopf, K. G. (1988) Integrated optics for sensors. *Proc SPIE,* **1011,** *Fiber Optic Sensors III,* 38-45.

6

Optical Fiber Chemical Sensors: Fundamentals and Applications

J. O. W. Norris

6.1 INTRODUCTION

Optical methods are some of the oldest and best established techniques for sensing chemical analytes, and have formed the basis for many chemical sensors. The development of inexpensive, high quality optical fibers for the communications industry has provided the essential component for the implementation of optical fiber sensors. There has been considerable research effort expended in developing sensors based on optical fibers for both physical and chemical analytes, with many interesting schemes having been proposed, since the late 1970s and continuing to expand since then.

Application areas for chemical analysis include process plant, environmental and pollution monitoring, military applications, laboratory-based analysis, clinical diagnosis and other medical applications. Each particular application area has its own requirements for sensitivity, precision, selectivity, sensor lifetime and unit cost. The physical state of the analytes studied has encompassed gases, dissolved gases, liquids, ions in solution and solids.

Physical sensors, most notably those for measuring pressure, temperature, fluid level and mass flow, are widely applied both for monitoring and controlling industrial processes. The chemical sensor most commonly used in industrial processes is the glass pH electrode. This system is relatively expensive, susceptible to electrical noise and not readily usable in the food industry because of the danger of breakage. As a consequence, a substantial interest has existed for many years in the development of miniaturized techniques that will overcome these problems and allow the measurement of a wide range of chemical parameters in an

K.T.V. Grattan and B.T. Meggitt (eds.), Optical Fiber Sensor Technology, 337–378.

ever increasing, diverse number of environments. Fiber optic techniques are continuing to make a significant contribution towards satisfying this need.

This chapter is intended to give the reader a clear and concise overview of the fundamentals and essential principles of optical fiber chemical sensors. It is structured so as to give a brief indication of the perceived advantages and disadvantages of optical fiber chemical sensors. There is a description of the major transduction principles, how the optical measurement is related to the analyte concentration. A convenient way of classifying these sensors is given and illustrated with some of the sensors that have been demonstrated. The examples given are intended to show an illustrative group of chemical sensing applications that have been addressed, and do not provide an exhaustive list, as a further discussion of applications has been given by MacCraith [1]. The sensing principles described can be, and will continue to be, extended to additional analytes. In addition to optical fiber sensors some mention is included of integrated optic devices, because this area, although currently not extensively developed, is a natural extension from optical fibers and will become increasingly significant in the future [2].

6.2 PERCEIVED ADVANTAGES AND DISADVANTAGES FOR CHEMICAL SENSING

6.2.1 Advantages

An essential component of any form of chemical analysis using optical fibers is the optical fiber itself. This is usually made from a silica-based glass, and less frequently from an organic polymer (e.g. polymethylmethacrylate). The development of fibers made from heavy metal halides or chalcogenides is extending the wavelength range at which optical fiber chemical sensors can operate. Several potential advantages arise from using an optical fiber as the basis of a chemical analysis technique:

- **Suitability for remote in-situ measurements.** The small size, often less than 1mm diameter, flexibility, chemical stability and high transmission efficiency of optical fibers enables remote *in-situ* chemical analyses to be undertaken using fiber optic sensors.
- **Electrical isolation and freedom from electromagnetic interference.** As optical fibers are made from insulating materials optical fiber sensors are, by their nature, electrically isolated from the interrogating

electronics. This is especially important for *in-vivo* medical sensors and application areas where flammable or explosive reagents are present. The associated immunity to electromagnetic interference (EMI) makes them suitable for use in electrically noisy environments.

- **Potential of distributed sensing.** In principle optical fiber chemical sensors can be made where the whole length of the fiber is sensitized to a particular analyte. The presence of the analyte at any point along the fiber can modify its optical properties, enabling a large area to be monitored simultaneously. Alternatively, the sensitized fiber can be interrogated by an optical equivalent of radar (optical time domain reflectometry, OTDR) to give a measure of the analyte concentration as a function of position over the extended area. Very few such devices have been developed.

6.2.2 Disadvantages

The major disadvantages of the use of this technology include:

- **Ambient light** may interfere with the optical signal of interest.
- **Sensor response time** may be long because it may be determined by mass transport to and in the reagent phase. These factors can be alleviated by appropriate sensor system design, for example by pulsing the interrogating light, and using phase sensitive detection to remove the optical background, and by using small probes or only thin films of the reagent phase.
- **Long term stability** In addition to the usual constraints on sensor accuracy caused by changes in the optical characteristics of the source, detector or transmitting optical fiber, chemical sensors may experience variable optical characteristics within the sensing transducer, e.g. due to fouling. For chemical sensors utilizing immobilized reagents, deterioration due to physical desorption or chemical degradation of the reagent may also limit the useful lifetime of the sensor.
- **Limited dynamic range** Many optical sensors utilize immobilized reagents, e.g. acid base indicators. These obey the law of chemical equilibrium (section 6.3.2). A plot of color intensity against species concentration, e.g. pH, follows a sigmoid shape, with the majority of the change occurring over a range of two to three pH units and the device being insensitive outside this range. Whilst this may be a disadvantage for applications requiring a large dynamic range, for some sensing applications, e.g. physiological monitoring, only a limited range is required, and with the appropriate choice of indicator, fiber optic chemical sensors provide an advantageous increased sensitivity over this limited range.

- **Components** Chemical analysis using optical fibers offers the potential of miniaturized equipment based on recently developed electro-optic components. However, the components available are principally determined by the needs of the major optoelectronics market, namely the communications industry, and are not necessarily optimized for sensing applications. The relatively limited market size for components for chemical analysis means that this requirement alone is usually insufficient to initiate the development of new devices, e.g. lasers operating at a particular wavelength.

6.3 UNDERLYING PRINCIPLES OF FIBER OPTIC CHEMICAL SENSORS

6.3.1 Optical effects

Optical fiber chemical sensing involves the probing of matter by photons. The photon may either be scattered or, if its energy is equal to that between the initially occupied and an excited state of the matter, it may be absorbed. Various parameters are associated with a photon flux, and changes in any one of these may give analytical information about the material being probed. These parameters, and some of the spectroscopic techniques used to measure their variation are shown in Table 6.1.

Table 6.1. Measurement techniques for parameters which vary in a photon flux

Parameter	Photon absorbed	Photon scattered
Intensity	Absorption and reflectance spectroscopy	Turbidity and nephelometry
Wavelength	Luminescence spectroscopies	Raman spectroscopy
Time characteristics	Luminescence lifetime	Photon correlation spectroscopy
Phase/polarization	Polarized absorbance, circular dichroism	Ellipsometry

6.3.1.1 Absorption spectroscopy

For the vast majority of absorbing (nonscattering) media the initial intensity of a beam of light, I_0, will be reduced to an intensity, I, given by the Beer-Lambert law:

$$\log_{10}\frac{I_0}{I} = \varepsilon lc \tag{6.1}$$

where ε is the molar absorptivity, and is a constant for a particular species at a particular wavelength, l is the pathlength (cm) and C is the species concentration (mol l^{-1}). The ratio $\log_{10} I_0 / I$ is, by definition, the absorptivity of the medium, and also, by definition, 10 times the medium loss in decibels. Occasionally, concentration-dependent reactions, e.g. polymerization or complexation with the solvent cause deviation from equation (6.1). However, this equation is generally applicable and is one of the fundamental tenets of many sensors because measuring the absorptivity of a medium allows the concentration of a species to be calculated given ε, the geometry of the system and no interfering absorbing species. More usually in fiber optic analyses, the product εl is found by calibrating the system with standard solutions.

6.3.1.2 Reflectance spectroscopy

Reflection occurs when light meets a refractive index discontinuity. Specular reflection and diffuse reflection occur from optically flat and optically rough interfaces, respectively. The former is most important for intensely absorptive, crystaline analytes, where the reflectivity of an absorptive sample, in air, is given by

$$R = \frac{(n-1)^2 + k^2}{(n+1)^2 + k^2} \tag{6.2}$$

where k is the material index of extinction, and n is its refractive index. The latter is wavelength dependent, particularly in the vicinity of an absorption band. Consequently the reflected light intensity decreases with increasing absorption coefficient. However, virtually no fiber optic sensors use this basic reflectance technique as the transduction mechanism.

Many optical fiber chemical sensors incorporate relatively weakly absorbing species in conjunction with an optically denser, *white* material, whose reflectance characteristics are virtually constant over the wavelength range interrogated. In this configuration the light passes through regions containing the absorbing species before and after being reflected by the second phase. Consequently the intensity of the reflected light decreases with increasing absorption coefficient.

The relationship between this reflected intensity and the concentration of the absorbing species is not linear. For diffuse reflectance the most widely used model is that of Kubelka-Munk, which assumes a semi-infinite scattering medium and relates the reflectance, R, to the concentration of the absorbing species on the scattering layer, C, by

$$F(R) = \frac{(1-R)^2}{2R} = \frac{\varepsilon C}{S} \tag{6.3}$$

where ε is the molar absorptivity as before, and S is a scattering coefficient. $F(R)$ is commonly referred to as the Kubelka-Munk, or remission, function.

This nonlinearity between reflectance, or absorptivity, and the analyte concentration means that sensors designed in this way require extensive calibration and characterization before they can be used analytically.

6.3.1.3 Luminescence spectroscopy

Following the absorption of an interrogating photon the resulting excited species may lose its energy by one, or more, of several processes:

- via a cascade of nonradiative deactivation steps, which could be intra- or inter- species, such that all the energy is converted into thermal energy;
- by emitting a second photon;
- by interspecies transfer with an acceptor species that subsequently emits a second photon.

The second and third of these processes usually occur in conjunction with the first, both competitively and as part of the deactivation route of a single moiety. Consequently the emitted photons are usually less energetic and shifted to the red end of the spectrum, relative to the initial excitation photons. This wavelength difference gives luminescence spectroscopy a sensitivity advantage relative to absorption spectroscopy: the former is measuring the appearance of photons of a particular wavelength relative to a low background, whereas the latter is measuring the reduction in intensity relative to the high background of the interrogating beam. Since noise levels are dependent on the square root of the background light level, it follows that luminescence spectroscopy has a fundamental signal/noise advantage. No distinction will be drawn between phosphorescence and fluorescence as the following generalities apply to both luminescence mechanisms, although lifetimes may range from around 1ns to 1ms.

Information about the concentration of an analyte can be obtained from intensity or lifetime measurements in one of several ways, as discussed below.

6.3.1.3.1 Intensity

The intensity of luminescence of an analyte is a function of the analyte concentration. This intensity is a fraction of the amount of light absorbed, and the latter is given by the Beer-Lambert law, equation (6.1), for the majority of species. Rearrangement of equation (6.1) gives

$$I = I_0 \exp(-\varepsilon l C \log_e 10) \tag{6.4}$$

where I_0 and I are the incident and transmitted light intensifies, respectively. If the analyte is only a weak absorber, i.e. $\varepsilon l C < 0.05$, then

$$I = I_0 \exp(1 - \varepsilon l C \log_e 10) \quad \text{to within } 0.7\% \tag{6.5}$$

whence

$$I_0 - I = I_0 \varepsilon l C \log_e 10 \tag{6.6}$$

where I_0 - I is the amount of light absorbed. Therefore, for weakly luminescing species the luminescent intensity is proportional to the luminophore concentration, the constant of proportionality being determined by factors including the quantum efficiency of the luminescence process, and the collection efficiency of the detector optics.

For more strongly absorbing species further factors become increasingly important:

- self-absorption of luminescence light by the analyte;
- self-quenching of the luminescing species;
- the variable volume illuminated by the exciting light and the variation in the collection efficiency with the distance from the end of the fiber. For example if the majority of the incident light were adsorbed within 1 μm of the end of the fiber this would result in a far higher detected luminescence signal than that obtained if the same amount of light were absorbed in a cone extending 1cm from the end of the fiber.

6.3.1.3.2 Quenching of intensity

Some of the complicating factors discussed above are significantly reduced if a constant concentration of luminescent species, L, is used with an analyte that quenches this luminescence, A. The mechanism may be depicted as follows. Under illumination with intensity I_0, the concentration of excited luminophore, $[L^*]$ may be expressed by the differential equation:

$$\frac{d[L^*]}{dt} = k_e I_0 [L] = k_1 [L^*] - k_q [L^*][A] \quad (6.7)$$

At equilibrium under steady illumination, the steady state luminescence intensity, $k_1[L^*]$, is given by:

$$k_1[L^*] = \frac{k_e I_0 [L]}{k_1 + k_q [A]} \quad (6.8)$$

The ratio of this intensity with a quenching analyte concentration of $[A]$, L_A, relative to the intensity when the concentration of A is zero, L_0, is the well known Stern-Volmer equation:

$$\frac{L_0}{L_A} = 1 + k_{SV}[A] \quad (6.9)$$

where k_{SV} is an amalgamation of constants known as the Stern-Volmer constant:

$$k_{SV} = \frac{k_q}{k_1} \quad (6.10)$$

6.3.1.3.3 Lifetime characteristics

An alternative method of assessing the degree of luminescence quenching is to use a pulsed source and to look at the temporal decay of the intensity. Solving the differential equation (6.7) gives

$$\frac{L_A(t)}{L_A(t=0)} = \exp(-(k_1 + k_q[A])t) \quad (6.11)$$

When the concentration of A is zero this gives the simple exponential decay of the luminescent species with its characteristic lifetime. For other concentrations, the luminescence half-life of the luminescing species is reduced from $\ln 2 / k_1$ to $\ln 2 / (k_1 + k_q[A])$.

This technique involves comparing the luminescence intensity at various times following excitation. Consequently, it is self-referencing, and compensates for factors like the fouling of windows or changes in the efficiency of the source, detector or electronics, which gives the technique a large advantage over simple intensity measurements for assessing the degree of quenching. However, it does of require an increased capital cost for the controlling electronics. Given the advances in electro-optic components and electronics (for pulsing the excitation me and for signal processing), luminescence lifetime measurements are becoming more widely used.

6.3.1.3.4 Energy transfer

Another form of luminescence spectroscopy involves energy transfer from the excited state of one species to the excited state of a second species, followed by luminescence from the second species. The luminescence intensity from this second species, *S*, is proportional to $k_{et}[L^*][S]$ where k_{et} is the rate of energy transfer between the two species, and is very dependent on the separation between the donor L^* and the acceptor *S*. Also, since the intensity depends on the concentration of L^*, the factors that determined the intensity of luminescent light detected from the species L^* are directly involved in determining the intensity of luminescence from the second species, *S*. The principal advantage conferred by this technique is that it differentiates between that portion of species S in very close proximity to L^* and unassociated *S*, and can therefore be used to characterize the amount of bound *S* in a competitive equilibrium, e.g. in an immunoassay.

6.3.1.4 Raman spectroscopy

The Raman process involves the inelastic scattering of photons from species with the loss (or gain) in energy being provided by the vibrational energy of the scattering moiety, to give the Stokes (or anti-Stokes) Raman features. This process has a much lower cross-section than has absorption spectroscopy, typically a factor of 10^9 less. However, because the scattered photon has a different energy to the incident photons, the Raman technique has a similar sensitivity advantage over absorptions spectroscopy to that of luminescence spectroscopy. Its principal advantage is that it allows analysis based on the vibrational fingerprint of an analyte to be carried out at visible wavelengths, a region where silica fibers transmit light efficiently. The low cross-section for the Raman process has several important consequences for analysis using this technique. These include:

- The interrogating light is not significantly attenuated due to the Raman process and therefore the intensity of a Raman band is directly proportional to the concentration of the analyte, assuming that the analyte absorbs neither the exciting laser nor the Raman scattered wavelength.
- Fluorescence from the analyte may swamp the Raman signal.
- Fluorescence or Raman scattering from the components composing the optical fiber may be more intense than the Raman signal from the

analyte. This can be significantly reduced by using different optical fiber to deliver the exciting light and to collect the Raman scattered signal.

Whilst the first factor is an advantage, the remaining two are drawbacks. However, these may be ameliorated by using the rapidly developing technique of Fourier Transform Raman Spectroscopy. This typically employs continuous wave 1.064 μm laser light, a wavelength which substantially reduces the number of samples that fluoresce, and makes this technique applicable to a larger variety of samples.

6.3.2 Chemical equilibria

The immobilized reagents, e.g. acid/base indicators, used in many optical fiber chemical sensors obey the law of chemical equilibria. Such indicators are typically a weak acid whose acid and conjugate base forms have different optical characteristics. In such a system an equilibrium is established between the reactant, e.g. the undissociated indicator acid HI, and the reaction products, e.g. the hydrogen ion, HI, and the conjugate base, I^-. This is shown below for an acid base indicator:

$$HI \leftrightarrow H^+ + I^-$$

The law of chemical equilibria states:

$$\frac{[H^+][I^-]}{[HI]} = K \tag{6.12}$$

where the constant K is the equilibrium constant of the reaction, and [HI] is the concentration of HI etc. (Strictly it is the activity of each component but in many circumstances replacing activity with concentration introduces little error.)

The amount of indicator present is usually a constant, c (i.e. $[I^-] + [HI] = c$) such that the concentration of, for example, the conjugate base can be expressed by:

$$[I^-] = \frac{cK}{[H^+] + K} \tag{6.13}$$

From the definition of pH it follows that:

$$[H^+] = 10^{-pH} \tag{6.14}$$

giving:

$$[\mathrm{I}^-] = \frac{cK}{10^{-\mathrm{pH}} + K} \tag{6.15}$$

Therefore, if I^- is optically distinct from HI, then by measuring the absorbance at a convenient wavelength, the concentration of I^- can be found, as discussed previously. At low pHs, where $10^{-pH} >> K$, equation (6.15) approximates to $[I^-]=0$, whilst at high pHs, when $10^{-pH} \approx 0$, equation (6.15) becomes $[I^-] = c$. Consequently a plot of $[I^-]$, related to the measurand absorbance, against pH follows a sigmoid shape, with the majority of the change occurring over a range of two to three pH units, as shown in fig. 6.1.

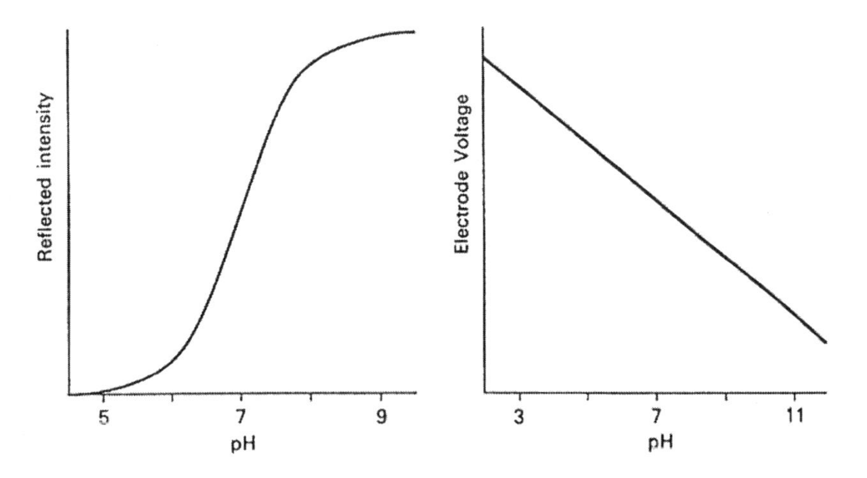

Figure 6.1. Comparison of signal / pH output from a typical optical fiber pH sensor and a glass pH electrode. (a) Typical reflected intensity / pH response of optical fiber pH sensor; (b) typical output voltage / pH response of glass pH electrode

The traditional method of measuring pH is to use a glass pH electrode. This potentiometric sensor follows the Nernst equation

$$E = E^0 + \frac{RT}{F}\log_e[\mathrm{H}^+] \tag{6.16}$$

(Again strictly it is the activity of H^+ and not its concentration but this causes little difference for $0 < pH < 14$.). Since $pH = -\log_{10}[H^+]$, then the Nernst equation can be expressed as

$$E = E_0 - k'\mathrm{pH} \tag{6.17}$$

illustrating that the output voltage of a potentiometric sensor is linear with pH, over the dynamic range $0 < pH < 14$, as shown in fig. 6.1

Whilst the different physical principles underlying potentiometric and optical sensors based on equilibria do limit the dynamic range of the latter, they also afford a high sensitivity over this limited dynamic range. Therefore, by choosing an indicator with the appropriate dissociation constant, a highly sensitive probe can be made for a specific pH range.

6.3.2.1 Immunological reactions

The immune system of animals contains cells that, after being in contact with a foreign molecule, known as an antigen (Ag), secrete proteins, known as antibodies (Ab), which are specifically shaped to bind only to that antigen. The antibody/antigen binding is similar to other chemical equilibria, and can be represented by

$$Ag + Ab \leftrightarrow Ag \cdot Ab$$

The equilibrium constant for this type of reaction, usually called the affinity by immunologists, is large, as can be anticipated, since usually both the antigen and antibody are present in low concentrations, $< 10^{-6}$ Ml^{-1}, and a small equilibrium constant would mean negligible bound complex was formed.

Such immune reactions from the basis of many clinical tests for the presence of antigens, e.g. human immunodeficiency virus and human chorionic gonadotrophin (the hormone used as a test for pregnancy), or antibodies, e.g. to rubella. These tests form the basis of a large commercial market, and consequently much research effort has concentrated on devising optical fiber and analytical techniques suitable for measuring the degree of immune reactions.

6.4 CLASSIFYING FIBER OPTIC SENSORS FOR CHEMICAL SENSING

Fiber optic sensors for chemical parameters can be classified in a similar way to physical sensors and thus may be subdivided into two major types, i.e.

- **Extrinsic sensors,** where the optical fiber merely acts as a light guiding link between the measurement point and the interrogating and display electronics;

- **Intrinsic sensors,** where the fiber, probably in some modified form, is the sensing transducer.

One type of extrinsic sensor, which will not be covered in this chapter, is the hybrid sensor, where the transduction mechanism produces a non-optical output which is then converted into an optical signal for transmission along an optical fiber to a receiver/display unit. An example of chemical analysis using such a device would be a glass pH electrode with associated electronics to produce a digitally, intensity or wavelength coded optical signal. Such a sensor would have some of the familiar advantages discussed earlier, and could be advantageously used to save weight, gain freedom from electromagnetic interference (EMI) or for safety reasons.

Optical fiber chemical sensors, and consequently chemical analysis techniques utilizing optical fibers, can further be classified as follows:

- **Species-specific sensors.** These comprise remote spectrometry, where the optical properties of the analyte are measured directly, and sensors using immobilized reagents, where the effect of the analyte on the optical properties of an added reagent are quantified.
- **Nonspecies-specific sensors.** These involve directly measuring some optical property that may be perturbed by any one of a number of analytes. An example is a fiber optic refractometer, for which a change in transmitted intensity merely indicates a change in the refractive index of the surrounding medium, not the specific species that caused it.
- **Indirect techniques.** These involve using an optical fiber sensor to measure some nonoptical parameter, e.g. strain or temperature, and relating the measurement to the analyte of interest.

Thc nonspecies-specific sensors and indirect techniques predominantly involve the intrinsic type of sensor.

6.5 DESCRIPTION OF SOME ILLUSTRATIVE SENSORS

Considerable research effort has been spent on demonstrating various fiber optic sensor concepts. The examples that follow are illustrative, demonstrating early illustrative and representative work on the range of sensors and sensing techniques that have been studied, and the advantages to be gained from using fiber optic sensors. More detailed sensor systems have been discussed by MacCraith [1]. The sensors are grouped according to the classification just described.

6.5.1 Extrinsic species-specific sensors

6.5.1.1 Remote spectroscopy

The characteristic absorption spectra of chemical species have formed the basis of many conventional chemical analyzers. It was therefore natural that the earliest fiber optic gas sensors used the optical fiber merely as a light guide, coupling the light source to an analysis cell, where it is partially absorbed by the analyte, and returning the remaining light to the detector. This scheme may be described as remote conventional spectroscopy, and is one of the most developed areas of fiber optic chemical sensing.

The simplest scheme is to measure the decrease in optical power, at an appropriate wavelength, that is transmitted through an absorption cell. This was demonstrated for methane [3]. The emission from light emitting diodes, whose peak wavelength was close to either the 1.33 μm or 1.66 μm absorption bands of methane, were focused into an optical fiber, through a gas cell, and returned to the detector via a second fiber. A wavelength absorbed by methane was selected using either a monochromator or narrowband (2-3 nm) interference filters. The LED was modulated with 90Hz current pulses at 50% duty factor, and that portion of the signal synchronous with the LED drive current was monitored. The detectable limit was quoted as 700 ppm (1.3% lower explosive limit, LEL) when using a 3 nm bandpass filter and 1km fiber links for both the transmitting and receiving optical fibers and around 2500 ppm (5% LEL) with two 5 km fiber links. A serious limitation of this scheme is the lack of any reference beam, such that an other factors that reduce the amount of power transmitted interfere, and give anomalously high measurements for the gas concentration.

This can be overcome by using two different wavelengths, e.g. two different lines from a laser system or two separate semiconductor sources, one of which is absorbed by the analyte of interest whilst the other is not. An example of this approach is the remote monitoring of NO_2 using a multiline argon ion laser. The 496.5 and 514.5 nm laser lines are strongly and weakly absorbed by NO_2, respectively. Appropriate narrowband interference filters placed before two photodiodes monitor the transmitted power at the two wavelengths. Taking the ratio of these two transmitted powers allows referencing against potential interferences such as particulates, steam or fouling of optical components.

Further referencing with respect to the input power at the two wavelengths can be achieved by placing a beam splitter in the laser beam, immediately after the laser, and using two further narrowband filters and photodiodes. Such a system has been described using a standard quartz halogen lamp and two interference filters, as shown in fig. 6.2 [4]. One filter had a central wavelength of 1.66 μm and a bandwidth of 2 nm, corresponding to the center of the intense Q-band of the $2\nu_3$ transition for methane, whilst the filter forming the reference beam was a broader band and at a wavelength away from the methane Q-band absorption. A single chopper was used, alternately to send light of 1.66μm or the reference wavelength down an optical fiber to the sensing head, where the remaining, unabsorbed, light was guided to the detector via a second fiber. A second detector monitored the input powers at the two wavelengths, and a microprocessor performed the appropriate calculations and displayed the gas concentration. With a total fiber length of around 600 m, and a gas sensing head which was a folded cavity of 80cm effective pathlength, calibration points between 0% and 40% LEL were measured with an accuracy of ±0.5% LEL [3].

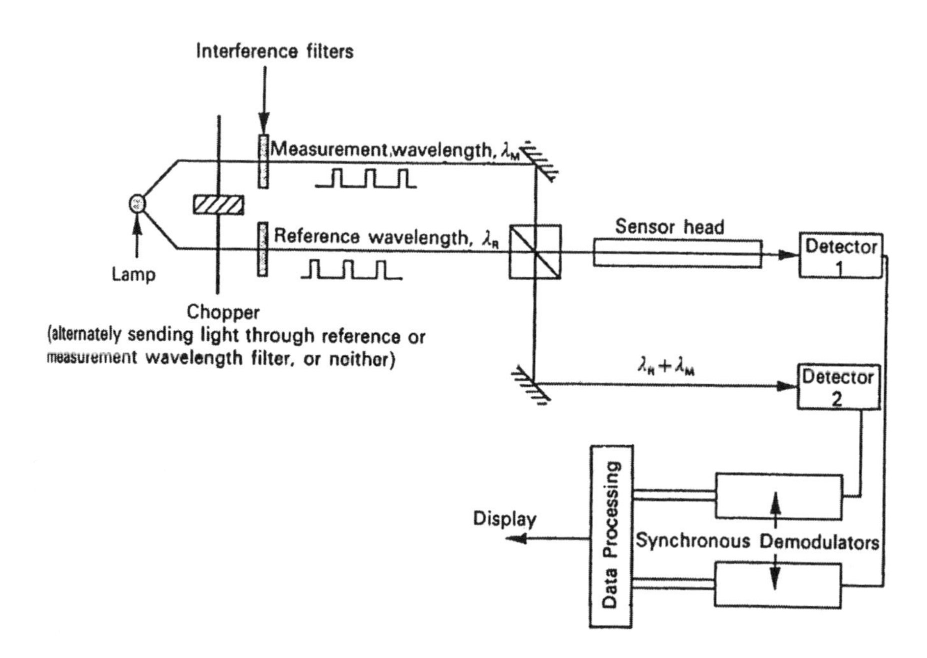

Figure 6.2. Schematic of two color fiber optic gas detection system. (From Hordvik *et al* [4])

Instead of using a small number of selected wavelengths a spectrum covering over several hundred nanometers can be measured, and subsequently examined to analyze for several components simultaneously. This was the approach adopted within the United Kingdom Atomic Energy Authority in 1986, where absorption spectroscopy is being used to analyze the composition of the various streams during fuel reprocessing. Used fuel from a nuclear reactor contains very radioactive fission products together with unused fuel: uranium and plutonium. The actinide elements have their own distinctive absorption spectra, and their concentration can be monitored using absorption spectroscopy. Optical fibers are useful for this application because of the extremely high radiation levels present, allowing the spectrometer to reside in an area where the radiation levels are low. Similar configurations, but using a dye laser, were described by workers at Karlsruhe where online measurements gave plutonium concentrations within 0.14 gl^{-1} for concentrations up to 50 gl^{-1}. In France the CEA had also undertaken similar research, developing several instruments which were manufactured under license for analysis principally of actinide species.

A further method of providing the different wavelengths for a differential absorption system that scans a narrow wavelength range is to use a laser diode and to vary cyclically its drive current. The output spectrum of a laser diode, operating in a single longitudinal mode, is a narrow line of width around 0.01 nm whose peak wavelength is temperature sensitive, $d\lambda/dT$ = 0.5 nm K^{-1}. By mounting the laser on a thermoelectric cooler and changing its drive current, its output wavelength can be changed from being coincident with, to being adjacent to, a sharp absorption line of the analyte gas.

An early demonstration of an elegant technique that both improves sensitivity and provides a reference against changes in sensitivity arises from work by Dakin *et al* [5]. Again the analyte of interest was methane, but instead of a narrowband interference filter, a scanning Fabry-Perot etalon – a tunable multilayer interference filter – was used, as shown schematically in fig. 6.3. The characteristics of this filter were a series of narrow (< 0.2 nm), evenly spaced lines whose separation is virtually constant whilst their absolute position is swept over a small range (3 nm). The 1.66 μm absorption band of methane consists of a central Q branch with well resolved rotational structure in the neighboring P and R branches. The important features of this rotational structure is that the absorption lines are very narrow, and are evenly spaced (there is little rotational anharmonicity). By tuning the separation of the Fabry-Perot etalon to be equal to the separation between adjacent rotational bands, a sensor specific to methane is obtained. As the Fabry-Perot etalon is scanned its output lines move relative to the methane absorption peaks such that it scans through these, simultaneously detecting the absorption of many bands. For this configuration any background absorption, scattering or obscuration merely appears as a constant offset on the output (plus a slight deterioration in the signal/noise ratio). The noise-limited resolution of the technique was reported as 100ppm, making this a sensitive and selective technique. Its principal drawbacks are that its use is limited to the few gases that have narrow, evenly spaced, rotational structure, and the cost of the scanning Fabry-Perot etalon.

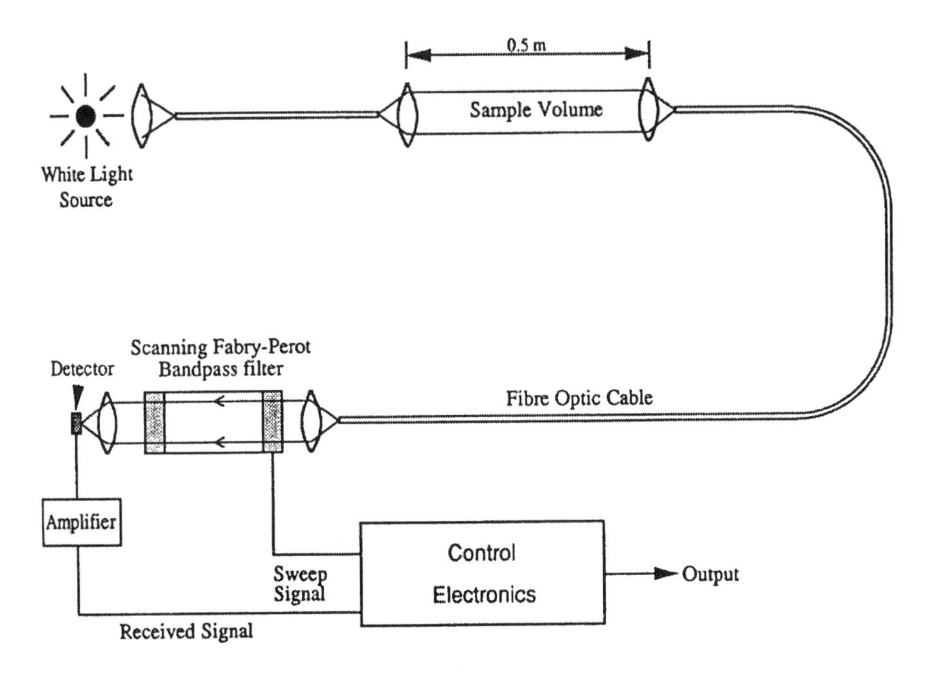

Figure 6.3. Schematic diagram of optical fiber methane sensor, using a Fabry-Perot filter. (From Dakin *et al* [5].)

For all the preceding examples, the principal advantage to be gained from using optical fibers is that the absorption cell can be placed in an awkward or hazardous location, for example on the top of a chimney, in an explosive atmosphere or in a radioactive environment, whilst the laser and the detector are kept in a suitable laboratory on the ground. Further, the instrumentation can be multiplexed to several absorption cells, allowing several locations to be monitored without the need to duplicate the transmitting and receiving apparatus, which may involve expensive components.

The principal disadvantage is that the species of interest must have an absorption band within the transmission window of the optical fiber. Figure 6.4 shows the transmission windows of different types of fiber, with some indication of the minimum loss achievable, since this determines the maximum length of fiber that can be used. The degree of development of these various types of fiber decreases from silica to chalcogenide, with a wider range becoming commercially available compared to the situation in early work.

Simple extrinsic species-specific sensors can also be used for remote luminescence measurements. This was demonstrated by Wolfbeis *et al* [6] for the determination of aluminum in the 1-800 ppm range by monitoring the fluorescence intensity of the aluminum-morin (2-2,4 - dihydroxyphenyl) -3, 5, 7 - trihydroxy - 4H - benzopyran - 4 - one complex on titrating with diaminohexanetetraacetic acid (DCTA). The use of a fiber optic configuration gave good precision even when the solutions are colored or turbid. This concept has been built upon extensively in subsequent work.

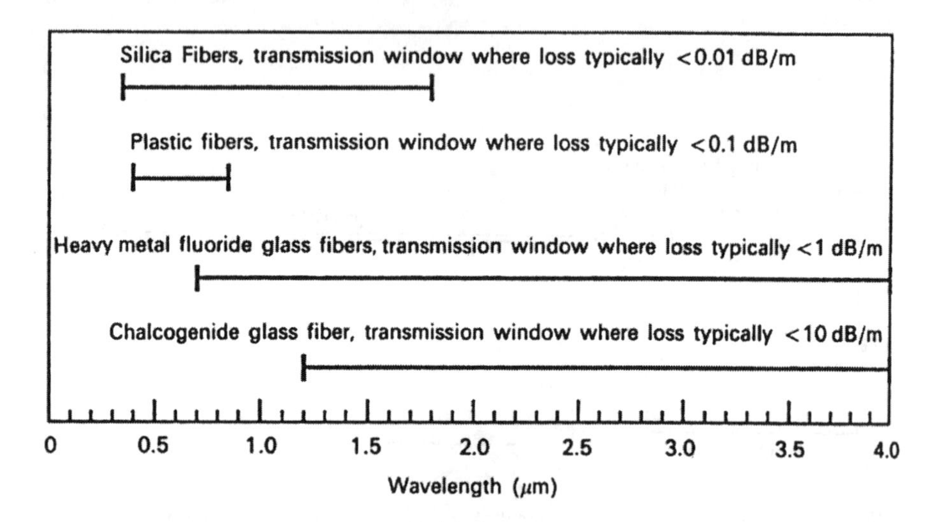

Figure 6.4. Transmission windows of typical optical fibers in the visible and near infrared regions

Another illustration of the use of extrinsic luminescence sensing in inaccessible areas was its use to establish whether a radioactive waste repository, in the USA, was leaking. It was known that the repository contained uranium, which fluoresces in its common chemical state, UO_2^{2+}. Therefore Hirschfeld and workers at the Lawrence Livermore Laboratory introduced optical fibers down very small bore holes, close to the repository. They shone blue light from an argon ion laser down the fibers and looked for the characteristic green fluorescence from any uranium.

In the biomedical field the thickening of arteries was first assessed by introducing an optical fiber, via a catheter, and examining the luminescence spectrum, because normal arteries and atherosclerotic arteries give different luminescence spectra [7]. The ratio of the luminescence intensity at 600 nm relative to 550 nm is particularly useful, with diseased arteries showing much lower intensity at 600 nm. Further work in biomedical fiber optic sensors has been reviewed by Thompson [8].

Remote absorption and luminescence spectroscopy may be employed to detect minor species in combustion products. A specially designed water-cooled probe was used to guide 5 ns pulses of 310 nm light (obtained from the doubled output of a Nd:YAG pumped dye laser) into the flame of a premixed air-methane burner and a combustion driven flow reactor [9]. The fluorescence from OH radicals present was collected by the probe and analyzed. The detectable limit was found to be 30 ppm OH radicals, allowing OH profiles to be measured within these high temperature, hostile environments. Such studies allow optical characterization of the combustion process, and could be used for scientific studies or as diagnostic equipment on production plant.

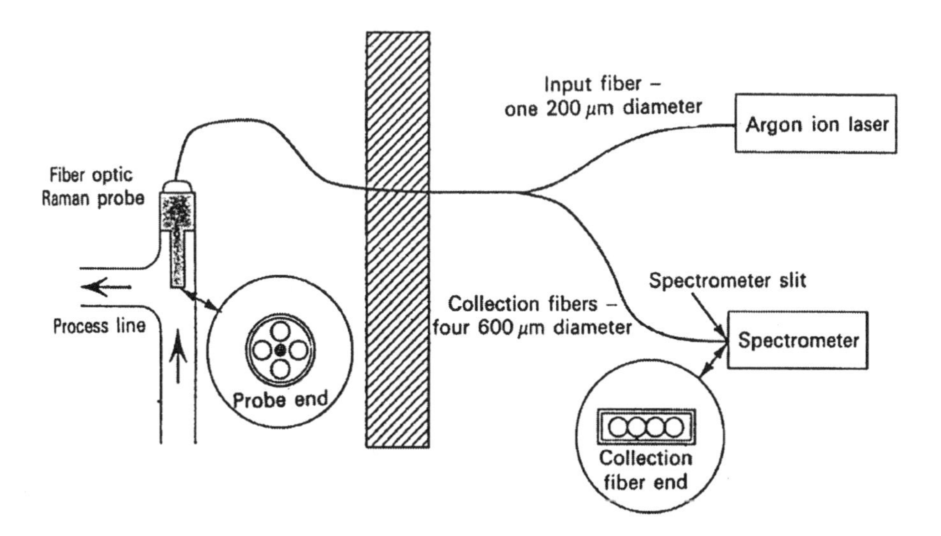

Figure 6.5. Schematic presentation of fiber optic Raman analysis of a process line. (From Gantner and Steinert [10].)

Remote Raman spectroscopy using optical fibers has also been demonstrated. A detailed study of the parameters governing the design of an optical fiber system for use in Raman spectroscopy concluded that such a configuration can significantly reduce background scattering and luminescence levels relative to conventional Raman spectroscopy. This technique has been used to monitor an alcoholic fermentation, giving information on ethanol, glucose and fructose concentrations. Higher levels of accuracy (~1g l^{-1}) have been reported for the analysis of uranyl species (UO_2^{2+}), in aqueous nitric acid [10]. This work achieved its objective of demonstrating the feasibility of using remote laser Raman spectroscopy in a nuclear process control environment. The optical arrangement, shown in fig. 6.5, provided an early illustration of some of the flexibility provided by optical fibers. The probe end consisted of a central 200 μm diameter fiber delivering the laser light, surrounded by four 600 μm diameter fibers to collect the Raman scattered light. At the detector, the four collection fibers were aligned linearly to match the shape of the entrance slit to the monochromator.

A further advance in the field of Raman spectroscopy has been the emergence of Fourier Transform Raman Spectroscopy, using 1.064 μm excitation from a Nd:YAG laser. Fiber optic bundles were used to deliver the laser light, and to collect the scattered light [11]. A good quality Raman spectrum of the bright yellow, naturally occurring antibiotic, Amphotericin A, illustrates a major advantage of this technique relative to conventional Raman spectroscopy using visible excitation when the spectrum is dominated by an overwhelming fluorescence background. Instrumentation for FT Raman Spectroscopy is now commercially available, and it is anticipated that this will become a further powerful tool in the armory of the spectroscopist.

6.5.1.2 Sensors utilizing immobilized reagents

Notwithstanding the range of spectroscopic techniques which may be applied remotely to analytes, there are many common species for which there is no convenient direct spectroscopic analytical technique. This may arise because there are no convenient electronic or vibrational transitions, e.g. as in the case of HI and aluminum ions, or due to a lack of sensitivity because the available absorption bands are too weak to be conveniently measured for typical analyte concentrations. For these species the approach is often to use an added reagent which has more appropriate optical characteristics that are dependent on the concentration of the analyte of interest. For the vast majority of optical fiber chemical sensors of this type the additional reagent is bound using either physi /chemi-sorption or via a chemical bond to either the optical fiber directly or to a substrate.

6.5.1.2.1 Absorption

One of the earliest and representative examples of this type of sensor is shown schematically in fig. 6.6, from Kirkbright *et al* [12]. The sensor described was made from a bundle of plastic optical fibers, of nominal diameter 1 mm, with a sensitive tip consisting of a styrene-divinyl benzene copolymer, onto which the pH indicator bromothymol blue was adsorbed. This was retained at the end of the probe by a PTFE membrane. The complete device had a diameter of about 2 mm. Changes in pH in the vicinity of the sensitive tip caused a variation in the absorption spectrum of bromothymol blue as it changed from its yellow acid tautomeric form to a blue basic form.

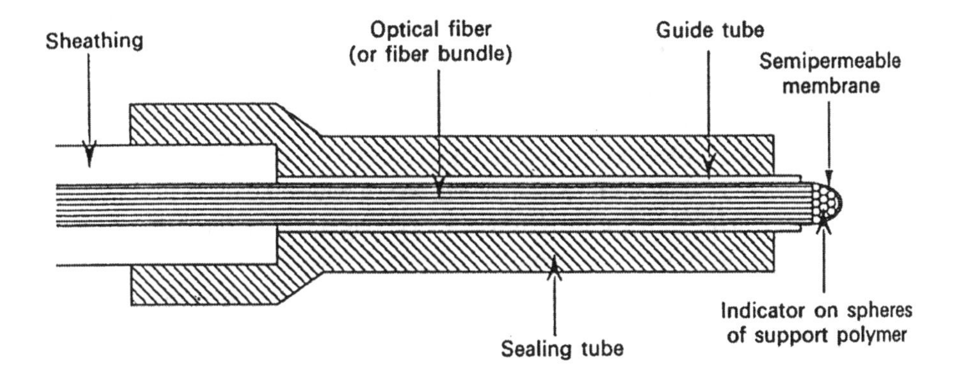

Figure 6.6. Cross-sectional diagram of fiber optic pH probe. (From Kirkbright *et al* [12].)

The measured response was the attenuation of the light reflected back up the optical fibers at 593 nm relative to that at 800 nm, a suitable reference wavelength. The bromothymol blue probe is useful over the pH range 7-9. When subjected to a step change of one pH unit this early probe took around 65 s to reach 63% of its final value (this corresponds to the $1/e$ characteristic lifetime assuming an exponential response function) and showed a stable response after 5 min. This relatively long response time was believed to be due to the rate of diffusion of solvent and ions across the PTFE membrane and through the reagent polymer sensing matrix.

By using a thin, 3 μm, porous cellulose acetate film, formed by spin coating, as the support for an immobilized pH indicator other pH sensors were developed from this concept which took 0.32 ± 0.03 s to reach 63% of their final value when subjected to a pH change of 8 units [13].

Application areas range from process plant, environmental sensing, as demonstrated by a system designed to measure the pH of rainwater [14], to biomedical uses. The first fiber optic miniaturized pH probe, whose description precedes the publication of Kirkbright *et al,* is shown in fig. 6.7 [15]. This probe was designed for *in-vivo* use, and so is configured to be sensitive over the pH range of physiological interest, $7.0 < pH < 7.4$. Its quoted sensitivity is 0.01 pH units. In many ways it is similar to the probe of Kirkbright *et al* [12], described earlier, however, one of the major differences is the use of 1 μm diameter polystyrene microspheres, which increase the amount of light back-scattered into the fiber. Fiber optic pH sensors based on this general design principle were designed for medical use.

Sensors of this type, utilizing immobilized pH indicators, can, in principle, analyze for any aqueous acid and for acidic and basic gases, e.g. SO_2, NO_x, HCI, CO_2 and NH_3. The pH range over which the sensor is sensitive (usually 2-3 pH units) can be tuned by choosing pH indicators having different dissociation constants, pKa values.

A variation on this concept was subsequently described where, for example, the enzyme urease has been incorporated with the pH sensitive dye. This enzyme selectively oxidizes urea, forming ammonia as one of the reaction products. The ammonia, in turn, being a base, causes a change in the local pH. The sensitivity of this device depends on the rate of mass transfer of urea to the sensing volume, its rate of oxidation, and the rate of mass transfer of the ammonia formed away from the sensing volume. Alternatively, if the product of the enzyme reaction is colored, an immobilized enzyme can be used alone.

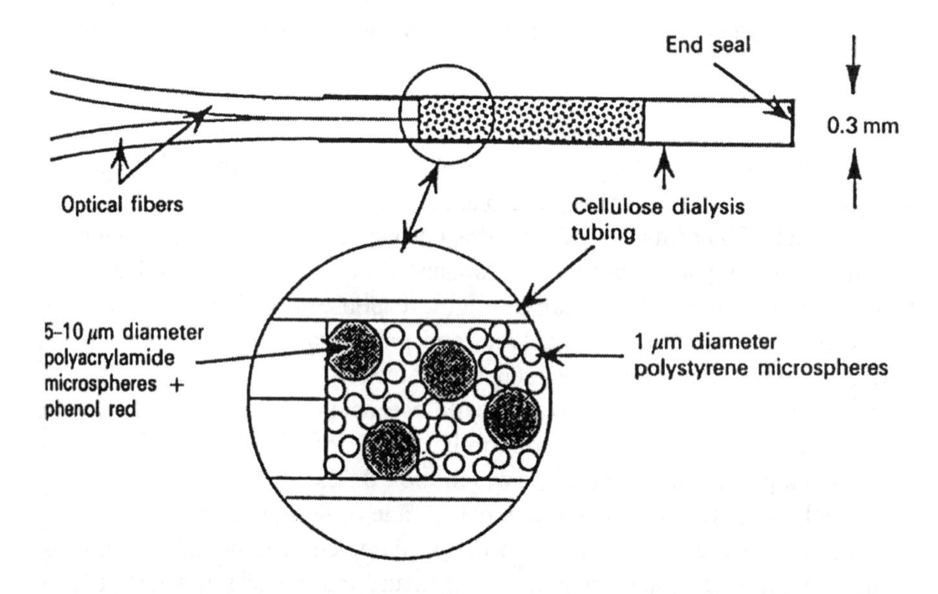

Figure 6.7. Schematic of fiber optic pH probe for physiological use

Reagents other than pH indicators have been immobilized at the distal end of a fiber to form sensors able to monitor other chemical species. One more complex example is the use of the Koenig reaction for the specific and quantitative detection of hydrogen cyanide gas. This involves the reaction scheme shown below:

$$CN^{-} + \textbf{oxidant} \rightarrow CN^{+}$$

$$CN^{+} + \textbf{pyridine} \rightarrow \text{gluconaldehyde}$$

$$\text{gluconaldehyle} + \textbf{primary amine} \rightarrow \text{colored species}$$

Consequently, three different reagents, the oxidant, a pyridine and the primary amine, have to be immobilized on a support matrix at the end of the fiber. After optimizing for sensitivity and speed of detection, probes were fabricated that could detect 1 ppm HCN in air within a minute (10 % of the Occupational Exposure Limit) [16]. One significant difference between this type of probe and those based on pH indicators is that this HCN sensor is irreversible, whereas the latter are reversible.

Many other reagents have been used to indicate the concentration of an analyte via the induced changes in the reagent's absorption spectrum. These include species-specific systems, e.g. the chromogenic crown ether (5-(4'-nitrophenyl-azo)-2-hydroxy- 1,3xylyl- 18-crown-5) for determining calcium ion concentrations and a lipophilic anionic dye (N-2,4-dinitro-6-octyloxylhenyl-2',4'-dinitro-6'-trifluoromethylphenylamine) for measuring potassium ions. The basic design of the probes described is similar to that shown in fig. 6.6. The potassium probe was sensitive over the range 10^{-6} to 1M at pH 7, measuring the absorbance at 513 nm, whilst the calcium ion probe covered the concentration range 5-50 mM.

Using immobilized triphenyl methane dyes fiber optic probes have been demonstrated that, when interrogated at 590 nm, respond to a range of organic species (tetrahydrofuran, ethylacetate, acetone, ethanol and methanol having been demonstrated) [17]. Hence these probes constituted a fiber optic sensor for organic solvents in waste water, with 30 ppm of organic solvent being analyzed in favorable circumstances.

6.5.1.2.2 Luminescence

Many research groups have described in their pioneering work, extrinsic species-specific sensors that used the luminescence of some immobilized reagent to obtain information about a chemical concentration. Two different transduction principles are used. In one the immobilized reagent undergoes a chemical reaction to form a second species whose luminescence characteristics differ from those of the original species. In the second, the analyte physically deactivates the luminophore, quenching the luminescence, by an extent which can be related to the analyte concentration by the Stern-Volmer equation (equation (6.9)).

An example of the first transduction principle is the pH dependence of fluorescein. This falls by around 90 % from pH 9 to pH 3, and consequently the pH of an unknown aqueous solution can be found by measuring the luminescence intensity of a previously calibrated fluorescein probe [18].

This early scheme was developed by Fuh *et al* where the chemical linking of fluorescein isothiocyanate to silanized, controlled porosity glass beads has been described [19]. The glass beads had an average pore diameter of 50 nm and an average bead diameter of 150 μm. A single bead was stuck to the end of a 105/125 μm all silica fiber using an ultraviolet curable epoxy resin. This miniature pH probe gave a change in fluorescence intensity at 525 nm over the pH range 2.5 to 7.5, with a response time of around 20 s. No claim was made regarding its accuracy but relative intensifies were quoted ± 0.025, corresponding to ± 0.25 pH at pH 4 and ± 0.1 pH at pH 7. The fluorescence intensity was also found to be temperature sensitive, changing by the equivalent of 0.8 pH units between 8°C and 33°C, and also dependent on the other ions present. Unfortunately each individual probe must be calibrated. Notwithstanding the limitations discussed above, the fact that the dye is covalently attached significantly improves long term stability relative to devices, where indicators are simply adsorbed onto a substrate, except at high pH when hydrolysis of the covalent linkage occurs.

Another physiologically important gas is carbon dioxide. This has been measured by luminescent techniques, making use of its equilibrium with water:

$$CO_2 + H_2O \leftrightarrow H^+ + HCO_3^-$$

The CO_2 generates a change in pH, which can be measured using a pH sensitive indicator. This technique has been used to measure the partial pressure of dissolved CO_2 using an isolated bicarbonate buffer solution, behind a hydrophilic, gas permeable silicone matrix [20]. Such measurements are important in medicine, and this probe represented an early design for *in vivo* use.

Similar sensors were reported for use in gaseous environments. One such uses fluorescein as the pH sensitive luminescent reagent, dispersed in a poly (ethyleneglycol) membrane. This sensor was found to have a detection limit of 0.1% (v/v) CO_2 in nitrogen, with a response range of 0 to 28% (v/v) [21].

As for sensors based on optical absorption, the pH range that can be measured using this type of probe can be extended by using other pH dependent luminescent indicators.

Sensors for many species in addition to pH have been constructed using changes in luminescence characteristics as the fundamental transduction principle. An example of a sensor using the quenching of an immobilized luminescent reagent is described by the same group of researchers who developed the miniaturized pH probe. This measured dissolved oxygen, another physiologically important species, using the dye perylene dibutyrate (the optimum of 70 dyes examined for this application [22]). Experimentally, it was found that a plot of fluorescence intensity versus dissolved oxygen partial pressure was close to that predicted by theory (the Stern-Volmer plot, equation (6.9)). This probe was an early development for use *in vivo*, particularly to measure dissolved oxygen partial pressure in the blood of patients under anesthesia. However, it had been characterized in both aqueous and gaseous environments, and was found to be sensitive over the range 0-20 % oxygen. Accuracy of around 0.13 % was claimed within this range for solutions and over most of the range in a dry gas stream. Significant, and progressively increasing, errors occurred after 80 min use in whole blood during an *in-vivo* experiment due to fouling, i.e. the formation of blood clots and other proteinaceous layers. The authors reported that the probe was removed, cleaned, re-inserted and found to be restored to its original sensitivity. This paper clearly demonstrates that such probes can be usefully used for physiological monitoring.

A similar sensor, using the quenching of luminescence from pyrene butyric acid and the Stern-Volmer relationship to relate the luminescence intensity to the oxygen partial pressure, was developed to monitor oxygen concentration in a gas stream between 300 and 500 K [23]. This device included a thermocouple to provide data for the temperature compensation.

An example of the use of lifetime measurements, rather than intensity measurements, to determine analyte concentrations is described by Lippitsch *et al* [24]. The reduction in the fluorescence lifetime of tris 2,2'-dipyridyl ruthenium (II) dichloride hydrate, caused by fluorescence quenching, was used to measure oxygen partial pressure. This was achieved experimentally using a relatively cheap blue light emitting diode, rather than a pulsed laser, modulated at 455 kHz and measuring the relative phase shift of the luminescence signal with respect to the driving current of the light emitting diode, to give a precision and reproducibility of ±2 ns. This corresponded to the determination of the lifetime to ± 1%, and the measurement of oxygen partial pressure to ± 0.37% in the range 0-20% oxygen. Measuring changes in lifetime is inherently a self-referencing technique, such that the measured lifetime is independent of the concentration of the luminophore. As anticipated, the sensor described showed negligible drift due to the leaching and bleaching of the indicator.

Another advantage that was illustrated from using time resolved luminescence studies is the significant reduction of interference in favorable circumstances. In biological analyses there is frequently a mixture of many luminescing species of which the majority have short, <1 μs, lifetimes, and give a large background obscuring the luminescence signal from the biological species of interest. Rare-earth metal chelates, which have long luminescent lifetimes, typically 0.1-1.0 ms, can be used as luminescent labels on the species of interest. By using a pulsed excitation source, and looking only at the luminescence that occurs some time, a few microseconds, after the excitation pulse, the background may be discriminated against. A pioneering report using this technique for a model immunoassay demonstrated an increase in the limit of detection of nearly three orders of magnitude, relative to an assay using a label with a short lifetime [25].

In the late 1980s, Wolfbeis *et al* [26] reported an ion selective optrode for the continuous determination of potassium. This used the fluorescence intensity of a lipid-soluble, modified rhodamine B dye, incorporated in a Langmuir-Blodgett deposited bilayer lipid membrane (BLM), to sense electrical potential. The fluorescent, hydrophilic end of the dye is in the central region of the BLM, insulated from the external solution. Specificity was obtained by incorporating valinomycin, an ionophore specific to potassium, into the BLM. A plot of the reduction in fluorescence intensity relative to the fluorescence intensity in the absence of potassium ions gave an approximately linear dependence against $\log_{10} [K^+]$ over the range 10 mM to 10 μM. Relative response to sodium ions, a potential interference, was shown to be around 2×10^{-3}, a 10 mM NACl solution giving a similar response to a 20 μM KCl solution, and the authors described how this may be further reduced using the output from a second reference sensor containing no valinomycin.

A further luminescence sensing concept based on competitive binding within a semipermeable membrane was first described and patented by Lui and Schultz [27]. The sensor described measures glucose concentration and is shown schematically in fig. 6.8. This is of particular importance in biomedical applications [8]. The sensor involved immobilizing a reagent that binds to sugars like glucose and dextran, concanavalin A, onto the inside walls of a short length of dialysis tubing. Dextran labeled with the fluorescent dye fluorescein isothiocyanate (FITC) was bound to the immobilized concanavalin A. This relatively large molecule could not diffuse through the walls of the dialysis membrane, whereas glucose could. When the sensor was placed into a solution containing glucose some of these molecules passed through the membrane and displaced some of the FITC-dextran from the coneanavalin A into solution. The sensor was illuminated with blue light, to coincide with the absorption spectrum of the FITC, and the intensity of the resulting green fluorescence was monitored. Because the walls of the dialysis membrane were substantially outside the volume illuminated by the fiber, in the absence of any glucose little fluorescence intensity was observed. The sensor was found to be linear in the range 2.8 to 22 mM glucose, and had a typical response time of 5-7 min. In this type of device the response time is determined not only by the diffusion and permeability characteristics of the membrane, but also by the forward and reverse rate constants of the binding reactions. Changes of the chemical equilibria, the membrane thickness and composition could be used to modify the time characteristics of the device.

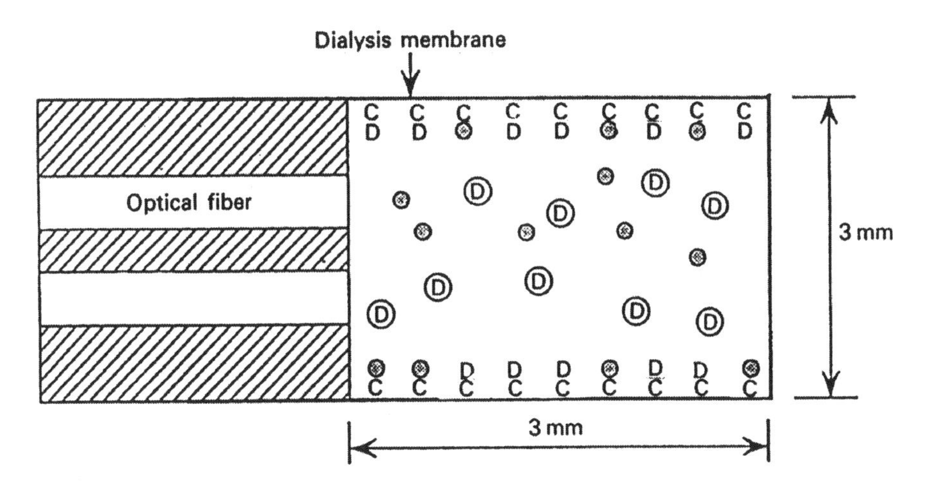

Figure 6.8. Schematic diagram of glucose sensor, using a competitive fluorescence assay. ⊛ = glucose; C = concanavalin A; D = FITC labeled dextran; Ⓓ = free FITC labeled dextran.

Fluorescence energy transfer was reported as a transduction mechanism in a modified version of the above sensor [28]. Labeled concanavalin was used in conjunction with labeled dextran, one as a donor and the other as an acceptor. In the absence of glucose, efficient energy transfer occurred between the bound labeled dextran and concanavalin A. In the presence of glucose some of the dextran was displaced from concanavalin A by glucose, and consequently there was a reduction in the luminescence intensity from the acceptor species.

The luminescence sensors described so far have were designed to be reversible. An irreversible fiber optic sensor for H_2S has been described which uses the reduction in the luminescence intensity of a fluorescein-lead acetate mixture on exposure to H_2S [29]. This sensor could detect concentrations of 0.5 ppm H_2S within 2 min: its irreversibility limited its potential applications, however.

Fiber optic probes have been demonstrated for chemiluminescence [30] since the late 1970s. Chemiluminescence reactions generate photons as a result of a chemical reaction, and consequently no interrogating photon source is required, but merely a detector. In this work, the probe demonstrated contained the immobilized enzyme peroxidase, in a polyacrylamide gel on the end of the fiber. The enzyme catalyses the oxidation of luminol by hydrogen peroxide, giving chemiluminescence with maximum intensity at 430 nm. The sensor was configured so as to provide an analysis for hydrogen peroxide. The experimental data and theoretical treatment showed that the emitted light intensity is proportional to the concentration of hydrogen peroxide when luminol is present in excess, and certain other conditions are met. The advantage of this type of sensor was its optical simplicity. The principal disadvantages were that there are only a very limited number of chemiluminescent reactions, and that in the configuration described excess reagent, luminol, has to be added.

Many other species have been sensed with fiber optic chemical sensors utilizing immobilized luminescent reagents. These include gases, e.g. SO_2, NO_2, HCl and Cl_2, ions, e.g. fluoride or aluminum ions and organic materials, e.g. ethanol, lactate and nicotinamide adenine dinucleotide (NADH). The examples listed above illustrated the principal transduction principles used, the wide range of applicability and some of the very ingenious sensing schemes first developed to tackle these problems. These basic concepts were then extended to other analytes when suitable reagents were identified [1].

6.5.1.2.3 Sol-gel based fiber optic sensors

The sol-gel process has been well known for some years and was extended to fiber optic chemical sensors first in the work of Badini *et al* in 1989 [31] and much of the detail of the process and its wider applications have been discussed by Klein [32]. Key issues have been studies of the properties of fibers and coatings, different from those in bulk materials, and the development of stable, crack-free thin films, important for fast response times and high sensitivities. Shahari has discussed the use of these techniques extensively [33] in their application to optical fiber sensors, and they provide one of the most effective means of providing analyte - specific coatings on fiber optic chemical sensors.

6.5.2 Intrinsic species-specific sensors

The interrogating light remains guided for intrinsic sensors. Therefore interaction with the analyte, or an immobilized, analyte-sensitive reagent, can only occur within the waveguide or in its vicinity by an evanescent wave interaction.

The idea of propagating light through the reagent phase was first proposed by Hardy *et al* [34], although in practice it suffers from significant technical challenges. It requires an optical guiding region of the reagent phase into which the analyte can penetrate. This invariably implies fabricating special cylindrical (fiber) or planar waveguides from nonstandard waveguide materials, which have a nonstandard microstructure, for example being quite porous instead of a solid glass. The physics of the waveguiding process, and the chemistry of the sensing process lead to conflicting requirements for the optimum microstructure for the intrinsic sensor as a whole. The ease with which the analyte can enter the sensitive region will depend on the size and number of pores or channels in the waveguiding region. Circumstances that increase the rate of mass transport will degrade the optical quality of the waveguide, principally through scattering losses from the pores.

Despite these disadvantages this type of sensor was demonstrated using a thymol-blue-doped polymethylmethacrylate fiber. The absorption spectrum of this indicator varied with pH, and the sensor produced was found to be reversibly sensitive to around 10ppm NH_3 and 5ppm HCl gases [32]. The same group of Japanese researchers had also detected moisture/relative humidity in air and soil, using changes in the absorbance characteristics (at 500 nm) of a polymethylmeth-aerylate fiber doped with phenyl red and a breathing monitor using the luminescence from an umbelliferon-dye-doped plastic fiber.

6.5.2.1 Evanescent wave sensors

A much more experimentally convenient configuration for an intrinsic species-specific chemical sensor is to have the analyte, or a reagent sensitive to the analyte, in close proximity to the waveguide, and to measure its interaction with the evanescent field to determine the presence of the analyte. Because the evanescent wave decays exponentially with distance from the interface, with a characteristic $1/e$ decay distance of typically 50-100 nm, the spatial volume probed is restricted to that within the vicinity of the waveguide.

This property was exploited in fiber optic chemical sensing, principally in the biomedical area to measure antibodies or antigens. Some of the earliest patents and papers were published in 1975, describing the use of evanescent absorption spectroscopy. Sodium picrate, codeposited on a 1mm silica rod with polyvinyl alcohol, was demonstrated to give a selective test for cyanide ions, detecting $< 0.1\mu g$ of cyanide [34]. Absorption spectroscopy, and the same experimental configuration, was used to detect ammonia, by its reaction with ferric sulphate (which is off-white) to produce violet ammonium ferric sulfate. One difficulty with these two reactions is that the equilibrium constant for each is very large, i.e. the reactions are essentially irreversible.

An early reversible optical waveguide sensor for ammonia was described [35] which used the pH sensitive dye oxazine perchlorate. Ammonia concentrations as low as 60 ppm were detectable, and the device had a time constant of less than a minute. One novel aspect of this device, shown in fig. 6.9, was the use of a coated capillary tube instead of a solid glass rod, to increase the number of reflections with the outer surface, and hence sensitivity.

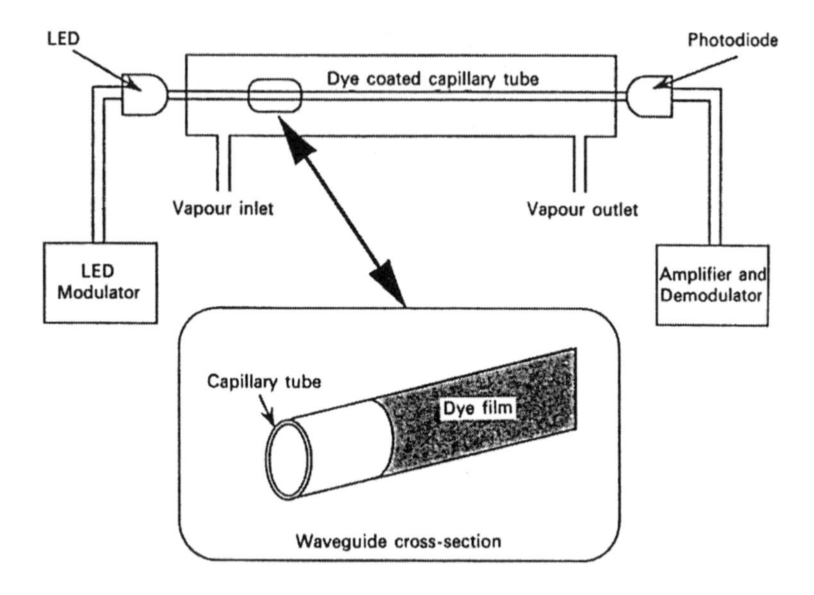

Figure 6.9. Schematic of the optical waveguide vapor-detection system using a dye-coated capillary tube

Further developments of this technology include work by Dress *et al* [36] using advanced low loss waveguides [37] for ultraviolet absorption in water quality monitoring.

A humidity sensor has been described, again using a coated capillary tube geometry, where the variation in absorbance characteristics of the reversible hydration of cobalt (II) chloride was used as the transduction mechanism. This device was most sensitive over the range 60-95% relative humidity.

Evanescent wave absorption sensors have been demonstrated for gaseous and liquid analytes, without the addition of any further reagent. For gases a relatively strong absorption band must be selected to provide adequate sensitivity. Methane gas has been detected by evanescent absorption spectroscopy using its strong absorption of the 3.392 μm line of a He-Ne laser, which corresponds to a fundamental C-H rotational vibrational absorption band. This has been demonstrated using an infrared transmitting fluoride fiber [38]. Three probe designs were discussed. In one the gas diffuses through the PTFE cladding of a multimode fiber, causing an attenuation of the evanescent field and thereby a reduction in transmittance through the fiber. In the second a short section of cladding is removed from a multimode fiber to access the evanescent field region. In the third a single mode fiber was used with its cladding thinned locally by polishing. Referencing was carried out at non-absorbing wavelengths in the 3.2 to 3.6 μm region.

The work of Paul and Kychakoff showed, using the bare silica core of a 200 mm plastic clad silica fiber and rhodamine dye as the absorber, that liquid analytes could be sensed [39]. However, the fiber in such a configuration became fragile when its protective polymer coating is removed. It has been demonstrated that non-polar solvents containing dissolved analytes can penetrate into the silicone coating of a plastic clad silica fiber [40] thereby allowing spectroscopic analysis without severe degradation of the mechanical properties of the fiber. The sensitivity measured was lower than for the bare fiber [39], due partially to the larger diameter of the coated fiber, and partially to the presence of the coating. An interesting observation was that coiling the sensing region increased the sensitivity by up to a factor of 2, due to a redistribution of the light into higher order modes (for which the penetration depth of the evanescent field is greater). The effect of coiling the fiber also provides a longer interaction length within a compact probe, further increasing the sensitivity without significantly increasing the size of the sensing volume.

In addition to absorption spectroscopy, scattering of the evanescent wave is a basic technique in chemical sensing. A silica fiber surrounded by a microporous silica cladding, comprising corpuscles up to 0.2mm, shows a transmitted power level that is dependent on the surrounding relative humidity. This sensor was demonstrated to respond to 20-95 % relative humidity. Several of these were joined via unmodified optical fiber to give a total length of 130 m. This was interrogated using optical time domain reflectometry to give a quasi-distributed sensor.

Various aspects of evanescent wave sensing have been described in the patent literature, reflecting protection of early ideas in the field. Whilst these cover many different aspects of evanescent wave sensors the Battelle patents in particular quote results based on absorption and luminescence spectroscopic determinations of immunological reactions. The experimental configuration described was principally to verify the feasibility of the concept rather than to form the basis of a commercially viable sensing system. Typically the antibody to the antigen of interest was immobilized on the outer surface of a bare 600 μm diameter optical fiber core. A known quantity of fluorescently labeled antigen was added to the analyte and passed through the surrounding flow cell. There was competition for the immobilized antibody binding sites between the added fluorescent antigen and that already present in the analyte solution. The higher the concentration in the original analyte the fewer the number of fluoreseently labeled moieties that bind.

An alternative configuration demonstrated was a sandwich immunofluorometric assay, a standard immunological assay technique. This was demonstrated using an antibody to Immunoglobulin G (IgG, a blood serum protein) immobilized on the fiber, adding a standard solution of IgG, and then incubating for 10 min. This solution was then flushed out, and a second fluorescein isothio-cyanate-labeled antibody to IgG was added, which selectively bound to those sites already containing an IgG molecule. Therefore, this assay gave an increase in fluorescence signal with an increase in analyte concentration.

The experimental arrangements described by the Battelle researchers were principally aimed at demonstrating the concept of evanescent wave sensing. Their significance was both technical, in the use of the spatial selectivity of the evanescent wave interaction to discriminate between bound and unbound analyte, and commercial, as there is a large potential market for simple, disposable clinical biosensors, recognized in the early 1980s. More commercially attractive configurations were patented in the mid-1980s, for example those based on an optical fiber within a surrounding capillary tube (fig. 6.10) [41]. The fiber had a portion of its cladding removed, and replaced with a covalently attached analyte sensitive coating, e.g. antibodies. This fragile fiber was protected by the surrounding rugged glass sleeve. When the open, lower, end of this device is placed in the analyte of interest, capillary action draws up the analyte to fill the space between the glass sleeve and the coated fiber. This results in a known volume of analyte, predetermined by the dimensions of the device, being sampled and analyzed. Consequently, this elegant design both protects the fragile fiber and makes the device much simpler to use. Similar 'capillary' fill devices using a planar waveguide geometry have been proposed and demonstrated for biomedical analysis.

A large potential market for optical fiber chemical sensors is for distributed sensors where a large area is simultaneously monitored. Since extrinsic sensors are, by nature, point sensors, the most likely form of such a sensor is an evanescent wave device. Early quasi-distributed pH sensors have been fabricated by Kavasnik and McGrath [42]. These comprised 25 cm long sections of plastic clad silica (PCS) fiber with their cladding removed and coated with a thin film containing cresol red, or the laser dye cryptocyanine. The spectral characteristics of these two dyes are pH sensitive. These sensing sections were joined by 5 m lengths of unmodified

PCS fiber, a distance which could easily be extended. A tunable dye laser of pulse width 5 ns, and operating at an appropriate wavelength (540 nm for cresol red and 640 nm for cryptocyanine) was used to interrogate the sensor. The intensity of the Rayleigh back-scattered signal was measured as a function of time (optical time domain reflectometry, OTDR) giving a measurement of the reflection from the individual sensing elements. Changes in pH caused a change in the form of the dye and a resulting change in the intensity of portions of the OTDR signal. This paper demonstrated the viability of distributed, or quasi-distributed, evanescent wave sensors, but little success has been seen in the development of subsequent sensor systems using this concept. Work continues, for example using plastic optical fiber coated with oxygen-sensitive dye [43].

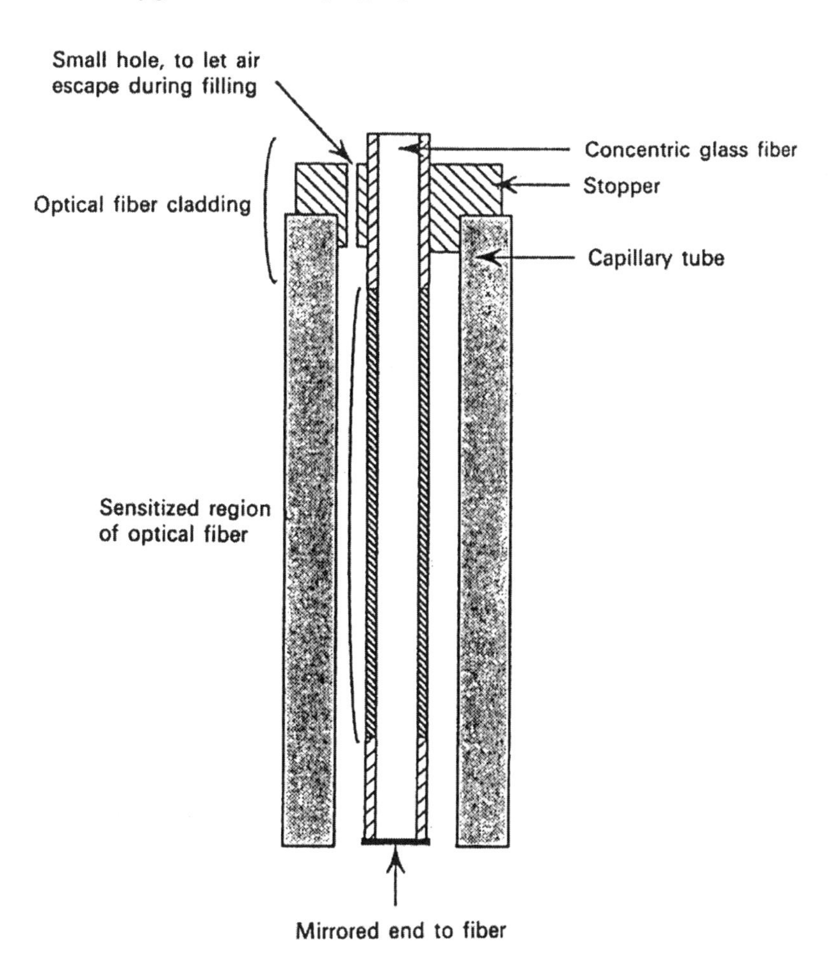

Figure 6.10. Schematic of capillary fill evanescent wave sensor [41]

Another evanescent wave sensor discussed was an integrated optical device designed to sense hydrogen, made and demonstrated by Nippon Sheet Glass. This comprised a single substrate, of $LiNbO_3$, which was masked and treated to give a 'Y' shaped optical splitter. One arm of the substrate was coated with a thin sensitive coating of tungsten trioxide, WO_3, and palladium whilst the other was untreated. If hydrogen is present it dissociated on the surface of the palladium and reduces the pale yellow tungsten trioxide to form a hydrogen tungsten bronze, H_xWO_3, which is deep blue. Detectors at the end of the two arms of the device measured the intensity of 1.3 μm light transmitted through each arm. In the presence of hydrogen the increased absorbance of the coated arm leads to a reduction in the transmitted intensity. The attenuation at 1.3 μm was found to obey the relationship

$$\text{Attenuation} \propto [H_2]^{0.588}$$

and levels of hydrogen as low as 20 ppm were detected, although the response time of the device was around several minutes.

6.5.3 Nonspecies-specific techniques

Changes in refractive index can be used to measure variations from an optimum composition, e.g. of a feedstock in a process plant. Whilst such measurements indicate a change in the chemical composition generally they do not reveal which species is involved - this has to be found by other techniques, or by a knowledge of the process involved. An early fiber optic refractometer was described by Harmer [44]. This used a series of curves with different radii of curvature, and in different directions, in preferably a step index optical fiber. The presence of the cladding does not fundamentally modify the phenomenon of light loss, and consequently polymer-coated fibers could be used without having to remove the protective polymer coating, obviating the problem of fragility which occurs when bare fibers are used. In an example using a step index, all plastic fiber, the coefficient of contrast varied by a factor of 120 when the refractive index changed from 1.40 to 1.45, indicating that this is a sensitive configuration. This concept has been developed over the years by several groups to achieve higher sensitivity and reliability.

Another simple optical fiber refractive index sensor has been described where two fibers, with their protective coatings intact, were twisted together. The fibers comprise a 200 μm fused silica core, a 100 μm silicone cladding and a 70 μm nylon jacket [45]. The amount of light coupled from one into the other was found to be dependent on the refractive index of the surrounding medium. There are some uncertainties about the exact mechanism involved, but the simplicity of the technique makes it attractive.

A simple and early approach to measuring changes in refractive index was to use a directional coupler, a typical device being illustrated in fig. 6.11. The relative intensifies from the two output ports are dependent on the coupling between the two waveguides which, in turn, depends on the refractive index of the surrounding medium. The device demonstrated acted as a pH sensor [46] and used a coaxial directional coupler which sensed a change in the refractive index of an outer polymer coating. In the configuration described the polymer contained the pH sensitive dye phenol red, whose refractive index at the wavelength of the interrogating HeNe laser, 633 nm, changes with pH due to anomalous dispersion. Although the accuracy of this device was less than that quoted for some end-on intrinsic, species-specific pH sensors, its speed of response was rapid, around 5 s, to changes in pH of a gas stream.

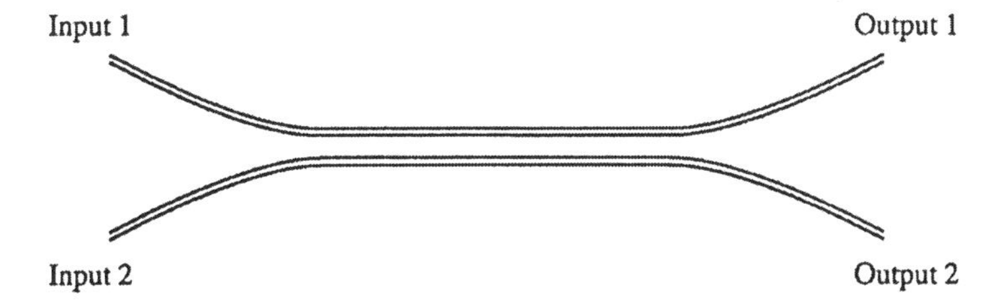

Figure 6.11. Schematic representation of an optical directional coupler

Surface plasmon resonance (SPR) has provided the basic transduction mechanism for nonspecies-specific chemical sensors for many years. A surface plasmon is a particular form of electromagnetic wave which propagates along the surface of a metal. It can be excited optically by light undergoing total internal reflection at the glass-metal interface. With the appropriate choice of metal, usually silver or gold, and its thickness, usually a few nanometers, excitation occurs at a particular angle of incidence, leading to a sharp reduction in the intensity of the reflected beam at that particular angle. This angle is related to the surface plasmon resonant frequency, and is very sensitive to variations in the refractive index of the medium immediately adjacent to the metal surface. The sensitivity of the resonant frequency as a function of distance from the metal surface falls experientially, having the same form and a similar decay constant as the evanescent field interaction discussed previously.

This analysis principle has been shown to be suitable for sensing anesthetic gas concentrations, and for detecting immune binding. In the latter case the spatial selectivity conferred by the technique makes it particularly suitable for this type of analysis. Most references to this technique describe experiments that used bulk optics, a metal-coated prism, and measured the angular dependence of the dip in reflectivity. An optical fiber nonspecies-specific chemical sensor, based on SPR, has been described using a metal-coated prism in contact with the analyte to give information about the analyte refractive index [47]. Light was transmitted to and from the sensing head using an optical fiber. A mirrored cylindrical surface on another facet of the prism provided the focusing required. The transmitted intensity, for a fixed position of the optical fiber with respect to the sensing head, was sensitive to the refractive index of the surrounding medium. This was investigated in a controlled manner using various concentrations of aqueous sucrose solution. The dynamic range of the sensor was found to be $\Delta n = 4 \times 10^{-3}$ (Δn is the change in analyte refractive index), equivalent to a 2.8% change in sucrose concentration, and the sensitivity was around $\Delta n = 4 \times 10^{-6}$.

Kreuwel *et al* [48] discussed an extension of this using a planar waveguide configuration which only supports a few modes. White light was selectively coupled into one mode. The different wavelengths have different propagation characteristics and only a narrow range of wavelengths is capable of exciting the surface plasmon, resulting in an absorbed narrow wavelength band. As the frequency of the surface plasmon altered, demonstrated by placing reagents having different refractive indices on the surface of the device, the wavelength absorbed altered. This invention has converted the angular dependence into a wavelength selective filter, which has been demonstrated on a planar waveguide, and could be adapted to an optical fiber.

6.5.4 Indirect techniques

These measure, optically, a physical parameter which varies as a consequence of the presence of a chemical. Interferometers can be used to measure changes in optical pathlength of one arm relative to a reference arm. These changes may result from a change in the physical length of the arm, or in the refractive index of the core. Such configurations have been used as the basis for chemical sensors. The first of these used a Mach-Zehnder interferometer to detect hydrogen, as is illustrated in fig. 6.12 [48]. A portion of one arm was coated with a thin film of palladium which, in the presence of hydrogen, underwent a small change in length, due to the formation of nonstoichiometric palladium hydride. This caused a small stress, and a resulting small change in the optical pathlength of the measurement arm, which is detected as a change in the fringe pattern formed by the interferometer. Consequently such a device forms the basis of a hydrogen-specific sensor. This concept has been extended over the years and a very elegant sensor, utilizing the same basic sensing principle was reported by Peng et al [49] in 1999. This used a palladium electroplated Bragg grating for hydrogen detection for applications in the US space shuttle program, using the expansion of the palladium to stretch the Bragg grating and change its wavelength of operation. The response is shown in fig. 6.13, showing also the effect of heat treatment on the sensor.

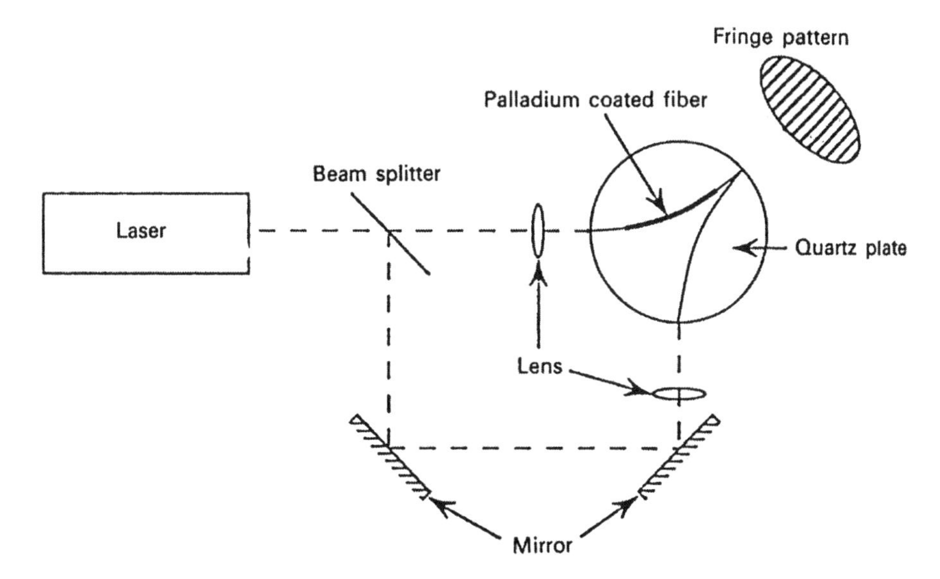

Figure 6.12. Schematic of the Mach-Zehnder interferometer used to detect hydrogen gas

A further demonstration of the value of fiber coatings in gas detection is in an optical fiber coated with a catalytic coating, to catalyze the exothermic oxidation of a combustible gas, combined with a technique to measure the temperature of the fiber, to produce a fiber optic combustible gas sensor [50]. Again an interferometer was used to sense, optically, the small increase in temperature. A change in temperature caused both a change in the refractive index of the fiber and its length. For silica the former effect was dominant. When a 100mm long sensing element was used, made by evaporating a 3 μm coating of platinum on this region, a temperature rise of 1°C was observed for 2% C_4H_{10}, and 7% CH_4, the different sensitivities reflecting the different rates of catalytically enhanced oxidation. The sensitivity was around 10^{-4}°C, indicating that this technique could detect these gases at concentrations << 1%.

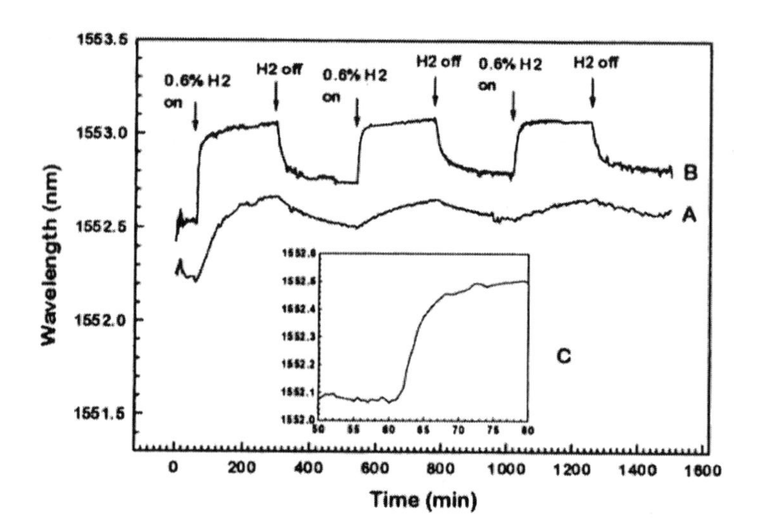

Figure 6.13. (A) Tests directly after the sensor was made; (B) after heat-treated in air; (C) a zoom-in at the first cycle in (B)

Another indirect nonspecies-specific chemical sensor was developed by British Gas in collaboration with Pilkington Security Equipment Ltd [51]. By the appropriate choice of materials it is possible to arrange for the core and cladding refractive indices of a special fiber to cross over at around –55 °C. Therefore if any part of the fiber is below this temperature the refractive index of the cladding exceeds that of the core, and previously guided core modes move out into the cladding where they are either absorbed or scattered by the fiber outer coating. This fiber was used as a distributed cryogenic leak detector system, where several kilometers can be distributed beneath a cryogen, e.g. a liquefied natural gas (LNG) store. In the event of a leak the LNG (boiling point -165°C) was detected by the fiber and triggered an alarm. This was a particularly elegant way of sensing a large area with one, intrinsically safe sensor, and demonstrates the potential benefit to be gained from distributed sensors [52].

6.6 CONCLUSIONS

Considerable research and development has been undertaken in the area of optical fiber chemical sensors. A large number of concepts have been illustrated in the laboratory and early principles developed, but currently only a limited number are commercially available in spite of the promise of early work. The principal emphasis in the future is likely to be the identification of appropriate markets for the sensors, and the development of some of the demonstrated concepts into commercially viable instruments. The success of this will depend on how the optical fiber chemical sensors perform relative to other, principally electrochemical, sensors. Therefore significant developments providing improved electrode systems will continue to have a direct detrimental effect on the attractiveness of fiber sensors.

For sensors using immobilized reagents the success of development work depends critically on being able to develop reagents and techniques that reproducibly give the sensors fabricated the required sensitivity and lifetime and sol-gel techniques have overcome some of the problems with early immobilized sensor schemes. Even so, the sensors that are most likely to become available are either those not using immobilized reagents, i.e. an increasing use of remote spectroscopy, or those for which optical fiber sensors offer very significant technical advantages over existing sensors, especially where safety considerations are important. These include *in-vivo* medical sensors, where safety is vital and a long sensor lifetime frequently is not important, immunological assays, where the spatial discrimination of evanescent wave spectroscopy provides a large technical advantage, and sensors for use within fire hazard environments. Also the commercial advantages to be gained from being able to monitor analytes continuously, and *in situ,* as distinct from single laboratory based measurements, will provide the impetus for further development of the already demonstrated concepts.

In summary, fiber optic chemical sensors have been demonstrated for many chemical analytes of importance to society and this chapter has reviewed the principles and early manifestations of the devices. They offer the potential of a valuable complement to existing chemical analytical techniques and sensors.

6.7 REFERENCES

1. MacCraith, B. D. (1999) Optical fiber chemical sensor systems and devices in *Optical Fiber Sensor Technology 4*. Eds. Grattan K. T. V. and Meggitt, B. T., Kluwer Academic Publishers, London, 15-46.
2. Magill, J. (1999) Integrated optic sensors in *Optical Fiber Sensor Technology 4*. Eds. Grattan K. T. V. and Meggitt, B. T., Kluwer Academic Publishers, London, 113-132.
3. Chan, K., Ito, H. and Inabe, H. (1984) An optical-fibre based gas sensor for remote absorption measurement of low-level methane gas in the near-infrared region. *J Lightwave Technol.,* **LT-2,** 234.
4. Hordvik, A., Berg, A. and Thingbo, D. (1983) A fibre optic gas detection system, in *Proceeding 9th European Conference on Optical Communications (Geneva),*p. 317.
5. Dakin, J. P., Wade C. A., Pinchbeck, D. and Wykes, J. S. (1987) A novel optical fibre methane sensor, in *Proceedings Fibre Optics '87,* London, Proc. SPIE, **734**, 194.

6. Wolfbeis, O. S., Schaffar, B. P. H. and Chalmers, R. A. (1986) Fibre-optic titrations IV: Direct compleometric titration of aluminium(III) with DCTA. *Talanta,* **33, 8**67.
7. Kittrel, C., Willett, R. L., de los Santos-Pancheo, C. *et al* (1985) Diagnosis of fibrous arterial atherosclerosis using fluorescence, *Appl. Optics,* **24,** 2280.
8. Thompson, R. B. (1999) Biomedical fiber optic sensors: problems & prospects in *Optical Fiber Sensor Technology 4.* eds. Grattan K. T. V. and Meggitt, B. T., Kluwer Academic Publishers, London, 67.
9. Kimball-Linne, M. A., Kychakoff, G. and Hanson, R. K. (1986) Fibre-optic absorption/fluorescence combustion diagnostics. *Combust. Sci. Technol.,* **50,** 307.
10. Gantner, E. and Steinert, D. (1990) Applications of laser Raman spectrometry in process control, using optical fibres. *Fresenius J. Anal. Chem.,* **338,** 2.
11. Lewis, E. N., Kalasinsky, V. F. and Levin, 1. W. (1988) Near-infrared Fourier transform Raman spectroscopy using fibre-optic assemblies. *Anal. Chem.,* **60,** 2658.
12. Kirkbright, G. F., Narayanaswamy, R. and Welti, N. A. (1984) Fibre-optic pH probe based on the use of an immobilised calorimetric reagent. *Analyst,* **109,** 1025.
13. Jones, T. P. and Porter, M. D. (1988) Optical pH sensor based on the chemical modification of a porous polymer film. *Anal. Chem.,* **60,** 404.
14. Woods, B. A., Ruzicka, **J.,** Christian, G. D. *et al* (1988) Measurement of rain water pH by optosensing flow injection analysis. *Analyst,* **113,** 301.
15. Peterson, J. I., Goldstein, S. R., Fitzgerald, R. V. and Buckhold, D. K. (1980) Fibre optic pH probe for physiological use. *Anal. Chem.*, **52,** 864.
16. Bentley, A. E. and Alder, **J.** F. (1989) Optical fibre sensor for detection of hydrogen cyanide in air. *Anal. Chim. Acta,* **222,** 63.
17. Dickert, F. L., Schreiner, S. K., Mages, G. R. and Kimmel, H. (1989) Fibre optic dipping sensor for organic solvents in waste water. *Anal. Chem.*, **61,** 2306.
18. Wolfbeis, O. S. and Posch, H. E. (1986) Fibre-optic fluorescing sensor for ammonia, *Anal. Chim. Acta,* **185,** 321.
19. Fuh, M.-R. S., Burgess, L. W., Hirschfeld, T. B. *et al* (1987) Single fibre optic fluorescence pH probe. *Analyst,* **112,** 1159.
20. Gehrich, J. L., Lübbers, D. W., Optiz, N. *et al* (1986) Optical fluorescence and its application to an intravascular blood gas monitoring system. *IEEE Trans. Biomed. Eng.,* **BME-33,** 117.
21. Kawabata, Y., Kamichika, T., Imasaka, T. and Ishibashi, N. (1989) Fibre-optic sensor for carbon dioxide with a pH indicator dispersed in a poly (ethene glycol) membrane. *Anal. Chim. Acta,* **219,** 223.
22. Peterson, J. I., Fitzgerald, R. V. and Buckhold, D. K. (1984) Fibre-optic probe for in vivo measurement of oxygen partial pressure. *Anal. Chem.,* **56,** 62.
23. Opitz, N., Graf, H.-J. and Lübbers, D. W. (1988) Oxygen sensor for the temperature range 300 to 500 K based on fluorescence quenching of indicator-treated silicone rubber membranes. *Sensors Actuators,* **13,** 159.
24. Lippitsch, M. E., Pusterhofer, J., Leiner, M. J. P. and Wolfbeis, O. S. (1988) Fibreoptic oxygen sensor with the fluorescence decay time as the information carrier. Anal. *Chim. Acta,* **205,** 1.
25. Petrea, R. D., Sepaniak, M. J. and Vo-Dinh, T. (1988) Fibre-optic time-resolved fluorimetry for immunoassays. *Talanta,* **35,** 139.
26. Wolfbeis, O. S. and Schaffer, B. P. H. (1987) Optical sensors: An ion-selective optrode for potassium. *Anal. Chim. Acta,* **198,** 1.
27. Lui, B. L., and Schultz, J. S. (1986) Equilibrium binding in immunosensors. *IEEE Trans. Biomed. Eng.,* **BME-33,** 133.

28. Meadows, D. and Schultz, J. S. (1988) Fibre-optic biosensors based on fluorescence energy transfer. *Talanta,* **35,** 145.
29. Roe, J. N. and Hirschfeld, T. (1988) Fibre-optic hydrogen sulphide detection. *Int. J. Optoelectron.,* **3,** 289.
30. Freeman, T. M. and Seitz, W. R. (1978) Chemilumineseence fibre optic probe for hydrogen peroxide based on the luminol reaction. *Anal. Chem.,* **50,** 1242.
31. Badini, G. E., Grattan, K. T. V., Palmer, A. W. and Tseung, A. C. C. (1989) Development of pH-sensitive substrates for optical sensor applications. *Springer Proc. In Physics*, Springer, Berlin, **44**, 436.
32. Klein, L. C. (1994) *Sol-gel optics*. Kluwer Academic Publishers, Boston.
33. Shahari, M. R. (1999) Sol-gel fiber optic chemical sensors in *Optical Fiber Sensor Technology 4*. eds. Grattan K. T. V. and Meggitt, B. T., Kluwer Academic Publishers, London, 47.
34. Hardy, E. E., David, D. J., Kapany, N. S. and Unterleitner, F. C. (1975) Coated optical guides for spectrophotometry of chemical reactions. *Nature (London),* **257,** 666.
35. Giuliani, J. F., Wohltjen, H. and Jarvis, N. L. (1983) Reversible optical waveguide sensor for ammonia vapours. *Optics Lett.,* **8,** 54.
36. Dress, P., Belz, M., Klein, K. F., Grattan, K. T. V. and Franke, H. (1998) Water-core for pollution measurements in the deep ultraviolet. *Appl. Opt.*, **37**, 4991.
37. Klein, K. F., Rode, H., Belz, M., Boyle, W. J. O. and Grattan, K. T. V. (1996) Water quality measurement using fiber optics at wavelengths below 230nm in Chemical, Biomedical and Environmental Fiber Sensors VIII, ed. Lieberman, R. A., *Proc. SPIE,* **2836**, 186.
38. Ruddy, V., MacCraith, B. and McCabe, S. (1990) Remote sensing using a fluoride fibre evanescent probe. *Proc. SPIE,* **1267,** 97.
39. Paul, P. H. and Kychakoff, G. (1987) Fibre optic evanescent field absorption sensor. *Appl. Phys. Lett,* **51,** 12.
40. DeGrandpre, M. D. and Burgess, L. W. (1988) Long path fibre-optic sensor for evanescent field absorbance measurements. *Anal. Chem.,* **60,** 2582.
41. Block, M. J. and Hirschfeld, T. B. (1986) Fluorescence immunoassay, *GB Patent 2,* 180 338A.
42. Kvasnik, F. and McGrath, A. D. (1989) Distributed chemical sensing utilising evanescent wave interactions. *Proceedings SPIE,* **1172,** 75.
43. Morisawa, M., Muto, S. and Vishno, G. (1998) POF sensors for detecting oxygen in air and in water. *Proc. POF98*, Berlin, 243-4.
44. Harmer, A. L. (1980) Refractive index responsive light-signal system, *US Patent 4,* 240 747.
45. Smela, E. and Santiago-Aviles, J. J. (1988) A versatile twisted optical fibre sensor. *Sensors Actuators,* **13,** 117.
46. Attridge, J. W., Leaver, K. D. and Cozens, J. R. (1987) Design of a fibre optic pH sensor with a rapid response. *J. Phys. E,* **20,** 548.
47. Villuendas, F. and Pelayo, J. (1990) Optical fibre device for chemical sensing based on surface plasmon excitation. *Sensors Actuators,* **A21-A23,** 1142.
48. Butler, M. A. (1984) Optical fibre hydrogen sensor. *Appl. Phys. Lett.,* **45,** 1007.
49. Peng, Y. T., Tang, Y. and Sirkis, J. S. (1999) Hydrogen sensors based on palladium electroplated Fiber Bragg Gratings (FBG). in 13th International Conference on Optical Fiber Sensors, eds. B.Y.Kim and K.Hotate, *Proc. SPIE* **3746**, 171-4.
50. Farahi, F., Akhavan, P., Jones, J. D. C. and Jackson, D. A. (1987) Optical fibre flammable gas sensor. *J. Phys, E,* **20,** 435.

51. Pinchbeck, D. (1986) The optical fibre cryogenic leak detection system. *Trans. Inst. Meas. Control (London),* **19,** 46.
52. Hartog, A. (1999) Distributed fiber optic sensors - principles and applications. In *Optical Fiber Sensor Technology 1*, eds. Grattan, K. T. V. and Meggitt, B. T., Kluwer Academic Publishers, London.

Index